THE AGE OF MAMMALS

INTERSECTIONS: HISTORIES OF ENVIRONMENT, SCIENCE, AND TECHNOLOGY IN THE ANTHROPOCENE

Sarah Elkind & Finn Arne Jørgensen, Editors

THE AGE OF MAMMALS

NATURE, DEVELOPMENT, & PALEONTOLOGY IN THE LONG NINETEENTH CENTURY

CHRIS MANIAS

University *of* Pittsburgh Press

Published by the University of Pittsburgh Press, Pittsburgh, Pa., 15260

Manufactured in the United States of America
Printed on acid-free paper
10 9 8 7 6 5 4 3 2 1

Cataloging-in-Publication data is available from the Library of Congress

ISBN 13: 978-0-8229-4780-6
ISBN 10: 0-8229-4780-3

Cover art: Painting of a *Megatherium* by Heinrich Harder from *Tiere der Urwelt*. Wandsbek-Hamburg: Verl. der Kakao Compagnie Theodor Reichhardt, ca. 1900.

Cover design: Alex Wolfe

CONTENTS

ACKNOWLEDGMENTS

I began this project in 2015, and spending a turbulent seven years (marked by the Covid-19 pandemic, continued environmental decline, large-scale industrial action, and a host of other things) at least partly in the company of giant sloths, mammoths, paleotheres, and other fossil mammals has been both a welcome opportunity and a spark to think about the world in a new light. This book would not have been possible without a range of support, help, and comradeship of various kinds, and I am delighted to acknowledge this here.

Over my academic career, Peter Mandler and Jan Rüger have provided tireless support since my PhD, for which I am extremely grateful; Robert Bickers, Kate Fisher, and Mark Jackson all provided crucial support for earlier iterations of the research. Colleagues at the University of Manchester and King's College London have been essential for helping the project develop, and for providing a stimulating and collegial work atmosphere. In particular, thanks are due to David Edgerton, Caitjan Gainty, Vincent Hiribarren, Anna Maerker, Julie Marie-Strange, Paul Readman, Adam Sutcliffe, and Abigail Woods for ideas, feedback, and conversation. At King's, the Centre for the History of Science, Technology and Medicine has proved a stimulating and supportive intellectual home, and I would particularly like to thank my colleagues here. My students, especially in the modules Defining Race and Culture, Humans and Nature, and Worlds in Objects, have always given me inspiration, new ideas, and encouragement to see things in new ways, and I am very grateful to them.

I am also extremely thankful to The British Academy (award: R117065), the University of Manchester, Princeton University Library, the American Philosophical Society, the Arts and Humanities Research Council, and King's College London for funding my research, which has allowed the project to take on such a wide-ranging and international character.

The book would also have been impossible without the support and help of staff in archives, libraries, museums, and other collections. I have particular gratitude to David Gelsthorpe and Henry McGhie (Manchester Museum); Greg Raml, Susan Bell, and Ruth O'Leary (American Museum of Natural History); Linda Oliveira (Princeton Library); Lisa Sisco (Carnegie Museum); Marie-Astrid Angel (Laboratoire de Paléontologie, Muséum National d'Histoire Naturelle); Franz Xaver Schmidt (Staatliches Museum für Naturkunde Stuttgart); Richard Hulser (Natural History Museum of Los Angeles County); Lea Gardam (South Australian Museum); Vanessa Finney (Australian Museum); Wendy Crawford (Western Australian Museum); and Josh Caster (archives of the University of Nebraska-Lincoln). I am also grateful to the staff at the British Library, Staatsbibliothek zu Berlin, Bibliothèque Centrale du Muséum National d'Histoire Naturelle, Archives Nationales de France, and London Natural History Museum Library and Archive.

My thanks are also due very much to the University of Pittsburgh Press. Abby Collier has steered the project magnificently and has persistently been helpful and encouraging. The anonymous readers of the manuscript also provided great help in sharpening and shaping the work, and giving it more coherence than it might otherwise have had.

One of the most enjoyable parts of this project is how it enabled me to build a community with other people interested in the cultural role of the deep-time sciences. The Popularizing Paleontology: Current and Historical Perspectives network, which grew alongside this book, has provided a great deal of stimulation and insight into the history of paleontology, and a genuinely friendly support group to deal with larger questions. Discussions with Paul Brinkman, Joe Cain, Pratik Charkrabarti, John Conway, Vicky Coules, Kirsty Douglas, Richard Fallon, Oliver Hochadel, Ellinor Michel (and the rest of the Friends of Crystal Palace Dinosaurs), Darren Naish, Bob Nicholls, Ilja Nieuwland, Elsa Panciroli, Irina Podgorny, Lydia Pyne, Sadiah Qureshi, Lukas Rieppel, Efram Sera-Shriar, Marco Tamborini, Will Tattersdill, Mareike Vennen, Mark Witton, and Rebecca Wragg-Sykes have been particularly helpful at various times, and I am especially thankful to those who were able to give comments on draft writing.

Of course the book has also only been possible with wider support. My parents have given me unflagging help throughout (and not least frequent queries of "How is the book going?," often at times when the book was not necessarily going). The book has also benefited from conversations about animals and history with Alex, Tom, Nick, and Emma. Jana and Andreas

Remy, and Heike Ewert have listened and helped with numerous things at numerous times. And it would be amiss in a book about connections between mammals and humans for me not to thank Belle and Yuri, my principal nonhuman mammalian friends. This book however is mainly for Elinor, who has been the best comrade and companion I could have hoped for in this and other journeys.

THE AGE OF MAMMALS

INTRODUCTION

Constructing an Age of Mammals

Over the nineteenth century visitors to metropolitan museums and readers of popular science books became familiar with a menagerie of extinct beasts. These included the mammoths, mastodons, dinotheres, and other ancient proboscideans, ancestors and relatives of modern elephants. The giant "toothless" edentates of South America, including the ground sloths and armadillo-like glyptodons, were known since the early nineteenth century. From North America came the horned herbivores like the dinocerata and titanotheres, and carnivores like the short-faced bear and saber-toothed cat. Fossil remains showed Europe had recently been inhabited by cave bears, hyenas, lions, woolly mammoths, aurochs, and hippopotamuses. More creatures were excavated in colonial territories, including great giraffids from India, and fossil marsupials from Australia. Evolutionary displays showed the step-by-step development of familiar creatures, such as horses and camels, from small early forms to the modern animals. The diversity of past life formed a narrative of the Tertiary and Quaternary periods—enshrined as "the Age of Mammals"—moving across the lush jungles of the Eocene, the bountiful forests and plains of the Miocene and Pliocene, the harsh Pleistocene glacial epoch, and the current "Age of Man," when most of the great beasts were lost.

Building this history for the mammals required diverse knowledge and expertise. Fossils were excavated in places connected to expanding nineteenth-century economies, including mines, quarries, agricultural fields, urban building sites, and territories surveyed for settlement and exploitation. Interpreting the remains required imagination and debate, as scholars compared ancient bones with modern animals and fashioned fragmentary fossils into workable specimens. Prehistoric animals were imagined by scholars, publics, and artists, and elaborated in novels, poems, and popular science works. These genres brought an unknown past to life and constructed a history for the animal world. The ancient history of the mammals could provoke wonder at the spectacles of creation, horror at fearsome beasts and their terrible relations (and their imagined terrible deaths), evocations of transcendent or perplexing mysteries, calls to action for research or preservation, or humorous musings on creatures that seemed strange and comical. The lost beasts of prehistory were discussed in various registers, building their relevance to the present. The mammals showed life and the earth had a history with moral messages. Like human history, this could be conceptualized as a story of confident progress, or of decline and fall and fears for the future. The sedimentary eras of the Age of Mammals undergirded understandings of nature and humanity, where faith in progress was tempered by uncertainty and trepidation.

This book is about how the deep history of the mammals was constructed across the nineteenth century, and its implications for the current world. Its starting point is that—contrary to current interest in dinosaurs—nineteenth-century scholars and public audiences seeking dramatic lessons on the history of life focused inordinately on mammals. Mammals were thought to represent the pinnacle of animal life and were crucial for understanding the natural world. Yet the assumed dominance of mammals combined with troubling notions: promising creatures had been swept aside in the "struggle for life," and modern nature was "impoverished" compared to previous eras. Why some ancient animals, such as the saber-toothed cat and ground sloth, became extinct, while others seemed to be precursors of familiar creatures like elephants and horses, were problems loaded with cultural assumptions and ambiguity. How humans related to deep developmental processes, and how the Age of Man differed from the Age of Mammals, provoked reflections on humanity's relationship to the natural world. Ancient mammals became crucial for engaging with nature and the environment, and the past, present and future of the world.

The Age of Mammals was constructed as the last era of earth's history, setting the foundation for the modern world. But this former world was not conceived as entirely lost. Sedimentary views of time and globalized visions of the natural world meant that remnants of ancient life were thought to still be present in places considered removed from progress. The "denial of coevalness" was a defining technique within nineteenth-century anthropology, colonial rule, and racial and cultural othering, as people and places around the world were defined by Western scholars as relics of past stages of development.[1] The modernity of many humans was denied, as was the modernity of many landscapes, animals, and plants. Debates over fossils, and reconstructions of organisms and ancient environments, marked particular creatures and locations as "progressive" or "primitive," "developing" or "decadent," "general" or "specialized," terms with strong ideological resonances. Studies of fossil mammals naturalized ideas of how familiar animals like the horse and the elephant came to be, how the life of South America and Australia was distinct from that of other places, and why Africa seemed to be the one place in the world where some great beasts survived. Fossil mammals linked nature with social, cultural, and economic values. The construction of hierarchy and order in the history of the mammals was inseparable from the construction of hierarchy and order in the modern world.

Paleontology across Boundaries and Borders

The growth of the concept of "deep time" is now regarded as one of the most significant shifts in human understandings of their place in the universe. Indeed, Martin Rudwick argues that the establishment of "the earth's deep history" should be regarded as one of the great conceptual revolutions, along with the Copernican, Darwinian, and Freudian.[2] Partly this was because of the tremendously long chronologies promoted by geology and paleontology, which constructed an earth history far beyond the time spans deduced from the Bible or the chronologies of India, China, and Pharaonic Egypt. Interest also derived from the sense of change and unfamiliarity. The earth and nature were reevaluated through ideas that landscapes had, in former ages, been radically different: covered in jungles filled with giant reptiles, or glaciated landmasses home to mammoths and great bears. How far back did this history go? How could it be made known? What forces drove change in the natural world? Was there a plan or order behind it? And what was the relationship between fossil organisms and the modern world? Behind these questions were sedimentary ideas of time and development. Geology and paleontology showed successions of environments layered on one another. There was no single timeless past or original state of nature, but a series of eras which stretched through the layers of the earth. Fossils and geological landscapes were gateways into these former worlds.

More recently Pratik Chakrabarti and others associated with a drive for "new earth histories" have examined the conceptual power of deep time obliterating other means of understanding the world and its pasts.[3] These studies argue that the establishment of deep time was intrinsically connected with structures of economic and political power. While recognition of connections between political ideology and evolutionary sciences is not new,[4] this more recent literature has taken a material and institutional focus, arguing that the construction of deep history was inseparable from control over territories and extraction of mineral resources. Katrin Yusoff has drawn attention to how "the sleight of hand of the Janus-faced discipline of geology (as extractive economy and deep-time paleontology of life-forms) is to naturalize (and thus neutralize) the theft of extraction through its grammars of extraction."[5] The extraction of fossils, the extraction of mineral resources, and the establishment of systems of authority are not separate stories, but deeply entwined. Similar observations have been made in the history of paleontology. Lukas Rieppel has argued that the paleontology of "the long Gilded Age" in the United States was predicated on the changing capitalist economy, drawing in people and techniques from mining, industry, corporate administration, and philanthropy, often in politicized and socially controlling ways.[6] The study of fossils was inscribed with power and linked with empire and economic dominance.

We therefore have a strong literature showing the conceptual importance of deep time, and how it was conditioned by power and control over the natural and human worlds. This book seeks to develop these perspectives, further linking nineteenth-century histories of the earth with the workings of cultural, political, and economic authority. The elaboration of the history of the mammals shows the connections between ideologies of progress and hierarchy, and how the natural world was inscribed with moral values. It also shows the wide ranges of people and places involved in these processes. In some respects it is a story of power and control. Mammal paleontology was connected with expanding nineteenth-century economies and was often furthered through empires and nationalizing states. Intellectually too the history of the mammals was constructed around ideas that living things could be arranged in scales of worth. However, the elaboration of the Age of Mammals also shows unevenness in both these areas. The geographic framing of the Age of Mammals reinforced the power and hierarchy of established centers, but could also displace them, by making areas regarded as strange, primitive, or unique essential for research. Similarly, these researches often raised more questions than they answered, reinforcing idioms of cyclical development, or doubts over the nature of "progress." In the construction of the Age of Mammals, valuations of progress and power were beset by uncertainty and threat.

The contested paleontological past rested on diverse perspectives. Indeed, rather than consider paleontology and the study of fossils as a single discipline, we should instead regard it, especially in its nineteenth-century form, as a linking field connecting many ways of knowing the natural world. While the nineteenth century has often been presented as a key period for the forging of modern disciplines, this was a difficult process.[7] Relations between different branches of knowledge were close, and fossils linked studies of the earth and minerals with the natural history of plants and animals. Since the inception of the field, there has been a constant tension around the extent paleontological research is a geological subject, a biological or natural history one, or something sui generis. These debates shifted across the nineteenth and twentieth centuries. David Sepkoski has discussed the rise of the field of paleobiology in the second half of the twentieth century, defined by researchers like Jack Sepkoski and Stephen Jay Gould, and the use of statistical methods to understand fossil records.[8] A different form of self-conscious paleobiology developed in the 1900s through figures like Henry Fairfield Osborn, Louis Dollo, and Othenio Abel, declaring (in often politicized manners) that "paleontology is the zoology of the past."[9] Early comparative anatomists like Georges Cuvier and scriptural geologists like William Buckland were also concerned with understanding fossil creatures as functioning organisms and communities.[10] These were all distinct projects, but showed similar attempts to relate the life of the past and present. Alongside these shifts, other scholars argued that the study of fossils should be primarily concerned with stratigraphy, with the presence of particular fossils allowing discernment of the age of rock strata—a process often connected with mineralogy and the search for resources.[11] These changing emphases were constantly negotiated, as fossils were used to debate the history of the earth and life.

Additionally, the study of fossils was not just an intellectual pursuit connected with what we might now call disciplines. Defining, locating, extracting, and analyzing fossils required practical knowledge and expertise. One very fruitful area of recent scholarship has examined how geology was enmeshed with the developing industrial economy and exploitation of coal, stone, metal ores, and oil, linking consciousness of deep time with economic processes.[12] Excavating fossils required skills around digging, preservation, and transportation, often relying on miners, quarry workers, and similar laborers.[13] Once specimens were taken to collections, considerable work was required to transform them into usable specimens. Paul Brinkman and Caitlin Wylie highlighted the importance of fossil preparators, whose technical expertise refashioned fragmentary, fragile fossils into workable scientific objects.[14] Knowing fossils often drew as much from the craft skills of manufacturing, casting, and preservation as it did erudite scholarly approaches to animals and nature.

The mixture of knowledge and expertise within paleontology raises a further point. As Claudine Cohen has argued, paleontological reasoning is based "not solely on observation and rationality, but sagacity and intuition, fiction and imagination, also play a necessary role in its hypotheses."[15] Partly this followed the trajectory of paleontology being entwined with literary modes of representation (as examined by Ralph O'Connor),[16] and artistic work, with art and imagination becoming necessary to reconstruct fossils and represent them as living organisms and "scenes from deep time."[17] Imagination also conditioned the practice of paleontological science itself. Understanding how fragmentary remains could connect with modern animals, and presenting assemblages of fossils not as masses of bone or rock, but as the relics of lost faunas and floras, depended on imagination and conjecture. The field certainly drew from detailed typological methods and claims of "objectivity," and some aspects showed the drive to mechanical reproduction and discipline in nineteenth-century science discussed by Daston and Galison.[18] However, given the gaps in the fossil record, distance in time, and strangeness of the paleontological past, studies of fossils required imaginative and speculative leaps. While the appropriate limits of speculation were a constant controversy, the field was persistently imbued with imagination and creativity.

As well as linking disciplines, people, and ways of knowing the natural world, the study of fossils also linked places. Paleontology was a self-consciously globalizing subject, and the construction of the earth's history was exactly that: the construction of a past which could accommodate the whole world. This global focus intermixed the ideological and material aspects of paleontology. Paleontologists sought to define the history of life across time and space, accumulating fossils and geological specimens from all over the world (while comparing them with the remains of modern animals and often humans). This built a vision of earth history defined by changes in life across different eras and between different places. Sometimes Indigenous, vernacular, or traditional knowledge of fossils, earth, and landscapes were engaged with, but more often these were subordinated, instrumentalized, or erased. The deep-time sciences worked and reworked a range of pasts and traditions around the earth, and incorporated them within their concepts.

Two types of location were particularly significant within the shifting geographies of fossil work. The first were field sites—the places where fossils were extracted. In the paleontological imagination, "the field" has a special status, often associated with remote and dangerous places—the badlands of the United States, the far reaches of Patagonia, and arid regions in continental interiors. These areas were certainly important for the study of fossils, and for ideologies around the paleontologist as a masculine field-worker, part scientist and part frontiersman.[19] But possibly more important were mines, agricultural fields, infrastructural cuttings, and urban digs. One core theme in this book is that exploited fossil sites tended to be in places being integrated into new industrial, commercial, and agricultural relations, rather than regions extremely remote to Western scholars. These sites could be difficult to work in, and the challenges of the modern environment were persistent features in excavation accounts. But fossil work almost invariably followed economic exploitation. Paleontology was a self-consciously "frontier science," although it was a medium- to late-stage entrant onto frontiers, using techniques and infrastructure set up by expanding political, economic, and colonial systems to assert conceptual and scholarly dominance over territories. Paleontology entered frontiers in a self-conscious manner, but did so when the balance had definitely shifted toward extraction.

The second major sites were central collections. The study of fossils depended on accumulation of material from across time and space. This was partly due to the rhetoric and practices around fossil work. An emphasis on analysis and comparison of specimens meant that centralization was critical to making sense of the past. As John Pickstone has argued, the museum collection, far from being secondary to laboratories and universities, developed over the eighteenth and nineteenth centuries as an important expression of new collecting and ordering modes of science.[20] Paleontology was a field centered on these institutions and ways of knowing. However, collections were sites of conflict and confusion as much as places of authoritarian dominance.[21] Who within the collection had authority to own, interpret, and display specimens was not an easy question to answer. And the role of other institutions in knowing the fossil past, in particular universities, private collections, and commercial operators, was often contentious. The world of collections was fractious, both among different collections and with other sites of knowledge.

Paleontology depended on relations between field sites and collections. Yet these two places were more clusters around which relationships could be consolidated, rather than binary poles or clearly identifiable centers and peripheries. Managing fieldwork, moving between the field site and collection, and negotiating for access and material contested the power of centers and built new ones. Much of this book examines the challenges and strategies of science being worked at a distance, whether through the organization of expeditions, managing collaborators, and transporting and preserving material. Control over field sites consolidated authority in particular places, especially as new scientific institutions developed in regions regarded as significant. Authority often varied depending on access, proximity, funds, and tradition. These relations conditioned how paleontology was undertaken and the concepts underlying the field.

The deep-time sciences therefore offer almost an ideal case study to understand how different forms of knowledge and claims to authority interacted and moved across the nineteenth and early twentieth centuries, a research problem most notably expressed through James Secord's contention that knowledge is produced through communication and circulation among different social and geographic contexts.[22] These transfers were not easy and depended on fierce debate and contestation. The history of paleontology allows us to get to grips with Fa-ti Fan's contention that "what is called 'circulation' may have been really a series of negotiations, pushes and pulls, struggles, and stops and starts."[23] It allows us to see movement across "lumpy" networks of power and exchange, and the connections of scholarship with hierarchies of knowledge and authority.[24] Global and totalizing messages were assertions of power from particular places, but also opened space for other voices—and blockage and conflict were just as important as circulation and exchange.

Understanding the Mammal Emphasis

My focus on the history of research on fossil mammals sets this book apart from most works on the history of nineteenth-century paleontology. Apart from broad studies examining the overall establishment and implications of geological time, most histories have taken dinosaur paleontology as the core focus of the field. A range of works have shown—in excellent detail—how dinosaurs and other Mesozoic reptiles captured public imaginations across American and European societies. For example, Paul Brinkman's *Second Jurassic Dinosaur Rush* discusses how dinosaur fossils were key to building scientific institutions in the United States, Ilja Nieuwland's *American Dinosaur Abroad* examines the transfer of dinosaur fossils across varied political and cultural contexts in Europe and the United States, Lukas Rieppel's *Assembling the Dinosaur* draws out the connections between dinosaur paleontology and Gilded Age American capitalism, and Richard Fallon's *Reimagining Dinosaurs* has shown how relations between science and literature were key to constructing the dinosaur as a transatlantic icon.[25] Meanwhile, the Dinosaurs in Berlin project has examined the early twentieth-century German-led excavations at Tendaguru in modern Tanzania to consider the links among paleontology, colonialism, and international politics.[26] We therefore have a large literature showing how dinosaurs became important icons of prehistory, especially in Anglo-American contexts in the late nineteenth and early twentieth centuries. This work emphasizes the relations between science and popular culture; transfer across national, local, and colonial contexts; and political and economic power.

While a great deal has been written about the impact of dinosaur paleontology, the equally prominent nineteenth-century focus on fossil mammals has been much less studied. The few exceptions are popular works[27] and books dealing with the reception of iconic creatures, particularly mastodons, mammoths, and giant sloths.[28] As well as missing a crucial focus within the history of paleontology, the relative lack of work on engagement with fossil mammals has obscured important aspects of the impact and role of the deep-time sciences. Presentations of dinosaurs tended to emphasize their strangeness and monstrosity, with accounts of their "grotesque," "ugly," and "ferocious" characters.[29] Similar terms were used for some extinct mammals, particularly early or large forms that seemed unrelated to modern organisms. Yet other prehistoric mammals were presented as comparable or ancestral to modern animals, explaining the origins of modern faunas and landscapes. Indeed, Rieppel's *Assembling the Dinosaur* includes an entire chapter implying that scientific interest in mammals, and valuation of "mammalian traits" of sociability and intelligence, was much greater than attention given to dinosaurs.[30] The history of the mammals gives us a deeper view of paleontology's cultural role: life's history was not just about weirdness, size, and monstrosity, but about empathy and linkages across the eras. Fossil mammals could show strange "extinct monsters," but also held the key to understanding the modern world and the forces driving life.

A focus on fossil mammals also gives new insights into the geographies of paleontological work. That histories of dinosaur paleontology orient around Britain, the United States, and to a lesser extent Germany does not just reflect the interests of historians, but the main places where dinosaur paleontology was conducted in the nineteenth and early twentieth centuries. Richard Fallon discusses the popularization of the term *dinosaur* as a decidedly US-British phenomenon.[31] Fossil work on dinosaurs was often unusual, requiring well-resourced collections with access to the rare sites containing well-preserved dinosaur fossils. This has therefore focused the history of paleontology around a few large museums in a few countries, which—while certainly important (and are indeed often key players within this book)—were not the only significant places. The history of other branches of paleontology—where fossils were more abundant, more easily worked, and spread more widely around the world—gives a broader vision of where and by whom paleontological work was undertaken. While dinosaur paleontology was geographically uneven, fossil mammals formed the basis for extensive collections across Europe, the Americas, Asia, Africa, and Australasia.

Studies of fossil mammals were also about recent history and made the "natural" past relevant to the present, conditioning understandings of modern environments and animals. In recent decades environmental history has become a wide-ranging project, with large historiographies examining human entanglements with nature and the construction of new hybrid environments, both metaphorically and materially.[32] As Simon Schama has influentially stated, landscapes were imbued with symbolic value and connected with variously imagined pasts, as "landscape is the work of the mind. Its scenery is built up as much from strata of memory as from layers of rock."[33] The history of the deep-time sciences allows us to think about how these layers of rock were themselves understood as representing deep and resonant memories. The importance of the deep past to engagement with modern environments has recently been drawn out in some nineteenth-century case studies, especially by Pratik Chakrabarti in the case of India and Daniel Zizzamia for the American West.[34] This book argues that these were not isolated incidents, but that the deep past permeated nineteenth-century engagement with the natural world. Where scientists and officials were concerned with making land "productive" through expanding agriculture, cutting through rocks to build roads and railways, or locating mineral resources like coal, knowledge of the deep past was critical to development. Long-term geological change was invoked to argue that modern environments were not static, but the latest phase of a much deeper series of eras. Past ages of lush forests, open oceans, or bountiful grasslands either laid down mineral resources or showed what the land could be like, if environmental conditions were managed.

The fossil mammals provided a history for the animal world, which raises a further point of intersection with the rapidly growing field of animal history, which contends that integrating nonhuman animals into historical processes allows us to see important issues in new lights.[35] Animals have been shown as essential for nineteenth-century economies and social systems, deeply tied to urbanization, economic change, and imperialism, and highly conceptually significant, with animals becoming symbolic of environments and places or thought to embody particular moral values.[36] Yet, strangely, histories of human–animal relations have rarely engaged with how the deep-time sciences affected engagement with modern creatures. Works in the field frequently refer to the impact of Darwinian evolution, theories of social development, and recent extinctions on human engagement with the animal world. But the construction of the long history of animal life is usually only obliquely touched on. Indeed, it is more common for works in both animal history and environmental history to discuss modern theories of the evolution of specific organisms or the paleoclimate of particular environments, rather than consider how many of these evolutionary and developmental narratives were themselves constructed in tandem with the nineteenth-century transformation of the environment and animal world. An underlying theme of this book is that engagement with fossils was a central means through which environments and animals were understood in the nineteenth century, and reflections on deep time were deeply entangled with changing knowledge of the current natural world.

We can see how modern creatures were defined through their assumed developmental past if we consider some of the major reasons why so much nineteenth-century attention focused on living and fossil mammals. Indeed, the originary work in the field of animal history, Harriet Ritvo's *The Animal Estate*, takes for granted that mammals were the main focus for Victorian observers, being the animals "with which people interacted most frequently and identified most readily."[37] More recently the excellent collection *Animalia: An Anti-Imperial Bestiary for Our Times*, examining entanglements between animals and the British Empire, devotes twenty-two of its twenty-six chapters to mammals, not only indicating historiographic emphasis, but the symbolic value of mammals.[38] This value was partly due to perceived utility and familiarity. An 1891 British text described mammals as "the best known and undoubtedly the most important group of the animal kingdom,"[39] and the French popular science writer Louis Figuier called them "the most important class of the vertebrates," who "interest us because they supply the animal auxiliaries who are most useful for our nourishment, work, and the needs of our industry."[40] As the history of human–animal relations has shown, mammals had crucial social and economic roles: cattle, pigs, and sheep were raised at increasing scales for meat, wool, and leather; horses powered cities and agriculture; dogs and cats were increasingly kept as companion animals; and exotic creatures like hippos, elephants, tigers, and bears became symbols of particular parts of the world and were hunted for commodities like ivory, hides, and fur. Mammals were pervasive, both as living creatures and as dead objects. Paleontology and the transportation of fossils were based on the same currents of global and imperial commerce as the movement of extant animals and their by-products. The life of the past was bound with the life of the present in tangible and material ways.

The prominence of mammals in the modern world was paralleled by their fossils. Mammal fossils were still rare, but considerably more common than the older remains of dinosaurs and other early reptiles and the usually fragile fossils of birds (with the notable exception of robust flightless birds like the moa of New Zealand).[41] Mammal fossils were also found throughout the world, from relatively recent geological periods. As a result, there were simply more mammal fossils in better states of preservation to be collected and studied than there were fossils of reptiles and birds. Large comparative collections were built up in numerous places. While it has been argued that this long knowledge of fossil mammals took away from their novelty and "by the end of the 1820s . . . hyaenas were old news, and a procession of bizarre extinct reptiles lurched into the limelight,"[42] it also meant fossil mammals could be used to engage with large problems, particularly those around development, variation, and distribution, at a time when scientific authority was often based around the accumulation of large amounts of material.

Of course, there were even larger collections of fossil invertebrates and fish, which were critical for forming ideas of development.[43] However, these never acquired the prestige of fossil mammals, for important cultural reasons. Nineteenth-century natural historians looked on mammals as the highest animals, at the summit of natural progress and exceeded only by humans (whose place within the mammals was itself debated). Histories of human–animal relations have often highlighted an overemphasis on charismatic mammals in animal studies, in contrast to the insights to be gained from studying human interaction with insects, fish, and microorganisms.[44] This book regards this mammal emphasis as an entry point rather than a problem. The privileging of the mammal derives from nineteenth-century views, where nature and human society were understood through hierarchy and progress.[45] Paleontology was crucial for this alignment, as life's history was used to show improvement up the scale of creation. While the regularity of progress was contested, the notion that animals could be arranged into a hierarchy of invertebrate, fish, amphibian, reptile, mammal, and human (with birds being difficult to place) was consistent. And the pervasiveness of scale-thinking made mammals crucial for defining natural progress.

Ideas of hierarchy within the mammals were complicated by older notions of the "chain of being." The idea that all creation could be ordered into a single schema, alternately called "the scale of nature" or "chain of being," was a long-standing one in European culture (and also the history of ideas, where Arthur Lovejoy's *The Great Chain of Being* is a founding work).[46] Nineteenth-century taxonomies had a variable relationship with this notion. The idea of a single scale was often criticized by naturalists as a holdover from classical thinking. Yet discussions of nature constantly referred to connection, linkage, order, "high," and "low." Museum displays, textbooks, and encyclopedias of natural history would consistently follow chain-of-being arrangements, either starting at a notional summit with humans or primates and then moving down to the "lower" creatures, or starting with the "simplest" organisms and then ascending. Often assumed rather than overtly stated, the scale maintained a continued grip, and as Harriet Ritvo has noted, "reports of the death, or even the displacement, of the chain were greatly exaggerated."[47]

The chain was complicated because mammals were not just defined as "high," but as incredibly diverse. Nineteenth-century scholars constantly stated how mammals had a unity of form, but varied lifestyles, including swimming whales, flying bats, burrowing rodents, large and small predators, and herd-living ungulates. The American paleontologist William Berryman Scott wrote, "It is as though a musician had taken a single theme and developed it into endless variations, preserving an unmistakable unity through all the changes."[48] Mammals in their diversity represented the widest flowering within the natural world. They became central to debates over comparative anatomy, Darwinian evolution and its branching patterns, and how animals formed communities, either as "ecologies" or as part of the "economy of nature." In reflections on mammals, progress and hierarchy were squared with diversity and variation.

A final point is that valuation of mammals rested on empathy and emotion. There was tremendous nineteenth-century debate over the relationships between humans and animals, but also persistent assumptions that mammals were close to humans. The notion of mammals as high in the scale of life was compounded by anthropomorphic characterizations, citing their intelligence, sociability, familial life, and complex emotions. The tremendous expansion of companion animals like dogs and cats and working connections with animals like horses, cattle, and sheep bolstered this perceived empathy.[49] Mammals seemed to presage human capacities, and were regarded as easier to understand than birds, reptiles, and other creatures. Dolly Jørgensen, in her study of valuing "lost" species, highlighted the need to pay attention to emotional engagement with animals and environments, as much as scientifically "rational" factors.[50] In the case of paleontology, the relationship between rationality and imagination, and the imposition of values on landscapes and creatures, was an emotive affair.

Mammals were therefore useful and good to think with for numerous reasons, with their assumed utility, abundance, hierarchy, diversity, and emotional resonances being particularly significant. Nineteenth-century paleontologists used mammals to create a deep history of progress and differentiation. Importantly, ancient mammals were not lost relics of a former age like the dinosaurs or trilobites. The Age of Mammals was recent enough to still be thought of as present in many parts of the world (even if often seen as under threat). Through focusing on the mammals, paleontology became not just about elaborating lost worlds, but understanding modern nature—even as it shifted, possibly into a new epoch.

Structure of the Work

This book therefore traces a large topic, examining how mammalian life was given a global history during the long nineteenth century. Selections must of course be made within this canvas. Geographically, the book has a center of gravity in Europe and North America, which (as work on the history of dinosaur paleontology has shown) were key locations for the elaboration of fossil worlds, the sites of large, often self-consciously universalizing collections, and core players within economic and colonial power structures. However, an emphasis on regions where fossils were found and how these were integrated into systems of knowledge brings in a wider geography. Examples from South America, Egypt, South Asia, and Australasia will be brought in as particularly important instances (although of course it must be noted that these were not the only places involved—further case studies on the Russian Empire, southeastern Europe, and eastern and central Asia would also be of great interest, but have mainly been omitted from this book due to limitations in my own linguistic abilities). Across these different places, we can see how the fossil world was elaborated across different geographies, the contestation between different places and actors, and how the fraught building of an Age of Mammals was linked to assertions of its importance to the present.

The book traces the elaboration of the Age of Mammals across four chronologically distinct sections. The first begins with the eighteenth-century redefinition of fossils and bones in the earth as the remains of lost creatures and indicators of ancient landscapes. This carries across three chapters examining particularly important systematizations of these ideas:

the definition of the mammals themselves (and how this was connected with studies of fossils and modern life); the construction of two particularly puzzling beasts; and the elaboration of lost faunas through the expansion of European power in regions regarded as "ancient," most notably India and Greece. These chapters see the building of a new fossil world, in which colonial and scholarly authority redefined the history and nature of life. This worked in complex ways with other means of knowing, and while new concepts of deep time were certainly important, they often reworked older mythic ideas as much as replaced them.

The second section sees how the Age of Mammals became increasingly ordered and conventionalized in the mid-nineteenth century. It first examines, across two chapters, how important institutions were founded in western Europe and North America, which were major centers of accumulation, but also field sites which reevaluated the modern territory. The remaining two chapters have a more conceptual focus, first tracing how the Age of Mammals was imagined as a series of eras, and then how paleontology became based around searches for origins and distribution. Paleontology was consolidated as a field in the years between 1850 and the 1880s, but in a contested way. Common values around progress, dominance, and links between the modern and ancient worlds were present, but often in a wary manner; uncertainty and calls to action were just as significant as confident pronouncements.

If the second section tells a story of increased consolidation of paleontological work, the third examines the heterogeneity within the field from the 1890s to the 1910s. This was certainly a period in which large institutions and particular models of the development of life were in the ascendant. The first two chapters of this section trace how institutions around the new museum movement and models of linear evolution (dramatically illustrated by the evolution of the horse) became powerful organizing principles. However, the next four chapters examine the messiness of these processes and the potential for contestation in places that could be regarded as peripheral, but that used their positions to become central to paleontological discourse. Fossil work in Argentina (especially Patagonia), Australia, Egypt, and the American West show how varied actors could claim authority within international networks and over the history of mammalian life.

The book concludes by examining a range of reflections on the natural world in the years around 1900, with the First World War being a natural break for this book, shattering the international and colonial links that paleontologists had grown to depend on across the nineteenth century. This period, marked by fin-de-siècle anxieties over the nature of development and the expansion of Western (and more generally, human) power throughout the world, saw melancholic reflection on change, and the possibility that the current era, perhaps a new Age of Man, was defined by loss and decline as well as human dominance.

Across the century paleontology linked different places and contexts, but not evenly or equally. The accumulation of fossils was frequently centered on a few institutions and localities, and these interacted and negotiated with counterparts across the world. Hierarchies of knowledge, access, and interpretation constantly shifted. Through these shifts, paleontology became a "world-building" project, constructing the modern environment, its manifold pasts, and its varied inhabitants. The created world was based on progress and hierarchy, although in unstable, contested, and variable forms. Fossil mammals became central for understanding nature, time, and the past, but showed that development did not move in a regular or inevitable manner, but was fragmentary and uncertain. These uncertainties applied both to the ancient history of the Age of Mammals and attempted human mastery of the modern world.

PART I

BUILDING A WORLD OF FOSSILS, 1700s–1840s

CHAPTER 1

The Cave and the Drift

Myth, Time, and Bones from the Earth

INTEREST IN FOSSILS WAS OF COURSE NOT NEW IN THE NINETEENTH CENTURY. BONES HAD been dug out from the ground for centuries prior to this and in many places. In European Christian contexts, these remains were understood through mythology and Scripture. Medieval and early modern records describe large bones taken to churches or aristocratic collections; they were understood as the remains of the Nephilim mentioned in Genesis or of dragons and giants. In the 1610s large bones found in a quarry in southeastern France were interpreted as the remains of a giant at least thirty feet tall, which some scholars linked to the Germanic king "giant Teutobochus," killed in battle against the Romans in 105 BCE.[1] In the seventeenth century peasants near Quedlinburg in central Germany extracted bones from the nearby "unicorn cave" to sell as medicinal objects, which attracted the attention of local antiquarians. The bones were famously represented as a unicorn in Leibniz's *Protogea*.[2] Both the Teutobochus case and the Quedlinburg fossils were controversial, sparking debate among scholars as to what they may have been. Discussing the bones was a means of understanding the history of the world and the place of humans and other beings within it.

Recent historiography has revised the idea that the search for fossils and new understandings of the earth are distinctly modern phenomena. Lydia Barnett has shown how the Noachian Flood was a crucial topic of debate across the early modern Republic of Letters, linking (but also separating) human and natural history.[3] Fossils and other "petrefactions" were dispersed around the world and found in layers of the earth that seemed marked by ancient water—hence the terms *drift* or *diluvium* for an upper layer of clay filled with pebbles. Caves were also important sources of bones. The idea that caves were the homes of witches, dragons, elves, and other entities was deeply rooted in many traditions. In late eighteenth- and early nineteenth-century Europe, folkloric motifs were reworked by Romantic writers, and caves were presented as sources of deep mysteries, while being exploited for minerals like saltpeter and quicklime.[4] Understandings of bones from caves and the drift as the remains of lost animals drew off older mythic traditions and was spurred by mining and commerce.

Knowledge of these remains also depended on engagement with the world beyond Europe. Europeans became more familiar with animals like rhinoceroses, elephants, lions, hyenas, and hippos at the same time that similar bones from the drift were collected in ever-increasing numbers. If the old bones were from ancient animals, they raised many questions. Were these the remains of creatures from historical periods, particularly the Roman era, where documents stated hippos, lions, and elephants were moved throughout the empire? Or were they from a deeper past? If the latter, had the animals actually lived in the territories where the remains were found? Or had the bones been transported by some great event? And were the bones from the drift the same as modern creatures, or were they distinct? These were difficult questions that required aligning ideas and materials to resolve.

Social relations were also crucial for understanding unearthed bones. Remains from caverns and the diluvium filled the cabinets of rulers, notables, and learned societies as part of displays on the wonders of nature.[5] The remains were also rooted in the places they were found. While scholarly communities in Paris, London, Saint Petersburg, and Philadelphia collected specimens, ascribing meaning to them was more diverse. Fossils from the caves and the drift interlinked new scholarly methods, Christian traditions of great floods, romantic nationalist usages of European folklore, and Indigenous beliefs in the Americas, Asia, and Australasia. As well as technical scholarly analysis, systems of "geomythology" (understanding landscapes and fossils through religion and the sacred) were constantly present in interpretations of ancient bones and "petrefactions."[6] The deep past was partly new but drew on older senses of wonder and mystery.

The Mamont and Incognitum across Siberia and North America

Some of the earliest specimens reevaluating ideas of the world's past and former inhabitants derived from the colonization of regions regarded as peripheral and remote. Northern Siberia and North America became particularly significant. As expanding polities like the United States and the Russian Empire annexed new territories, political authority expanded alongside economic exploitation and subordination of Indigenous inhabitants.[7] Trade in fur and the search for minerals and areas for settlement and conquest motivated expansion into new territories. During this colonial expansion, fossil remains were found and ordered into new conceptual schemas. Particularly striking were large bones exceeding the size of these territories' living creatures, and were often the subject of Indigenous traditions. Several works have argued that reflections on what would become the mammoth and the mastodon were central to establishing ideas of deep time.[8] These large fossils were given meanings which tied them to local contexts, while also illustrating global development.

Animal remains from the earth were features of life for Indigenous peoples in northern Siberia. The bones and tusks of a creature called the "Mamont" were traded for hundreds, if not thousands, of years, and were used by Siberian people like the Sakha and Evenki for making tools.[9] We are reliant on reports from traders and geographic expeditions working alongside Russian conquest to reconstruct these traditions. Evert Ysbrants Ides wrote in 1706 of seeing "Mammuts Tongues and Legs" and occasionally entire heads in Siberia, noting that "the Heathens of Jakuti, Tungusi, and Ostiacki" believe that the animal lived underground and if it "comes so near to the surface of the frozen Earth as to smell or discern the Air, he immediately dies." Meanwhile the "old Siberian Russians" settling in the region "affirm that the Mammuth is very like the Elephant," meaning "there were Elephants in this Country before the Deluge, when this Climate was warmer, and that their drowned bodies floating on the surface of the Water of that Flood, were at last wash'd and forced into Subterranean Cavities."[10] Western travelers tended to present Indigenous Siberian stories of the Mamont as picturesque folklore, and followed the Russian colonial view that the tusks and teeth were the remains of some sort of elephant. However, whether they indicated that Siberia had always been cold and these remains were carried from warmer regions by a great cataclysm (like the Noachian

Deluge), or whether they instead showed that Siberia had once been warm enough to support elephants was uncertain and reinforced discussions of climate and diversity in earth's history.

As Russian exploitation of Siberia intensified, there was increased interest in these remains. Particularly noteworthy was what became known as the "Adams Mammoth," which shows relations among Indigenous societies, geographic surveying, and political and economic power.[11] Following the narrative presented by Michael Friedrich Adams himself, these remains were found by the Evenki hunter Ossip Shumachov around 1800 near the "Frozen Sea." Shumachov saw tusks, feet, and the flank of a huge animal encased in ice. He initially avoided the remains, partly because they were very frozen but also because he was warned by elder Evenki that interference with similar remains had led to the deaths of whole families in the past. Yet Shumachov continued to visit the site over the years. By the fifth year tusks of "extraordinary size and beauty"[12] were sufficiently exposed to be extracted. Shumachov took them to the town of Yakutsk, the administrative, commercial, and military center of Russian power in the region, and sold them to a merchant named Roman Boltunov for 50 rubles. Boltunov also ventured to the remains with Shumachov and sketched them as a gigantic bristled piglike creature.

In 1806 naturalists attached to a Russian diplomatic mission traveling overland to China were diverted to Yakutia. One of these scholars was Adams, who heard of the frozen animal from the Russian merchant community, and traveled with Shumachov and a group of Evenki and Cossacks to the remains. Adams's account veered between the picturesque and the sublime, describing mountains "of a brilliant whiteness, and of a savage and horrid aspect," seeming like "the mutilated remains of grotesque and gigantic figures."[13] The Evenki were discussed in patronizing manners as Rousseauesque noble savages, "innocent children of Nature,"[14] which made Adams "convinced . . . that the inhabitants of the North enjoy happiness even in the midst of the frozen regions."[15]

Adams's report on the remains continued to emphasize wonder and horror. The carcass was "completely mutilated" after years of defrosting and refreezing, and partially eaten by "ferocious animals—white bears of the north pole, gluttons, wolves and foxes."[16] Adams enlisted ten men to strip the hide and boil away the flesh, while preserving some "bristles." Two crosses were raised to mark the locality, placing it under Christian Russian control. On returning to Yakutsk, Adams purchased a pair of mammoth tusks he was sure were the ones sold by Shumachov (although this was queried by later scholars). The remains were then shipped to Wilhelm Gottlieb Tilesius von Tilenau at the Saint Petersburg Kunstkamera. Tilesius wrote that "the skin when first brought to the museum was offensive,"[17] but he supervised the mounting of the bones, based on the skeleton of an elephant presented to Peter the Great and comparisons with drawings of African elephants (which led Tilesius to conclude the Siberian beast was more like an Asian than an African elephant). He further wrote that the extent of mammoth remains in Siberia meant "the number of Elephants now living on the globe is greatly inferior to the number of those whose bones are remaining in Siberia."[18] Tusks from the earth were potentially a greater source of wealth than living animals encountered in Africa and southern Asia. Bristles and hair, either from Adams's specimen or other sites in Siberia, and copies of Boltunov's drawing were circulated across scholarly networks. Johann Friedrich Blumenbach acquired one copy, and gave "the so-called mammoth" the name of *Elephas primigenius*, using the Linnean taxonomic system dominating European natural history.[19]

Shumachov's specimen was unusual, consisting of hide and frozen flesh rather than just bone. However, the story demonstrates several issues around how the remains of lost animals were extracted across the nineteenth century. The specimen was found by chance by someone familiar with the region, in this case an Evenki hunter. It was then secured by local elites and finally transported by a visiting scholar to a major collection. In the course of this journey, the remains were inscribed with senses of threat and myth, wealth and opportunity, and promoted engagement with a landscape subjected to surveying and domination by colonial powers. The remains connected contexts and ways of knowing the world. During this process, local connections were often occluded, or at best became instances of a picturesque background, in the eventual assessment of the specimen.

Large bones were also found in the Americas. In the Spanish Empire, Indigenous people and colonists observed the bones of "giants" across South and Central America.[20] In North America, Mark Catesby reported enslaved Africans identifying fossil remains as "the Grinders of an Elephant," an animal they knew from their homelands, an identification regarded as credible.[21] Other large teeth were not elephant "grinders," but cusped and deeply rooted, almost like human molars. These teeth were often found alongside large limb bones and were thought of as a new animal, the "American

FIGURE I.I. The Adams Mammoth on display in Saint Petersburg, next to two extant elephant specimens. Figuier, *Terre avant le déluge*, 363. Author's collection.

Incognitum." Amy Morris has described how scholars like Cotton Mather linked these remains with stories of a giant creature called "Maushop" by the Wampanoag, and biblical stories of the Nephilim who lived on the earth before the Flood.[22] The remains were given a hybrid status, mixing two mythic traditions. Meanwhile at Big Bone Lick, a salt quarry and way station along the Ohio River, Indigenous people, traders, soldiers, and surveyors found huge bones, which were taken to East Coast and European scholarly institutions.[23]

In 1801 the Philadelphia artist and museum builder Charles Willson Peale learned that more giant bones were found in the Hudson Valley by farm workers digging marl pits for fertilizer after a particularly dry year. After negotiations with, and payments to, the landowner, Peale set up a pumping and pulley system to drain the ditches, dramatized several years later in his painting *The Exhumation of the Mastodon*.[24] The excavation was a local attraction, and Peale's son Rembrandt wrote, "Every farmer with his wife and children, for twenty miles round . . . flocked to see the operation; and see a swamp always noted for being the solitary and dismal abode of snakes and frogs, became the active scene of curiosity and bustle."[25] A nearly complete skeleton was exhumed, but it lacked a mandible, leading to a scouring of neighboring fields for more bones. When a lower jaw was found, Rembrandt recalled, "the woods echoed with repeated huzzas. 'Gracious God, what a jaw! How many animals have been crushed between it!' . . . [A] fresh supply of grog went round, and the hearty fellows, covered with mud, continued their search."[26] Excavating the beast was a social occasion, involving crowds of people in the elaboration of the deep past.

Two skeletons were mounted from the bones, with missing pieces worked up in wood and papier-mâché. One was placed in Peale's Philadelphia Museum, and a second was taken on tour by his son Rembrandt. Paul Semonin argues that the beast became an icon of American identity, used to counter concepts that American nature was degenerate or inferior compared to the Old World.[27] The Incognitum showed the bounty and extensiveness of the land, and allowed American scholars to participate in international networks. Imaginings of the creature as a fearsome predator were reinforced into these ideas. The traveling showman Albert Koch, discussing his own Incognitum specimen as the "Missouri Leviathan," made similar links to Cotton Mather's: the creature was simultaneously redolent of the biblical Leviathan and Native American legends of the "Great Buffalo."[28]

The international reception of American specimens could be variable, however. Rembrandt regarded his exhibit of the Incognitum in Britain as a failure and never managed to reach Paris.[29] He also changed his view on the animal during the London exhibit. He described it in his first exhibit guide as "A Non-Descript Carnivorous Animal of Immense Size," judging from the teeth that it was a predator. However, the guide's second edition of 1803 noted "the first crude ideas from an imperfect examination were hastily given,"[30] and it was more likely a ponderous lake-dwelling animal, eating fish, turtles, and marine vegetation. The beast was still huge and strange, but not quite the terror of the ancient forests. Albert Koch meanwhile managed to sell his specimen to the British Museum, although it was rather humiliatingly revised by Richard Owen, who classified it as a proboscidean of some sort and noted so many errors in Koch's reconstruction that "the necessary reform in the juxtaposition of other parts of the skeleton could be effected only at a great expense."[31]

Interpretations of the great beast of the Americas shifted over the nineteenth century, but the mastodon created North America as a land of wonders, as understood through Christian and Indigenous antiquity. While Thomas Jefferson and others famously believed that the Incognitum might still be living in the US interior, this view was increasingly rejected. Rembrandt Peale quoted William Hunter's reflections from 1768 on American fossil remains that this was a great boon for modern humanity, as "if this animal was indeed carnivorous, which I believe cannot be doubted, though we may as philosophers regret it, as men we cannot but thank Heaven that its whole generation is probably extinct."[32] The great animals had disappeared, making the world suitable for humans.

Mining and Folklore in Britain and the German Lands

Interest in the Mamont of Siberia and Incognitum of North America mixed economic exploitation, learned debate, and myth. This combination was not limited to North America and northern Asia, but was also intermixed in places taken as central for new concepts of "deep time." In Britain the developing center of a world imperial system, the remains of what were interpreted as elephants, rhinos, hippos, and hyenas were found in mines, canal beds, and roadworks. Meanwhile, in the German lands, interventionist states, scholarly societies, and popular writers investigated the landscape and its deep past. In both Britain and Germany the fossils of the earth's last ages provided a new antiquity, tied to changing economic and state structures.

The establishment of the deep-time sciences in Britain in the late eighteenth and early nineteenth centuries is one of the most well-trodden areas in the history of science. The pioneering works of Martin Rudwick illustrated how institutions like the Geological Society of London and Geological Survey promoted an unfathomable antiquity for earth history and new social relationships for gentleman scholars and workers.[33] Geological texts and lectures reached large audiences, with William Buckland's contribution to the *Bridgewater Treatises* reinvigorating debates over natural theology by linking new geological finds with Scripture,[34] and Charles Lyell's *Principles of Geology* becoming a manual for those holding more gradualist perspectives on the earth's development.[35] The construction of the deep past was a wide-ranging process. Antiquarian and natural history journals were littered with reports on fossil bones. Buckland wrote of English and Welsh sites containing "the teeth, tusks, and bones of elephants of prodigious size," often "mixed with great numbers of the teeth, bones, and horns of elk, stag, ox, horse, hippopotamus and other diluvial animals."[36] So while scholars could pronounce on the immensity of the past, access to material was more widespread. The deep past pervaded metropolitan and provincial cultures as a thing of immediacy as well as antiquity.

Fossil excavations were the work of many people. While the famous caricature by William Conybeare depicted Buckland's investigation of the Kirkdale "hyena den" as the personal journey of a gentleman scientist into a former world, other images illustrate more collective work. Buckland's *Reliquiæ diluvianæ* depicted the Dream lead mine on the Derbyshire estates of the Whig MP Philip Gell (a keen antiquarian and collector of fossils), where miners in 1822 reached a pit containing the remains of a rhinoceros.[37] Geological surveying, infrastructural projects, and mining were the most extensive acts of digging in the nineteenth century, and a frequent source of fossils. While miners and quarry workers were often occluded in scholarly texts, they were often the people most responsible for excavating and finding fossils.

The deep past also connected geographic regions. It was not simply that British fossil hyenas and hippos were compared with modern specimens; understandings of past and present animal lifestyles and anatomy fed into one another. The often repeated story of how William Buckland acquired a hyena, named Billy, who "cracked the marrow-bones of oxen . . . exactly as did his ancestors ages before him in the wilds of Yorkshire," so that "it was impossible to say which bone had been cracked by Billy and which by the aboriginal hyena of Kirkdale," linked modern and ancient animals. The bones chewed by living hyenas like Billy and "his ancient British forefathers," as well as the "*album graceum*" left by both animals, made the ancient beasts comprehensible and gave modern creatures an ancient lineage.[38] And when the first living hippopotamuses were exhibited in Britain, most notably the one named Obaysch in the London Zoological Gardens in the early 1850s, "hippomania" was filled with the language of antiquity and myth.[39] Richard Owen populated an ancient British landscape with these animals, describing "gigantic elephants of nearly twice the bulk of the largest individuals that now exist in Ceylon and Africa," "the lakes and rivers were tenanted by Hippopotamuses as bulky and with as formidable tusks as those of Africa,"[40] and "troops of Hyaenas, larger than the fierce Crocuta of South Africa, which they most resembled, crunched the bones of the carcasses relinquished by

FIGURE 1.2. Buckland entering the hyena den. Reprinted in Gordon, *Life and Correspondence of William Buckland*, 61. Courtesy of the Wellcome Collection.

the nobler beasts of prey."[41] The deep past of Britain was as exotic as distant regions of the world, and chronological and geographic remoteness were brought into the same frame.

Similar processes occurred in the German lands, although in more localized manners. While scientific institutions in Berlin and Vienna, the capitals of Prussia, and the Austrian Empire often claimed status analogous to those of London and Paris, they faced difficulties in doing so. Germany—divided into thirty-nine states after the Napoleonic Wars—did not possess a single scholarly and political center.[42] While contemporaries would bemoan the splintering of German scholarly life, it did have advantages, as the different states promoted scholarship in their territories. While funding varied between states, patronizing art and natural history was often regarded as a governmental duty.[43] Meanwhile, the learned and educational culture of the German lands and their rich urban life meant that (male) German

FIGURE 1.3. Dream Lead Mine in Derbyshire. W. Buckland, *Reliquiæ diluvianæ*, plate 20. Reproduced by kind permission of the Syndics of Cambridge University Library.

notables formed associations, with antiquities and natural history societies being particularly widespread.[44] These two forces—princely institutions and middle-class associations—were sometimes rivals and sometimes worked synergistically. Over the nineteenth century royal collections were increasingly opened to the public, employed bourgeois scholars, and amalgamated with university and learned society collections.

Riesengebeine (giant bones) and *unicornu fossili* (fossil unicorn) like those at Quedlinburg, and petrified shells and plants were widely located, especially in southern Germany. Miners and artisans developed interpretations of these fossils and sold them to collectors.[45] German scholarly fossil collectors often attempted to gain aristocratic and state patronage for their work or—if this was not forthcoming—appealed to commercial culture or international backers. These scholars tended to focus on cave sites, often located in regions marked by developing currents in German Romanticism as sublime or picturesque. Walking through the mountains and valleys, and then into the caves, was a journey into antiquity, especially when bones were found. While hyenas were the most iconic British cave animals, German caves were known for giant bears, big cats, horses, great deer, and wild bovids. How long ago these creatures had lived was unclear, but they connected with traditions of German forests being the haunts of powerful animals.

As a case in point, Georg August Goldfuß, simultaneously professor of zoology and mineralogy at the University of Bonn and director of the city's Naturhistorisches Museum, wrote a travel account and guidebook *The Surroundings of Muggendorf*, a region in southern Germany with large numbers of caves (Muggendorf itself also contained a sizable number of taverns, which facilitated travel there). Goldfuß described several caves as well suited for "troubled minds to warm their hearts, by gazing at the glorious sublime nature," visiting the "rock valleys, grottos and springs, where nature speaks to us with magical powers."[46] The caves led to mythic worlds. The interior of Rosenmüller's cave was "a shimmering fairy temple"[47] with glittering crystal formations, while "Oswald's cave" was named after a nobleman who lived there as a hermit. The most famous bone cave, Gailenreuth, was like a "crypt," filled with the remains of "many hundreds of monstrous animals. In a horrible confusion, we saw heads, terrible teeth, long bones, vertebrae and claws, either lying loose on the floor or cemented on the walls with sinter."[48] The caves had also been examined by earlier generations of scholars. Johann Esper thought the bear skulls were the remains of polar bears and that their disarray indicated they had been swept there by a great flood. Meanwhile Johann Rosenmüller interpreted the bears as a specifically indigenous German species with the Linnaean binomial of *Ursus spelaeus* (the bear of the caves), and—as shown in an excellent study by Patrick Anthony—worked them into a primordial past where they may have lived alongside humans.[49]

Goldfuß's account of the caves fulfilled desires for journeys into mythic former worlds. However, his text illustrates that matters were quite organized. Early in the account Goldfuß described how the state government had appointed a "cave inspector" named Ludwig Wunder, who guided travelers, made sure the caves were safe, and ensured "the caves are not robbed of their ornaments by greedy or indolent hands." Goldfuß did not explain how the cave inspector judged who was a valid visitor and who was a potential risk, although this judgment clearly privileged educated scholars like himself.

Wunder was from a family of artisans who collected and sold local fossils as curiosities; his father had been the first appointed cave inspector. The Wunders' knowledge of the territory and social connections gave them status and income despite their lower-class backgrounds. Ludwig Wunder stipulated numerous regulations. Visitors needed to pause to adjust to the temperature of the cave before venturing in, and they should secure themselves with ropes and bring lanterns. Goldfuß stated that "several of the most beautiful and remarkable [caves] can now be visited without any danger, even by ladies."[50] The description of Gailenreuth indicated the exploitation of the site. It was now like a "heap of rubble" and unrecognizable compared to the first accounts of the cave.[51] Wunder himself claimed to have collected 150 cave bear skulls in the three years prior to Goldfuß's visit, and stated "there is still an inexhaustible supply."[52] The caves may have been interpreted through Romantic sensibility, but they were also bound to commerce and extraction.

The early nineteenth century saw increased interest in human antiquities as well as natural history ones—although whether humans had lived alongside the ancient beasts was a fraught question. In the German lands these studies frequently connected antiquities with nationalist discourse and desires to promote visions of Germanic antiquity.[53] One of the core texts within this revival was the *Nibelungenlied*, a thirteenth-century epic enshrined as a record of the ancient Germanic past, filled with giants, dwarves, dragons, and Teutonic heroes (which provided the basis for Richard Wagner's Ring cycle, with Bayreuth being close to Muggendorf). One reference in the *Nibelungenlied* stood out for natural history–inclined scholars, describing a hunt by the hero Siegfried:

> darnach schlüch er schiere, einen wisent und einen elch,
> starcher üre viere, und einen grimmen schelch.[54]
>
> Then he slew a wisent and an elk,
> four strong *üre* and a fierce *schelch*.

Most of the animals slain by Siegfried were quite knowable. The wisent was a bison, and the *ur* was the aurochs, the great cattle described by Caesar as terrorizing the primeval forests, whose last representatives were hunted in the early modern period (and whose remains were increasingly found in the drift and soil of central Europe). However, the next lines mentioned a

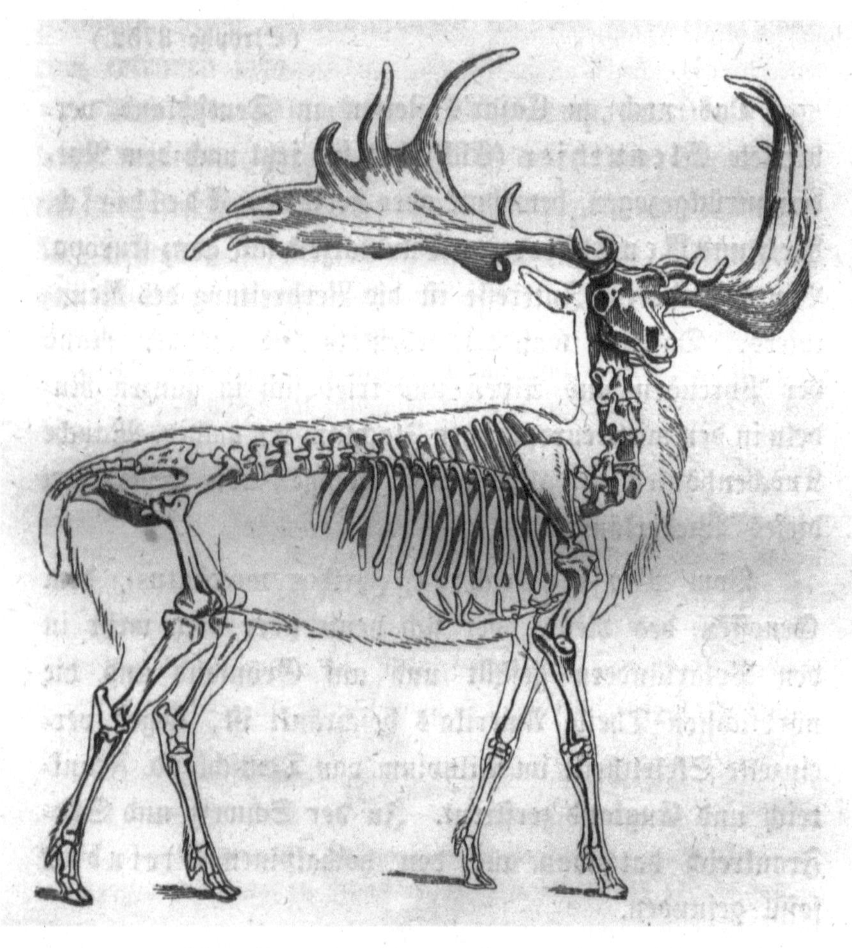

FIGURE 1.4. A later illustration of the giant deer, and the possibility of it being the schelch. Zittel, *Aus der Urzeit*, 507. Author's collection.

mysterious animal—the schelch. What this could be was debated. While some scholars saw it as a mythological animal, or a literary confabulation to enable a good rhyme scheme, a trope developed in paleontological literature that this was a surviving prehistoric beast. Just as the ur was a more primordial bison, the schelch was most likely a grander primordial elk.

There was a prime candidate for the schelch: the Irish elk or *Riesenhirsch*, literally the "giant stag." These were immense deer, whose males had gigantic antlers, excavated from peat bogs in the British Isles. The size of the animal, and cultural values of stags as hunting trophies, meant male giant deer, or at least their skull and antlers, often became museum centerpieces (such as in the Hunterian Museum in London, whose halls were dominated by a giant deer skull). Similar remains were found in Germany, with one specimen described by Goldfuß. His correspondent, Nees von Esenbeck, wrote how the writer of the *Nibelungenlied* had made "a freely-created image of a recently surviving, but nevertheless practically lost, picture of ancient Germanic love of hunting and prey."[55] While the alignment was initially tentative, the idea that the schelch was a memory of the giant deer grew to become a cliché in popular science works in Germany and farther afield. In the *Edinburgh New Journal* of 1830, Samuel Hibbert referred to the idea without having read Goldfuß's paper (and connected it with an animal mentioned in early Irish accounts named the *segh*).[56] From the 1840s German scholars would frequently describe Siegfried's hunting of the schelch to show the persistence of ancient animals. Germanic heroes had hunted the beasts of prehistory and were naturalized into an antiquity of myth, legend, and fossils. A primal and heroic past linked the human and the animal.

Antipodal Nature and Aligning Unknown Bones

The excavation of caves and drift layers occurred all over the world, mixing economics, local traditions, and developing scholarly norms and comparisons. Reflections on the ancient past showed that the whole world possessed relics of former ages. However, there was always tension between the local and the "universal," especially in how far the history of the ancient earth was uniform and how far there was local variation. Finds in Australia became important for engaging with these issues. Despite (or perhaps because of) its distance and difference from Europe, Australia became central to many debates. Political and economic control by Britain was coupled with intense efforts to catalog the natural products of the continent. The fauna of "New Holland" was constantly cited as strange and unfamiliar, with extensive debates on how Australian animals could be classified.[57] Australia's human population was discussed in similar terms, as their land, bodies, and artifacts were seized by European settlers. Those with control over Australian specimens—whether in Australia itself or in European collections accumulating Australian material—could garner great authority.

Bones found in Australian caves drew interpretations of the Australian landscape and its inhabitants (whether human or animal) with questions on the overall development of nature. The Bathurst settler George Ranken identified similarities in the limestone landscape of New South Wales to the caves excavated by Buckland. In 1830 he broke into the Wellington Caves on the land of the Wiradjuri and found some fossil bones. Using these finds to build his position within colonial society, he showed them to the local politician and minister John Dunmore Lang, who wrote a long letter to the *Sydney Gazette*, which was reprinted in the *Asiatic Review* and the *Edinburgh New Philosophical Journal*. Lang's letter rooted the finds within the Australian context, while interpreting them according to northern norms assumed as universal. He wrote how Ranken entered a chamber "into which no mortal man had ever entered before," citing Aboriginal beliefs that a spirit called Koppa dwelt in the caverns and forbade entry. Upon finding bones, Ranken initially thought they "might have belonged to some bush-ranger who had attempted to hide himself in the cave, and had subsequently died," but recognized they were deeply embedded in the rock, and were different sizes than human bones.[58]

Lang regarded the Wellington Caves as so similar to Kirkdale that they must have been inhabited by similar creatures—hyenas, rhinoceroses, and elephants—and date to before the Deluge. He concluded the letter: this was "another convincing proof of the reality and the universality of the deluge," and that it "supplies us also with a powerful motive of gratitude to Divine Providence." The region was now more suitable for exploitation and settlement, as "the tiger or hyaena would have been a much more formidable enemy to the Bathurst settler than the despicable native dog, though indeed they would certainly have afforded a much nobler game to the gentlemen of the Bathurst Hunt."[59] As with Hunter's reflection on the demise of the mastodon, the disappearance of ancient creatures was a boon for modern colonial settlement.

After these initial reports, the site was monopolized by Thomas Mitchell, surveyor general of New South Wales, who led militarized surveys across Australia between the 1820s and 1840s. Mitchell's book series, *Three Expeditions into the Interior of Eastern Australia*,[60] described his excursions, collecting efforts, and interaction with Aboriginal people—who were sometimes employed as collaborators and informants, but also treated with brutal violence, such as in 1838 when Mitchell's party attacked and killed a large number of Kureinji and Barkindji people at a site subsequently euphemistically named "Mount Dispersion." Mitchell's visit to the caves was

FIGURE I.5. The colonial explorer venturing into the Wellington Caves. Pentland, "Observations on a Collection of Fossil Bones," plate 5. © The British Library Board: P.P.1433.b.

a further exercise in scholarly and conceptual control over the territory. He described a journey into a cathedral-like structure, much like caves "well-known from the descriptions of Dr. Buckland and other writers."[61] Excavations required strenuous activity, descending with candles and ropes "six fathoms at one stage," and "crawling through narrow crevices we penetrated to several recesses, until Mr. Rankin [*sic*] found some masses of osseous breccia beneath the limestone rock, but so wedged in, that they could be extracted only by digging."[62] This was a controllable space to be entered and dominated.

Interaction with Aboriginal people was crucial for developing European understandings of the Australian environment.[63] In an excellent study Kirsty Douglas has discussed how European scholars engaged with Aboriginal peoples to locate sites and create different ways of knowing fossils.[64] Some caves served as habitation sites, while others were sacred or forbidden as the homes of ancestors or spiritual beings. Aboriginal traditions on the formation of landscapes, and mineral and geological products, including fossils, were followed by European prospectors. However, this should not be taken as an indication of respect—their knowledge often meant that Aboriginal people were naturalized within the landscape. An early picture of the Wellington Caves by Augustus Earle (before the discovery of the fossils) starkly dramatized this, with several Wiradjuri depicted, in faceless manners, as guardians of the caves. Ranken's discussion of Aboriginal fears of "Koppa" contrasted them with the British researcher, who was freed from "superstitions" and could fully exploit the caves. For other writers, however, Aboriginal knowledge could be practical and empirical. Lang noted that "the aborigines are very good authority . . . in the absence of such men as Professor Jameson, or Professor Buckland, or Baron Cuvier," and when shown bones "and asked if they belonged to any of the species at present inhabiting the territory, they uniformly replied, *Bail that belongit to Kangaroo, Bail that belongit to emu*, and c. and c.,"[65] implying these were animals unknown today. Other scholars emphasized myth, with Aboriginal Dreaming stories of beings in caves and watercourses like the Bunyip, possibly being interpretations of fossils as ancestral memories of engagement with ancient animals or unknown creatures that still lived in the Australian interior.[66] Understandings of fossils drew on but subordinated Aboriginal knowledge and cast Australia's landscapes, people, and animals as strange and threatening.

FIGURE 1.6. Caves in the Wellington Valley. Solander Box A33 #T79 NK12/41. Courtesy of the National Library of Australia.

Much of the Aboriginal contribution was occluded when the fossils were brought to European collections. Moving quite strongly against Lang's interpretation and use of Aboriginal knowledge, British writers generally agreed the Wellington Caves fossils paralleled modern Australian animals. They included remains of several types of kangaroo (some of which could be new species, owing to their "gigantic stature"[67]), wombat, something akin to a Tasmanian devil, and kangaroo rats. However, there were more confusing specimens, in particular a large robust limb bone. Jameson and Clift compared it with bones of animals in European collections, arguing it "bears a great resemblance to the radius of Hippopotamus."[68] Pentland meanwhile stated it looked like the bone of an elephant about "one-third smaller than the ordinary Asiatic elephant."[69] Owen in 1843 described similar specimens and believed they "incontestably establish the former existence of a huge proboscidean Pachyderm in the Australian continent."[70]

The idea of a primordial Australian elephant had puzzling implications for European observers. It indicated that Australia had possessed animals similar to those of the Old World in the past, rather than its current distinctive fauna. Clift regarded the ancient presence of gigantic animals as "a fact of high importance, when we recollect that the quadruped population of New Holland is at present but meagre, the largest species being the kangaroo."[71] Owen took this further, stating the bones "tell us plainly that the time was when Australia's arid plains were trodden by the hoofs of heavy Pachyderms; but could the land then have been, as now, parched by long-continued droughts, with dry river-courses containing here and there a pond of water?"[72] The rhetorical question implied Australia had a wetter climate in the deep past, and the current imagining of the land by Europeans as dry and impoverished was only a temporary state. Through management and control, the country could again potentially support large animals.

It is easy to ridicule these theories of ancient Australia being inhabited by large placental mammals, and they certainly represent an imposition of European understandings of life onto a very different environment. But a striking point is how quickly these suggestions that Australian bones may have belonged to mammoths and hippos were revised. More Australian fossil bones were shipped to London, and increasingly to Richard Owen, who built up a powerful base at the Royal College of Surgeons (Mitchell increasingly used him as the main conduit for fossil bones). After receiving bones from the German naturalist Ludwig Leichhardt, purchased from Aboriginal Australians near Darling Downs, Owen revised his opinion. The large bones were not the remains of a pachyderm, but an unknown giant marsupial, to be named *Diprotodon*. While the name was fairly unassuming (simply referring to the animal's two protruding front teeth), the creature was of central importance. In an 1844 paper Owen wrote how for "extinct as with existing Mammalia, particular forms were assigned to particular provinces."[73] Australia had therefore always had a local marsupial fauna, and the ancient world was as locally varied as the present. The ancient fossil world could be known through central collections, but life's history needed to be understood through specific places.

The investigation of the bones of fossil mammals through aligning different forms of knowledge and reconciliations between the local and the global would continue throughout the nineteenth century. The world building of paleontology simultaneously marked particular places through local histories, but they were understood through global comparisons. Fossils were excavated through industry, agriculture, exploration, and settlement, and defined the modern landscape through universalized ideas of ancient creation and concerns for local specificity. Caves in Germany and southeastern Australia, mines in Britain, agricultural and mineralogical sites in North America, and melting tundra in northern Asia all became entry points to the former world. These sites gave a deep antiquity to particular regions (often inflected by cultural stereotypes, modern economic and political engagement, and invocations of local myth and folklore) and reinforced ideas of transformation in the deep past. A world of fossils was constructed in the years around 1800, connecting the universal and the local, the mythic and the scholarly. The deep past became a way to think about the specificity of places and transformations across large geographies and stretches of time.

CHAPTER 2

Defining the Mammals

Hierarchy and Diversity in the Natural World

HISTORIANS OF PALEONTOLOGY HAVE OFTEN FOCUSED ON THE BEASTS OF THE DRIFT AS representing the initial flowering of interest in the fossil world and highlighted the *Megatherium* (discussed in the next chapter) as an originary case in paleontology.[1] However, contemporaries cited some less dramatic creatures: the beasts of Montmartre, fossils excavated by gypsum miners and restored by the figure heralded as the "founder" of the discipline, Georges Cuvier, chair of comparative anatomy at Paris's Muséum d'Histoire Naturelle. Cuvier worked these fossils into two new genera of mammals: "the unarmed beast" *Anoplotherium* and "the old beast" *Palaeotherium*. Louis Figuier wrote that these were "the first fossil mammals to be restored by our immortal naturalist," who "gave the signal, even the model, of innumerable researches which were soon undertaken across all Europe for the restoration of the animals of the ancient world."[2]

The beasts of Montmartre were unquestionably ancient and completely extinct. However, they were not strange primeval monsters, but functioning animals like those observed in the Muséum's workrooms and menagerie. In Britain Hugh Miller wrote, "The extinct Pachydermata of the Paris basin . . . are now as familiar to the geologist as any of the forms of the existing animals."[3] They were conceptualized as lost links in the chain of life and were crucial for constructing animal taxonomies and a new history for the earth. Knowing them integrated different forms of knowledge. Bringing the beasts to life linked two preoccupations in natural history. The *Palaeotherium* and *Anoplotherium* represented a lost world of organisms in the depths of the past. Understanding them linked modern mammals with this past, placing them on scales and systems of life that stretched through the animals "up" to humans.

The gushing statement from Figuier heralded the role of "the great Cuvier." From the French-speaking town of Montbéliard, then in the Duchy of Württemberg, Cuvier conducted research in Normandy after his home was annexed into France during the Revolutionary Wars, arriving in Paris in 1795. In a career traced by Deborah Outram, Martin Rudwick, and others, Cuvier played scientific authority; local, national and international politics; and appeals to public audiences to become a dominant figure in natural history.[4] In 1802 he was appointed chair of comparative anatomy at the Muséum d'Histoire Naturelle, a position he maintained until his death in 1832 (surviving the rise and fall of both the Napoleonic Empire and the Restoration monarchy).

Gaining a position at what French scholars simply called "le Muséum" placed Cuvier in an institution that epitomized contemporary drives toward centralized collecting and the reordering of natural history knowledge.[5] The Muséum originated in the royal gardens and cabinet of the Bourbon monarchs, which were transformed over the eighteenth century into a huge scholarly institution. While royal, state, and municipal collections of natural history specimens and live animals and plants were widespread at this time, the extent of French royal support and France's European and global entanglements—and the prestige of the French language—meant these collections could assume an important international role.[6] Scholars such as the Comte de Buffon and Antoine de Jussieu used control over the royal collections and networks of scientific patronage to promote new interpretations of nature. As the Jardin du Roi and Cabinet du Roi were seized for the French people early in the Revolution and transformed into the Jardin des Plantes and Muséum d'Histoire Naturelle, it provided a base where different branches of knowledge could be directed. Muséum scholars aimed to collect, order, and display the whole of nature in laboratories, exhibition spaces, outdoor gardens, monuments, and a menagerie of live animals (which were "liberated" from the royal menagerie at Versailles), forming the archetypal Latourian "center of calculation."[7]

Cuvier built up a formidable power base through controlling of institutions and gathering specimens. Some specimens were inherited from the old royal collections, some were donated from collectors in France and beyond, and others were seized from other collections by French soldiers during the Revolutionary and Napoleonic Wars. Some specimens were part of the "paper museum" of drawings, notes, and illustrations, others were entire skins and skeletons, and some were live animals (who would become skins, skeletons, and measurements upon their deaths), forming a combination of work and material which Rudwick describes as an assembly line of analysis, reconstruction, and tabulation.[8] The specimens within the Muséum linked past and present life. Cuvier's first major paper in 1796, "Species of Elephants," compared specimens of African and Asian elephants taken from Dutch collections with "the bones of enormous animals . . . found under the earth in Siberia, Germany, France, Canada, and even Peru"[9] to argue that fossil bones represented a species of elephant that no longer lived. Questions of "What was this primitive earth? What was this nature that was not under the authority of man?"[10] carried across Cuvier's career. The beasts of former ages were lost species distinct from modern animals; they had been destroyed by a succession of cataclysms that had annihilated swaths of life and radically altered the earth's surface.

Paris was also an important center for industry as well as scholarship, and this further promoted fossil work. Gypsum mines in the north of the city had been exploited since the Middle Ages to provide material for "plaster of Paris." A large working-class mining community developed around the hill of Montmartre, long before the area became renowned as a pleasure ground. Like the Muséum, the mines were reorganized and expanded during the Revolution, as a Benedictine monastery was dissolved and its lands opened for excavation.[11] Gypsum was excavated in great bivouacs of earth and stone, often fifteen meters tall, likened by contemporaries to great "cathedrals," but vulnerable to collapse (killing miners and causing landslides). Disused tunnels were also reputed to be the homes of vagabonds and thieves.

As well as economically valuable gypsum and rumored bandits, the mines contained strange bones. Cuvier wrote how fossils were so abundant "that there is never a day when the laborers who work in the quarries at Montmartre, Mesnil-montant, Pantin, Argenteuil, and other nearby villages fail to find some in the blocks that they shape into building stones," and "several connoisseurs [*curieux*] of this city have long collected the bones for their museums [*cabinets*]."[12] While sometimes complaining about miners damaging fossils, Cuvier also showed appreciation for their work, talking about the "care" with which they preserved fossils, and described incidents where he himself ventured into the mines to see particularly noteworthy (and difficult to extract) specimens.[13]

The fossils of the gypsum quarries were exceptionally ancient. Cuvier noted that "most remains of quadrupeds hitherto found occur in very loose deposits" in caves and drift formations. While these bones were relatively well preserved, dating them precisely was difficult. The Paris fauna, however, was "embedded right in the interior of the stone"[14] and had to be as old as the rock itself. Work across specialisms in the Muséum was a boon for Cuvier, especially through his collaboration with Alexandre Brongniart, who managed a large state-run porcelain manufactory in Sèvres (taking economic advantage of the local minerals) and was to become professor of mineralogy at the Muséum in 1822. Cuvier and Brongniart arranged the rocks of the Paris basin into a geological column, with distinct strata of progressively older ages.[15] The layering and alternation of the remains of marine and freshwater mollusks showed sedimented time and radical shifts as the land was connected to the sea, then under freshwater lake deposits, and then fully inland. This demonstrated Cuvier's view of "revolutions" in earth's history, with periods having defined characteristics and forms of life. The layers in the "gypseous formation" where the mammal fossils were found also contained fossils of freshwater mollusks and palm trees, which indicated periods of freshwater inundation and a tropical land of shallow lakes and marshes.[16]

The antiquity of the fossils from the gypsum, their embeddedness in the rock, and their fragmentary nature made interpreting them difficult. Cuvier presented numerous studies in the *Annales du Muséum* between 1804 and 1807, which were collected and revised in the third volume of his *Recherches sur les ossemens fossiles de quadrupèdes* in 1812. These works aimed to "follow my researches precisely in the order, or more often in the disorder in which I have found them."[17] This reference to "disorder" is significant. As examined by Gowan Dawson and Claudine Cohen, Cuvier's work on fossils was often interpreted through senses of wizardry or deduction, of bringing organisms to life from fragments of bone, imprints, and suppositions of laws.[18] He also often emphasized the technical and frustrating nature of the work. The restitution of antique beings was the painstaking solving of a puzzle whose pieces gradually revealed a coherent form, rather than a virtuosic

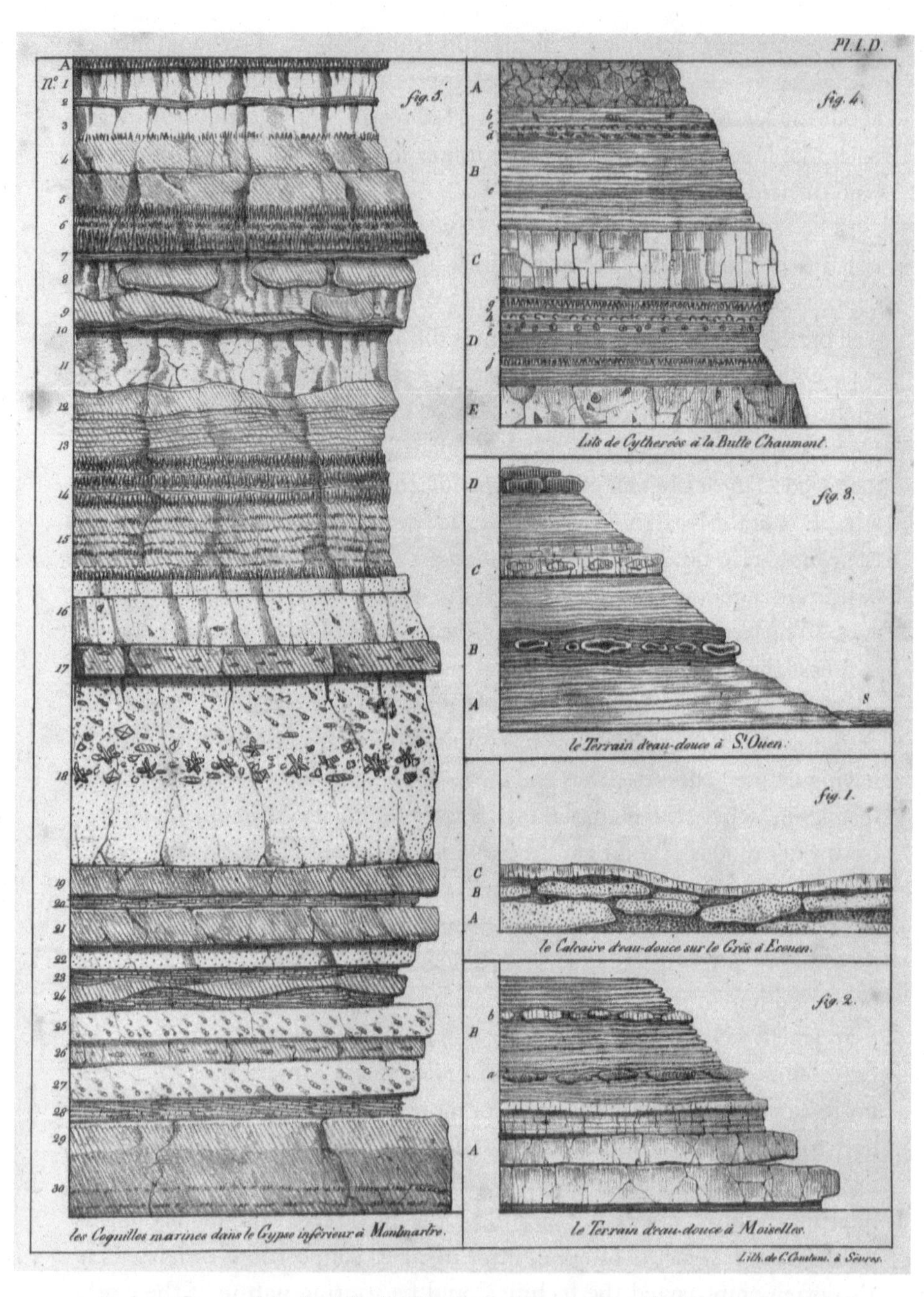

FIGURE 2.1. Strata of the Paris basin. Cuvier and Brongniart, *Essai sur la géographie minéralogique*, plate 1D. Reproduced by kind permission of the Syndics of Cambridge University Library.

conjuring trick. He wrote how he "had been given pell-mell the mutilated and incomplete debris of several hundred skeletons belonging to twenty types of animal," and needed to solve their relationships:

> I did not have an almighty trumpet at my disposal; but supplied with the immutable laws prescribed to all living beings and the voice of comparative anatomy, each bone, each portion of bone, assumed its place. I have no expression to describe the pleasure I felt in seeing the extent to which I could discover a characteristic, and all the more or less foreseen consequences of that characteristic would develop successively; the feet could be found which heralded the teeth; the teeth which heralded the feet; the leg bones, the thighs, all those which could reunite these two extreme parts . . . each of these species was reborn, so to speak, from these elements.[19]

Comparison and ordering of the fragmentary fossils from the gypsum linked bones of similar size and conformation and presented them as parts of a single organism. Comparisons with living animals were also crucial, especially around two features of anatomy: teeth and limb bones. Cuvier wrote, "The first thing to do when studying a fossil animal is recognize the form of the molar teeth; we can determine through this if it is a carnivore or a herbivore, and in the latter case, it can to a certain extent indicate the order of herbivores to which it belongs."[20] The limbs meanwhile showed the animal's locomotion and means of manipulating the world. Following these, other elements could be worked out, such as the presence of a tail and overall body size. All of these features needed to be in harmony, with herbivorous teeth corresponding to hooves or padded feet, and carnivorous teeth corresponding to claws and talons.

Naming was also an important scholarly act. The late eighteenth and early nineteenth centuries were a great period of renaming all the creatures of the earth.[21] The gradual acceptance of Linnean taxonomy gave opportunities to not just systemize terms, but to override earlier nomenclature. New Latinate and Greek-derived binomials drove out vernacular and Indigenous names for animals and plants, even if this was often an uneven process. New names also allowed fossil beasts to be defined in the same way as modern animals, including the *Ursus spelaeus* named by Rosenmüller, Blumenbach's *Elephas primigenius*, and *Megatherium americanum* described by Cuvier. For fossil beasts, no older names held precedence, and the names chosen

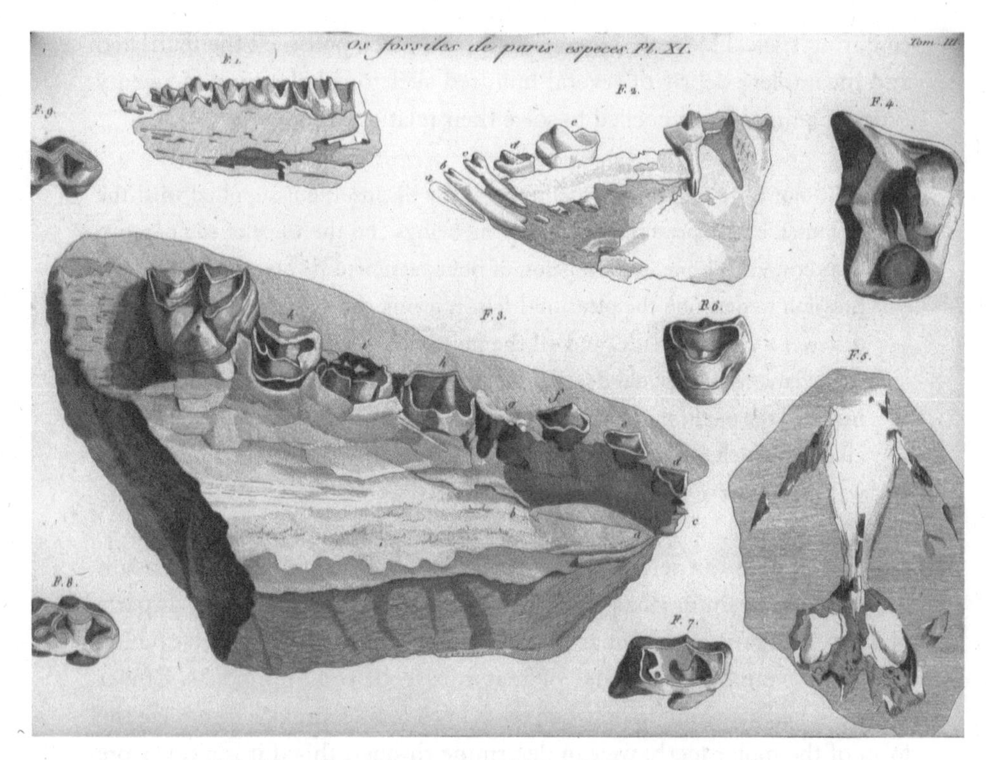

FIGURE 2.2. Fossil jaws from Montmartre. Cuvier, "Sur les espèces d'animaux: I. Mémoire," plate II. Reproduced by kind permission of the Syndics of Cambridge University Library.

inscribed the creatures with meaning. Cuvier classed one set of fossils as the *Palaeotherium* (the old beast), highlighting its great antiquity, and divided it into five species according to overall size and body shape: *magnum*, *medium*, *crassum*, *curtum*, *and minus*. Another set of fossils formed the *Anoplotherium* (unarmed beast), which lacked obvious teeth, claws, horns, or other weapons. Five species of *Anoplotherium* were also identified: *commune*, *secundarium*, *medium*, *minus*, and *minimum*.[22]

The Paris beasts were not just assemblages of bones, but functioning creatures. To reconstruct them, fossils were compared with specimens across the Muséum. Analogies were based on form, but were also full of cultural and symbolic values. There was a continuous point of comparison for the *Palaeotherium*. Cuvier wrote of the largest of these, *Palaeotherium magnum*, that "nothing is easier than representing this animal in its living state, for

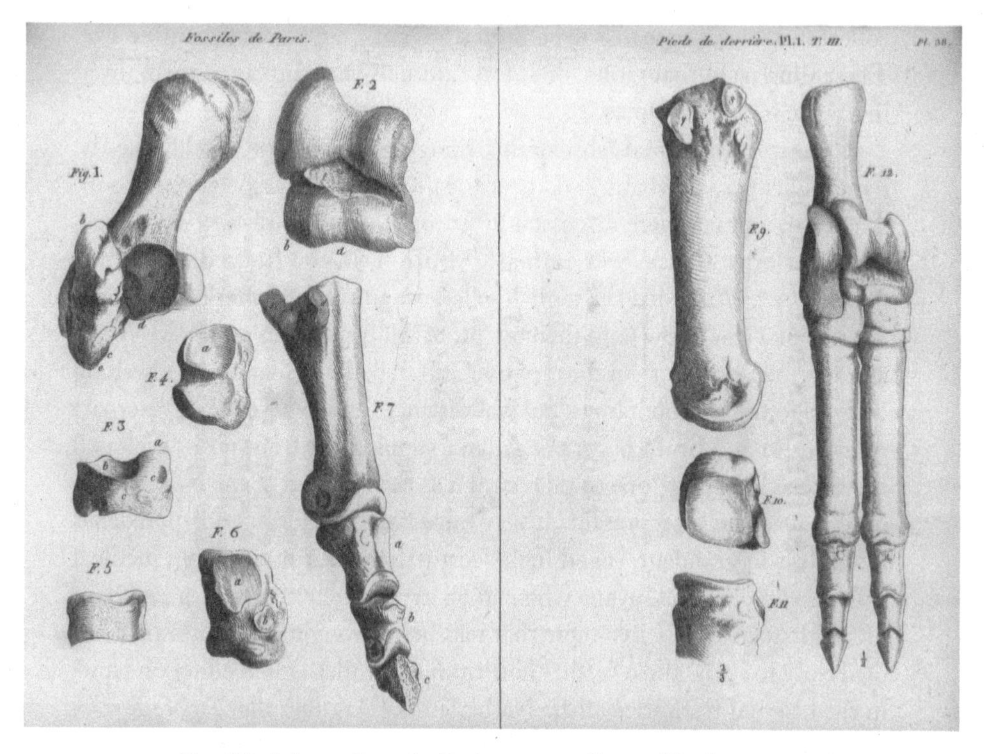

FIGURE 2.3. Fossil limb bones from the Paris gypsum. Cuvier, "Sur les espèces d'animaux: 1. Mémoire," plate 38. Reproduced by kind permission of the Syndics of Cambridge University Library.

one must only imagine a tapir as big as a horse."[23] The analogy with the tapir had important connotations. First, this was an exotic animal under Cuvier's conceptual authority. In 1804 he wrote "Osteological Description of the Tapir," which argued that the creature, while "one of the most interesting species,"[24] had been little studied by naturalists, to the extent that no one even agreed on the number of teeth it had. Cuvier's account, based on consultation of numerous texts and observations of three tapir skulls in the Muséum's collections, demonstrated that the creature had forty-two teeth, "canines which resemble those of carnivorous animals," and a "striking" arrangement of the skull and nose bones which "recall that of the elephant, indicating the presence of a mobile trunk."[25] This was followed immediately by an article ("On Several Teeth and Bones Found in France, which Appear to Belong to Animals of the Tapir Genus") describing some teeth in private collections which appeared to be from a "gigantic tapir" of elephantine size. These illustrated that tapirs, or at least animals very similar to them, lived in France in ancient times.[26]

The tapir, an animal "absolutely belonging to the New World, like the llamas, vicuñas, capybaras, and peccaries,"[27] had other associations. South American animals were discussed in European natural history as strange, and inferior to Old World animals.[28] Buffon, in the first edition of his *Histoire naturelle*, called the tapir "the largest animal of America," but only the size of a small cow (and thereby "proof" of his dictum that New World animals were smaller than those of the Old). It had a similar marsh-dwelling lifestyle to the hippopotamus, but was distinct from its African counterpart in its "mild and timid nature."[29] A later supplement from 1782 developed these ideas by saying "one could regard it as the elephant of the New World, only it nevertheless represents it very imperfectly by form, and approaches it still less in grandeur,"[30] and included further descriptions from medical officers stationed in Guyana who had observed the live animal, an account of the dissection of a live tapir that had been brought back to France but died close to Paris, and a further note from the work's Dutch editor on tapirs in the colonial territories of the Netherlands.[31] Arguing that *Palaeotherium* was tapir-like cemented an idea that creatures living in distant eras were like animals in regions now physically and climatically distant from Europe.

The other Paris beast, *Anoplotherium*, had no obvious single parallel. Not all the bones of the different species were available, but missing parts could be plotted following "the example of geographers," who indicated conjectural features on maps through dotted lines.[32] The creature was restored through varied analogies with modern animals and imaginings of the ancient environment. The largest species, *Anoplotherium commune*, was defined by "its heavy build, its short stout limbs, and above all its enormous tail" (only matched in proportion to its body by the kangaroo). It seemed somewhat like an otter, but its teeth indicated it was a plant eater "like the water rat, the hippopotamus, and all types of boar and rhinoceros," and probably "sought out the roots and succulent stalks of aquatic plants." Being "a swimmer and diver," it might have had "the sleek hair of the otter," or perhaps was hairless like the hippopotamus. Meanwhile, the *Anopolotherium medium* was "light like the gazelle or roe deer, it must have run rapidly around the marshes and ponds," and was "probably a timid animal, with large mobile ears like those of deer, to alert itself to the least danger." Finally, the tiny *Anopolotherium*

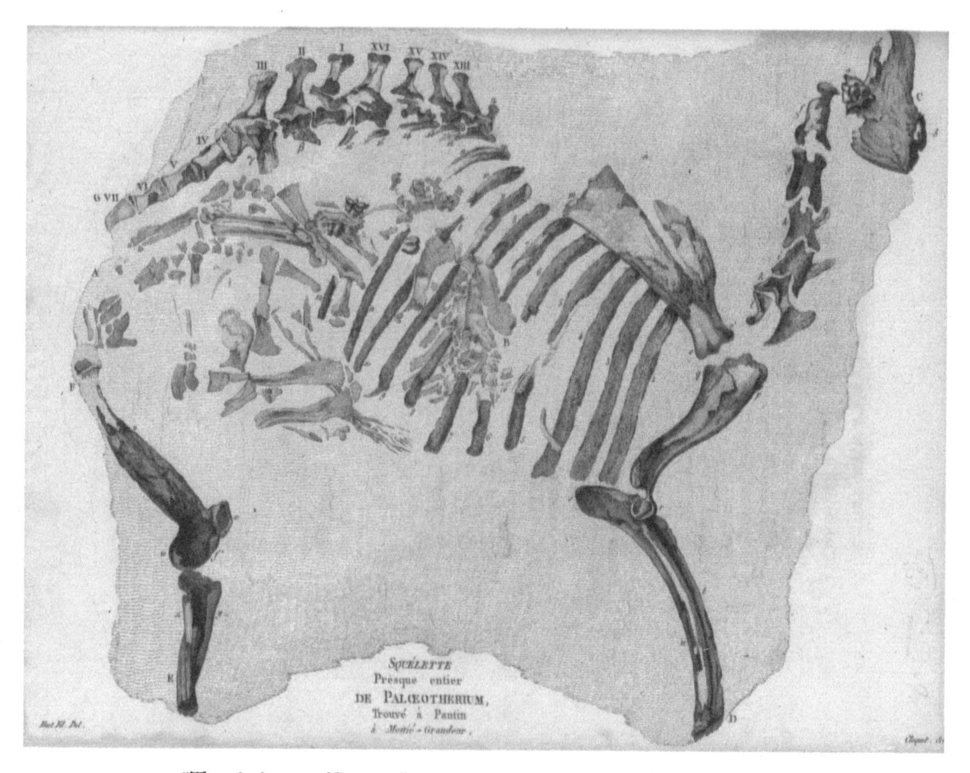

FIGURE 2.4. "The skeleton of Pantin," a complete *Palaeotherium* fossil, initially interpreted by gypsum miners as being that of a goat. Cuvier, "7. Mémoire" in *Recherches sur les ossemens fossiles*, vol. 3, final plate. Courtesy of Staatsbibliothek zu Berlin/ Preußischer Kulturbesitz/Abteilung Historische Drucke, Shelfmark: 4'Mi 5555–3:R.

minus was like "the hare—the same size, the same proportion of its limbs must have given it the same strength and speed, and the same type of movement."[33] A whole land of creatures, with analogies to modern animals, but still unique, was imagined from the fossils of the gypsum.

Cuvier's papers were filled with high-quality engravings of bones from the mines, arranged by type and size. In later papers these were bolstered by prize specimens backing up his earlier conjectures, complete skeletons of *Anoplotherium commune* and *Palaeotherium* found in the gypsum quarries. Finally there was a series of drawn reconstructions. The first were outlines of different species of *Anoplotherium*, apparently by Cuvier himself, attaching musculature appropriate to a water-dwelling animal onto the bones. The second was a cruder, and somewhat more fanciful, sketch by Cuvier's

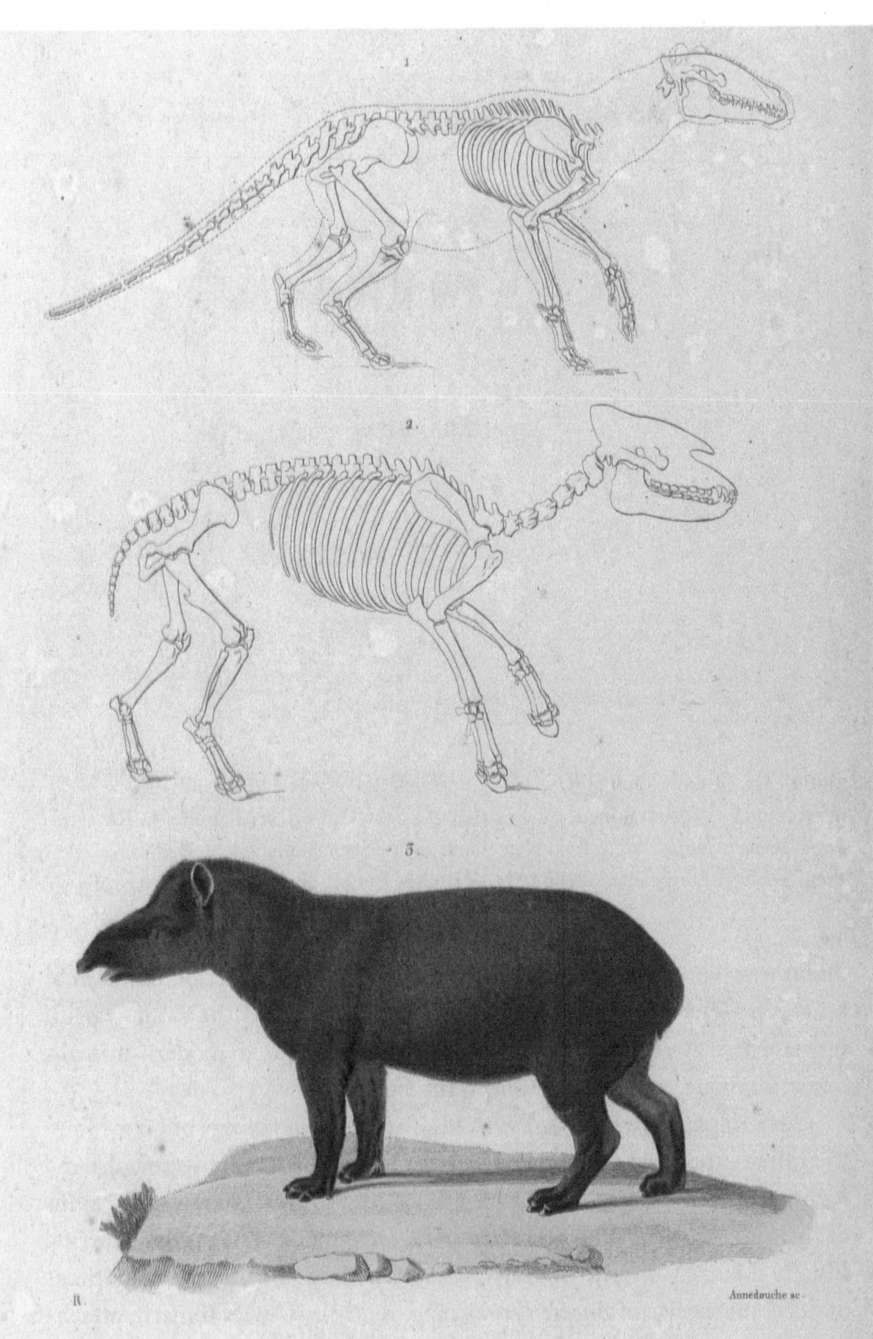

R. Annedouche sc.

1. *ANOPLOTHERIUM COMMUNE.* 2. *PALÆOTHERIUM MAGNUM.* 3. *TAPIR COMMUN D'AMÉRIQUE.*

◂ FIGURE 2.5. "Ordinary Pachyderms," depicting *Anoplotherium commune*, *Palaeotherium magnum*, and an American tapir as linked forms. Cuvier, *Règne animale distribué d'après son organisation* (1836–1849), vol. 1, plate 82. Courtesy of the Wellcome Collection.

FIGURE 2.6 (*above*). Laurillard's images of the Paris fauna, placed in a lakeshore environment in Page, *Past and Present Life*, 156. Author's collection.

secretary Charles Laurillard, showing the Paris animals in outline. Interestingly, it was Laurillard's reconstructions of the Paris animals that were most widely circulated, sometimes reprinted directly and at other times depicted within ancient landscapes. The illustrations rendered the fossils into visible animals.

Yet the Paris basin creatures, for all their lively reconstruction, were now gone. The idea that the world had formerly been inhabited by now disappeared animals has been highlighted as one of the principal conceptual contributions of the study of fossils—even though resistance to the idea of extinction persisted.[34] Paleontology depended on former worlds having vanished, and new forms arising. Those claiming expertise over fossils could restore lost creatures to life and see echoes of them in modern animals and landscapes. These acts were fraught with difficulty, requiring technical analysis, deduction of mysteries, and imaginative leaps (as well as manual labor disinterring, preserving, and arranging the specimens). They allowed the ancient animals and their lost world to be understood and related to the present. The *Anoplotherium*, *Palaeotherium*, and American tapir could all be seen as "Ordinary Pachyderms," interpreted in related manners and understood under the same framework.

Taxonomy as Superiority among the Mammalia

The beasts of Montmartre were part of wider discussions in the years around 1800, as the category of "mammal" was itself constructed by European scholars. This was a slow process, as seen in Cuvier's own writings. In the 1800s he still referred to the Paris creatures as "quadrupeds," an earlier term for four-footed animals. This and other terms were slowly, if unevenly, edged out of scholarly classifications and popular discussions over the following decades. These taxonomic reorderings represented control over the natural world, as products, specimens, and creatures were sorted into new systems, and animals, plants, minerals, and humans were assigned value and position. However, the regularity of this process should not be overstated. Harriet Ritvo in particular has explained how new scholarly divisions of the animal world frequently drew on popular and vernacular ideas, even as they attempted to dislodge them.[35]

Definitions of *mammals* reworked traditions that hairy, four-footed animals that gave birth to live young were connected to one another and were the closest animals to humans. In vernacular usage, *beasts* in English, *bêtes* in French, and *Tiere* in German were all terms used to discuss these creatures, while in scholarly networks the term *quadrupeds* derived from Aristotelian systems, characterizing animals by their four-footed locomotion and then dividing them further by whether they laid eggs or gave birth to live young.[36] Folk and classical taxonomies were revised across the early modern period. For example, John Ray's *Synopsis Methodica Animalium* of 1693 divided the viviparous quadrupeds into two large groupings: the hoofed "ungulates" (including horses, ruminants, and rhinos) and the digited "unguiculates" (including carnivores, primates, and "anomalous" creatures, mainly insectivores).[37] Ray's work followed an important principle in natural history: the subordination of characteristics. This identified particular anatomical features as "dominant" and used them to class and divide groups—although of course the selection of which characteristics were dominant was strongly dependent on cultural assumptions.

The Mammalia were named by the Swedish naturalist Carl Linnaeus in the 1758 edition of his *Systema Naturæ*, positioning them as the highest class in a general table of nature, above the Aves (birds), Amphibia (mainly comprising what today would be classed as amphibians and reptiles), Pisces (fish), Insecta (insects), and Vermes (worms).[38] The name, literally "the breasted ones," referred to their suckling of young. The gendered implications of this are discussed by Londa Schiebinger in one of the few studies on the definitions of mammals, who argues that this classification reinforced contemporary gender ideologies. Mammalia signified the group's abundance and fertility, and naturalized visions of maternal females.[39] In addition, humans were expressly included within the group, with *Homo sapiens* at the summit of Mammalia. This was an inclusion that other scholars, notably Buffon, argued against vociferously, but it had important implications: it placed humans within the same categories as the rest of nature; and the subdivision of humans into four main "varieties" (plus a further range of "monstrous" types) connected the ordering of the natural world to definitions of humanity.[40]

These new taxonomic systems can be taken as a movement away from classical and religiously derived visions of the natural world, although—as will be seen throughout this book—Christian and classical notions had long, if often subterranean, persistence in the valuation and comprehension of animals. Another explanation of changing taxonomic mindsets highlights the expansion of global trade and empire in this period, as huge varieties of plants and animals were observed around the world and taken to European collections and trade ports.[41] This mass of unfamiliar material required new systems of classification to understand it and assess how it might be exploited. And these systems needed to be fluid enough to incorporate any new discoveries. Animals and plants were stamped into categories where Europeans were dominant, as "one by one the planet's life forms were to be drawn out of the tangled threads of their life surroundings and rewoven into European-based patterns of global unity and order."[42] Naming and systematization was as much about control as about understanding, and control required flexibility.

New animals were not just found living in territories beyond Europe. They were also excavated as fossil remains from the ground—and these remains were ordered into the same classificatory systems as living creatures. The mammals of the Paris basin and creatures of the drift were studied and named in exactly the same institutions—scholarly collections like the Jardin des Plantes—that studied and named modern creatures. They were understood according to the same principles: body shape, locomotion, and means of acquiring food. Taxonomies needed to be flexible enough to accommodate living animals like the tapir, orangutan, and pangolin, and newly constructed fossil species like *Palaeotherium* and *Anoplotherium*.

The connections between fossil and living mammals can be seen in Cuvier's taxonomies of mammalian life. Marshaling his position at the Jardin des Plantes, Cuvier ordered the whole of the animal world in *Le règne animal*. The 1817 edition defined *Mammifères* through their superiority: "The mammals must be placed at the head of the animal kingdom, not only because it is the class to which we ourselves belong, but even more so because they enjoy the most multiple faculties, the most delicate sensations, the most varied movements, and are where the combination of all the properties appear combined to produce a more perfect intelligence, more fertile in resources, less slave to instinct and more susceptible to perfectibility."[43] Cuvier emphasized some features of Linnaeus's definition, citing their "fertility" and place at the summit of animal life, and incorporating humans within the group. He also added other features, praising their bodily harmony, intelligence, and improvability. Mammals more than any other group displayed order within nature.

Finer classifications of mammals into subcategories often focused on limbs and teeth, reflecting primary functions of movement and diet. The distinctive, but still apparently homologous, limb bones of mammals enabled different groups to be linked, while illustrating whether they were swimming or flying beasts, gracile runners, ponderous walkers, or powerful predators. Teeth also showed divergence and unity, ranging from the shearing bites of meat eaters, to the grinding teeth of herbivores, to the reduced teeth of insect and fruit eaters. Yet beyond this, mammals had similar types of teeth, with varying combinations of incisors, canines, premolars, and molars. This split the mammals from the reptiles, fish, and amphibians, which generally had one type of tooth repeating in the jaw. Moreover, each mammal species seemed to differ in the number, combination, and structure of their teeth, and could be classed through numerical dental formulae. Teeth simultaneously split the mammals into groups, reflected their diets and lifestyles, and showed the unity of the class.

Studies of limbs and teeth also linked extant and fossil mammals. Robust limb bones and hard teeth were by far the most common fossils. Indeed, most vertebrate fossils collected before the nineteenth century had been isolated limb bones or teeth of large mammalian herbivores like hippo, elephant, mastodon, and rhino. Richard Owen, in his *Odontography* wrote how teeth were "so intimately related to the food and habits of the animal as to become important if not essential aids to the classification of existing species," and for fossil species "the teeth are not unfrequently the sole remains."[44] Fossil teeth were durable and were regarded as distinctive enough to define animals, ascertain their diets and lifestyles, and position them on the scale of life.

Ideas of "perfectibility" meant that those interested in understanding the mammals focused on another aspect of their anatomy: the brain and nervous system. The assumed superiority of mammals rested on intelligence. Analyzing the behavior and learning abilities of living mammals was one way of investigating this, and was undertaken within the Paris menagerie and similar institutions, linked with studies of human intelligence.[45] However, investigations of intelligence increasingly depended on studies of the brain. According to the British naturalist William Charles Martin, the brain was crucial "as it is in this mysterious laboratory that all mental operations are conducted—moreover, as according to the degree of excellence of this organ, so is the grade of the animal in the chain of being . . . we find the brain the most completely developed in the Mammalia—and, among the Mammalia, in Man."[46] The brains of animals were arranged into series, along with the brains of humans divided by these same scholars into "races." Defining the raw size, but also the conformation, of brains of humans and animals legitimized numerous valuations of superiority and inferiority.

The focus on brains led to another link between the comparative anatomy of modern animals, extinct beasts, and humans—a privileging of the skull. There is a large literature on how human skulls were accumulated within eighteenth- and nineteenth-century scientific institutions, stretching across phrenology and craniology and reinforcing ideologies of race.[47] This was connected to a similar rush for extant and fossil animal skulls, which linked racial studies and paleontology. The driving forces of human craniology also studied fossils. Johann Friedrich Blumenbach, who gave the fossil elephant of Siberia its Linnaean name, is most famous for his "Golgotha" of skulls and racial taxonomy. Cuvier also developed his own racial typology. The study of skulls therefore linked attempts to know and control humans and animals.

Unlike teeth and limbs, which were relatively common, fossil skulls were exceedingly rare. When skulls were located, they became prize specimens. Skulls of emblematic organisms became some of the most reproduced and discussed fossils, with much attention devoted to their retrieval, study, and comprehension (and they also often required a great deal of remodeling and reconstruction). On an anatomical level, skulls allowed the entire dentition and sensory apparatus of the animal to be appreciated and understood. Skulls were also given aesthetic value, being seen as anatomical gateways into more cerebral qualities. The skull could show the brain cavity and potentially the nerves, and confronting the skull of a creature was a way of understanding it as a type. Taking the skull of the *Palaeotherium*, mastodon, and *Anoplotherium* or the tapir, elephant, and wolf enabled communing with the essential characteristics of the beast across the ages.

The Difficult Chain of Beasts

The category of "mammal" was hierarchical, marking the group as the summit of animal creation. However, mammals were also defined by diversity, with different subdivisions of the Mammalia suited for particular lifestyles. This mixture, of hierarchy and diversity, meant that attempts to devise subdivisions of mammals were often tense. While scholars took certain features as markers of "highness" and "lowness," especially intelligence, the complexity of the nervous or circulatory systems, and the development of dentition, it was unclear which (if any) had precedence. Likewise, whether it made sense to class plant-eating cattle and horses as superior or inferior to meat-eating wolves or bears, much less flying bats, was not assured. Mammal taxonomy could simultaneously reinforce, but also raised doubts in, understandings of natural hierarchies.

The way uncertainty combined with hierarchical views can be seen in how two scholars at the Jardin des Plantes attempted to subdivide the Mammalia. In successive editions of *Le règne animal*, Cuvier organized the mammals according to "dominant" features of limbs, constructing a scale moving from animals with highly differentiated digits and opposable thumbs (starting with humans and moving on to primates), then to "ongicules" with separate digits (like carnivores and rodents), intermediate types (the pachyderms, a large category of thick-skinned herbivores), ungulates (moving through goats and cows, to the "solipedes" or horses), and finally swimming seals, dolphins, and whales. While arranged as a scale, Cuvier's vision was nonhierarchical. With the exception of humans, there was no indication that any mammal groups were intrinsically superior or inferior due to their limb conformation. They were simply distinctly formed to move and live in particular ways. Elements of Cuvier's model were deployed throughout the nineteenth century, such as in Étienne Geoffroy Saint-Hilaire's *Cours de l'histoire naturelle des mammifères* (despite their differences over transformism)[48] and Demarest's *Mammalogie* of 1820–1822, which organized the mammals according to Cuvier's schema, and included fossil organisms like *Anoplotherium* and *Palaeotherium* in the same categories as still living animals.[49]

Despite Cuvier's authority, his nonhierarchical arrangement of the Mammalia was often contested. Many commentators preferred the system of another French naturalist, Henri de Blainville. While most of the literature on French natural history has focused on Cuvier, Blainville was in many respects equally significant—and a distinct rival.[50] Cuvier patronized Blainville's early career, but the two soon had a disagreement, after which Cuvier blocked Blainville's access to specimens at the Muséum. It was only through intercession by the minister of public instruction that Blainville was eventually granted the newly established chair of mollusks and worms following the death of Lamarck, and he eventually succeeded Cuvier as chair of comparative anatomy in 1832. For much of his career, Blainville attempted to undo Cuvier's legacy, opposing his interpretations (something which will be seen in future chapters, where Blainville's reconstructions of fossil animals almost always rejected Cuvier's earlier studies).

Blainville's taxonomy of mammals was based on superiority and inferiority. He divided mammals into two groups based on their reproductive systems: the marsupial Didelphida, mainly found in Australia; and the Monodelphida, the placental mammals, with the marsupials presented as lower than the placentals. Toby Appel, in one of the few works on Blainville, describes his taxonomy as an attempt to work against the Cuvierian orthodoxy, reviving a more traditional concept of nature as a single scale (potentially deriving from Blainville's Catholicism).[51] Cuvier placed mammals at the summit of the natural world, but their subdivisions were based on function rather than progress, and all were equally suited to their conditions. Blainville's taxonomy meanwhile had a clear progression and revived the scale of nature.

While Blainville and Cuvier were rivals and presented different taxonomies, there were important similarities between them. Both were opposed to evolution and transformation, both saw research on mammals as crucial to understanding the "summit" of the natural world, and both believed that extinct and living animals needed to be understood together, according to the same principles. These similarities meant it was common for scholars to combine their ordering systems. The division of the mammals into superior monodelphs and inferior didelphs was common in nineteenth-century natural history. So while many of Cuvier's categories were retained, they were mixed with Blainville's overall schema.

Whole groups were judged as "superior" and "inferior," or "normal" and "abnormal," according to how they reflected the core qualities of the mammals: complexity of teeth; development of the brain, nervous, and circulatory systems; and their degree of maternal care and sociability. Assessments were partly derived from the comparative anatomy of living and fossil organisms, but equally significant were emotional and even spiritual dimensions. Reports from travelers and animal keepers were crucial (as were reports from Guyana for explaining the tapir), and animals were linked with valuations of the environments where they lived. Scholarly taxonomies were also deeply entangled with popular and traditional ideas, with Christian and classical notions retaining a strong influence.

Interpretations of particular mammals as high or low drew from observations of modern creatures and assumptions of their history. Animals like the tapir that seemed to predominate in earlier periods of earth's development were regarded as lower on the scale of life. While evolutionism was hotly debated in these years, the idea that life's history showed progress through different orders of creation was widespread (and lurked in the background, if not often overtly stated, in Cuvier's, Owen's, and Blainville's models). However, the presence of particular animals in the past was not always a mark of inferiority. It could also demonstrate the importance of certain animal groups in earlier periods, marking them as creatures with the potential to dominate the world. In particular, the extent of elephants and carnivores in the ages of the drift potentially showed their superiority.

Connections between the present and past of mammalian life meant the lowest groups were generally agreed on, orienting around the animals of Australia, as Australian nature, flora, and fauna, was marked by European scholars as strange and unfamiliar. Taxonomic systems developed these valuations further, arguing that Australian fauna represented earlier (and potentially more primitive) phases in life's history. The platypus and echidna were particularly debated in these years.[52] Their seemingly mixed forms, and controversies over whether or not they laid eggs, threatened to destabilize clear divisions between "kinds," linking mammals with birds and reptiles. Marsupials meanwhile had a range of associations. One of the most influential aspects of Blainville's taxonomy was the break between marsupials and placental mammals, with marsupials being lower than their placental counterparts. While Cuvier had initially included Australian marsupials within his carnivore group, in later decades they were marked as distinct and inferior.[53] The discovery of what Cuvier identified as an opossum within the Montmartre fossils also marked marsupials as a group belonging to the distant past.[54] They were seen as deficient in two core features of mammalian existence—their brains, argued as less complex, and their reproductive systems. Marsupials giving birth to "premature" and undeveloped young illustrated their inferiority. However, they were not always defined as completely deficient. Richard Owen argued that marsupial reproduction reflected the arid Australian landscape, as "by the marsupial modification the mother is enabled to carry her offspring with her in the long migrations necessitated by the scarcity of water."[55] While Owen still placed marsupials in a low position (especially through analyses of brain structure), this was due to their environment.

The lowest rung of placental mammals was also conventional. This was a large group named the edentates (toothless ones), including sloths, anteaters, armadillos, pangolins, and aardvarks—a grouping which Natalie Lawrence has described as having roots in early modern taxonomies.[56] Living across a wide territory encompassing South America, Africa, and southern Asia, edentates were linked by two "inferior" features. The first was heralded by the name: edentates had reduced dentition compared to other mammals, demonstrating their lack of development. In addition, their brains and circulatory systems were often classed as undeveloped. That they lived in regions regarded as strange and primitive by European scholars was a further issue, marking them as aberrations from tropical environments. However, as with the marsupials, the absolute inferiority of the edentates was often contested, and as will be seen in the next chapter, the edentates were seen as having a more powerful history.

Discussions of the highest mammals tended to cluster around two major groupings. The first should be unsurprising; these were the primates, who were taken even in nonevolutionary contexts as the closest animals to humans. Their grasping hands, large brains, and complex social relations heralded human features. Hunting for primates was a major preoccupation among natural historians, with orangutan specimens reaching an incredibly high price, and dead gorillas acquired by the Franco-American travel writer Paul Du Chaillu being of sensational popular interest and scholarly discussion (with several being dissected by Richard Owen).[57] Primate fossils were also searched for. While Cuvier was skeptical that they could be located, a number were reported from the 1830s, meaning primates could also be positioned within the tracks of deep time.

The second "superior" group was more idiosyncratic. These were the elephants, which—while not placed at the crown of creation—were nevertheless cited as important animals. Nigel Rothfels has recently investigated how elephants played a key role in Western culture, simultaneously regarded as wise and sociable, but also dangerous and threatening to humans.[58] Elephants were seen to manifest intelligence and sociability, with grasping trunks allowing engagement with the world and their fellows. Moral assumptions around elephants often drew uncritically on classical writers like Pliny the Elder, who wrote the elephant was "the nearest to man in intelligence" and "possesses virtues rare even in man, honesty, wisdom, justice, also respect for the stars and reverence for the sun and moon."[59] The eighteenth and early nineteenth centuries saw tremendous discussion of elephant behavior, including experiments at the Muséum in Paris on whether they appreciated music.[60] Valuations of modern elephants connected with the study of their fossil counterparts. The remains of the drift showed that elephants, and their possible relatives the mastodons, had a wider former range, inhabiting Europe, northern Asia, North America, and possibly Australia (depending on how the bones at the Wellington Caves were interpreted). They were not just a powerful group in the present, but a dominant one in the past.

While naturalistic hierarchies positioned marsupials and edentates at the bottom, and elephants and primates at the top, the stages in between were subject to more debate. Bats and whales moved unpredictably in different taxonomies, with whales sometimes placed low owing to their similarity (and sometimes contested position) with fish, while bats could be placed near the top due to their mastery of flight (and some anatomical analogies with primates). However, these were tendentious. Despite tremendous interest in marine mammals, their features were regarded as perplexing, and they were often dealt with separately from other mammals.[61] Whale fossils were fairly common, but certain superficial resemblances to reptile anatomy (especially in the dentition) made them difficult to analyze. Albert Koch's *Hydrarchos*, envisioned as a great sea serpent, was eventually reinterpreted as being made up of several whale fossils, while another specimen named *Basilosaurus* (king reptile) from the southern United States, was also later redefined as an early whale (along with attempts to evade the embarrassing misnomer by renaming it as *Zeuglodon*).[62]

Whether assertions of highness and lowness were even appropriate for the masses of rodents, herbivores, and carnivores was an open question. So instead of working out granular hierarchies, scholarly attention focused on two large groups: ungulates and carnivores. The medium- and large-size herbivores showed a dazzling array of subdivisions and were potentially all grouped as ungulates, or divided into solipedes, ruminants, and pachyderms; or artiodactyls and perissodactyls. These herbivores were valued for cultural and economic reasons. Herd-dwelling creatures like sheep, horses, and bovids seemed to epitomize the social characteristics of mammals. Animals within this group were also the "most useful" to humans, either as domesticates or wild creatures to be hunted. The ungulates also demonstrated extensive variation and suitability for particular environments. Figuier wrote how "with the pachyderms and ruminants we enter an already perfected organic structure."[63] Their anatomies showed variations on a single design, with foot structures, teeth, body plans, and (in some varieties) antlers and horns, illustrating how variation could manifest. As large-bodied and robust creatures, they could be used as striking taxidermized specimens or skeletal mounts. Collecting large numbers of herbivores and comparing minute details of anatomy, dentition, and horn structure became an important project.

Carnivorous mammals meanwhile were observed with interest, but also trepidation. The power and intelligence of big cats, bears, and canids was frequently a source of admiration, but their lifestyles of killing and eating other animals reflected a low moral status. Figuier demonstrated this ambivalence, calling carnivores "the strongest and most formidable of terrestrial mammals," "organized for murder and carnage," that "spread terror around them." However, they nevertheless "have a providential role in nature to limit the number of herbivorous species," and "strange as it may appear at first sight, their disappearance from the surface of the earth would cause real disorders."[64] That carnivores played a key—if distasteful—role in the "economy of nature" was acknowledged. However, they were often regarded with fear or even disgust and seen as competitors with humans. Owen wrote, "As the order Carnivora includes the most noxious and dangerous quadrupeds . . . it has suffered the greatest diminution through the hostility of man, wherever the arts of civilization, and especially those of agriculture, have made progress."[65] Understandings of fossil carnivores interacted with these assumptions. The remains of hyena, cave bear, lion, and wolf in the drift created an image of a dangerous and frightening former world. However, the decline of large carnivores over recent geological history indicated providence in natural history, as the world became safe for humans.

These ordering systems could incorporate the entire natural world in the past and present. They connected scholarly valuations of teeth, limbs, and brains, moral arguments around the virtues of particular lifestyles, and assessments of the qualities of different parts of the world. These systems drew together varied ways of knowing nature—especially comparative anatomy and modern animal behavior, but also symbolism and empathy. Mammal orderings simultaneously stressed progress and hierarchy, variability and diversity of animal lifestyles. Ordering the world reinforced conceptual mastery over nature, but this was a slippery mastery, resting on diverse valuations of the animals themselves.

Taxonomy also linked modern and ancient nature. Modern animals could be understood only through the history of life, and the features used to define modern mammals were also the features looked for in fossils, especially limbs, skulls, and teeth. Nature was a single system, across time and space. The British writer David Page stated that there were benefits "of a moral kind that spring indirectly from the study of paleontology. There is no other science, perhaps, that tends to engender so much the feeling of community; none that connects more closely the whole of animated nature into one inseparable system."[66] Reconstructing fossil beasts like *Palaeotherium* and *Anoplotherium*, positioning them within their ancient world and connecting them with modern animals, enabled the whole of Creation to be categorized and mastered. The old beasts showed that life in the present and the past existed according to the same principles.

The Lost Animal Community of Sansan

The visions of life presented at the Jardin des Plantes and through new orderings of mammalian life connected the living and fossil animal worlds. Networks of interest could stretch far—to Guyana, Siberia, and the interior of North America. However, it could also take more localized forms, examining fossils not just as representing individual creatures or positions on a scale of creation, but as assemblages of animals living together. These ideas of animal communities were also an important way in which the history of the mammals became understood, and to examine this, it is useful to trace the example of Edouard Lartet, a lawyer in the southern French department of Gers.[67] After studying in Paris, Lartet set up a legal practice and became a notable within small-town society, dispensing legal advice to local peasants in exchange for unearthed objects, including "stone axes, ancient medals, and even bones and shells." A no doubt embellished account published after Lartet's death, in the idiom of a folktale, described one particularly important offering: "One day, a peasant took a large fossil tooth to Lartet, which had been found on the hillsides bordering the valley of Gers. Other (and possibly most) collectors would have contented themselves to place this curious thing in a cabinet to be seen by visitors. But Lartet, who had become interested in scientific matters during his stay in Paris, to whom long and fruitful readings had opened the most varied horizons . . . wanted to draw all that this mysterious tooth could teach an investigative spirit."[68] Lartet's knowledge and "zeal" had been opened by the learned centers of Paris. However, gaining material required connections with the locality.

Following this, Lartet organized excavations in several southern French sites, the most productive of which was at Sansan, close to Gers. Lartet recalled being informed about the "camp de las Hossos" (noting the name, meaning either "field of bones" or "field of ditches" in the local Occitan language) and reported on stories around the site. He wrote how "fragments of bone brought to the surface of the soil must have been noticed in the past by the inhabitants of the country," and local peasants regarded these bones and shells as "the work of the devil," who, being "jealous of the creations of the good Lord, also wished to make animal forms, but could not succeed in giving them life and movement; the debris of these imperfect sketches were buried in the depths of the ground."[69] While Lartet did not endorse these stories, he also did not refute them: fossil legends were part of the picturesque locality and gave knowledge for excavations.

Lartet employed peasants and quarriers to dig and excavate fossils of proboscideans, rhino, paleotheres, the claws of what appeared to be edentates, and most strikingly of all, the jaws of fossil apes. While there were some large and dramatic specimens, Sansan's significance rested on the diverse assemblage of creatures that had lived together in the deep past. The "animals buried in Sansan" mixed species from different climatic zones. Some had "living analogues only found in the hottest regions of our continents," some would be "considered today as essentially northern," while "most approach species which still live in a wild state in our temperate climates."[70] This mixed fauna showed the richness and abundance of ancient France.

While excavations depended on local knowledge, Lartet gained support from Parisian institutions and sold a large collection of mammal fossils to the Muséum in 1845. Laurillard described these as a crucial acquisition, as it "would be infinitely regrettable if Lartet's collection passed to a foreign country, or was dispersed in various provincial cabinets."[71] The acquisition was only the start of the Muséum's work at Sansan. When the hill came under threat from foreign researchers wanting to excavate fossils, and from commercial operators wanting to dig for minerals, the Muséum and Academy of Sciences in Paris lobbied the Ministry of Public Instruction to purchase the site for the French state, preserving its paleontological riches. Excavations continued, largely organized by Laurillard, retrieving the remains of a mastodon, which was exhibited in the galleries of comparative anatomy in 1853 after a year of restoration.[72]

The diversity of ancient and modern life was discussed in Paul Gervais's *Zoologie et paléontologie françaises*, which aimed to define all creatures that had lived on the territory of modern France, compared with material from across Europe, Asia, the Americas, and Australasia. Gervais noted that while the land mammals of modern France "appear to have little variety if we compare them to such parts of the globe as Africa and India, and even more so in America and New Holland," matters were different if fossil creatures were included. The history of life in France had been more varied, involving "other orders which we completely lack today in Europe, such as the quadrumanu, marsupials, sloths, and edentates."[73] Bringing the fossil beasts into the assemblage of life showed that the whole of nature was much greater than could currently be seen. A related point was made by Hugh Miller in Scotland: "Our present mammaliferous fauna is rather poor; but the contents of the later deposits show that we must regard it as but a mere fragment of a very noble one."[74] The life of the present was a shadow of the life of the past, and to fully appreciate the extent, grandeur, and variety of mammalian life, it needed to be regarded historically and as a community, not just as a linear scale.

CHAPTER 3

Great and Terrible Beasts

Wonder, Accumulation, and Transmission

> *Pupil*: Good Lord, what an enormous skeleton! It is undoubtedly that of an elephant.
> *Master*: Examine it a little closer, and try to discern it.[1]

JUAN MIEG'S 1818 GUIDE TO THE CABINET OF NATURAL HISTORY IN MADRID WAS WRITTEN as a dialogue between an unnamed "Master" and "Pupil," who on entering the Sala de Petrificaciones encountered a huge skeleton. On closer inspection the student's expectation that it was an elephant—even "one of those elephants no longer alive today" he had read about—were confounded. The Master explained it was in fact the *Megatherium*, literally "the Large Beast," which had been found in South America and described by "the wise Cuvier" as "between the sloths and the armadillos." Its powerful jaws and claws showed that it "broke and crushed the branches of tree-roots like the elephant and the rhinoceros, after unearthing them with its claws." And its huge size showed the life of the so-called New World was not "small and degenerate," and possessed the same antiquity and "splendor of nature" as the Old.[2]

Eighteen years later an advertisement announced that "managers of collections of natural history" could purchase oil-colored plaster casts of the skull of another remarkable creature, *Dinotherium giganteum*, "the gigantic terrible beast."[3] The skull was decidedly strange, with a large nasal opening, teeth somewhat like those of the mastodon, and huge tusks descending from the lower jaw. Copies of this specimen could be ordered from the

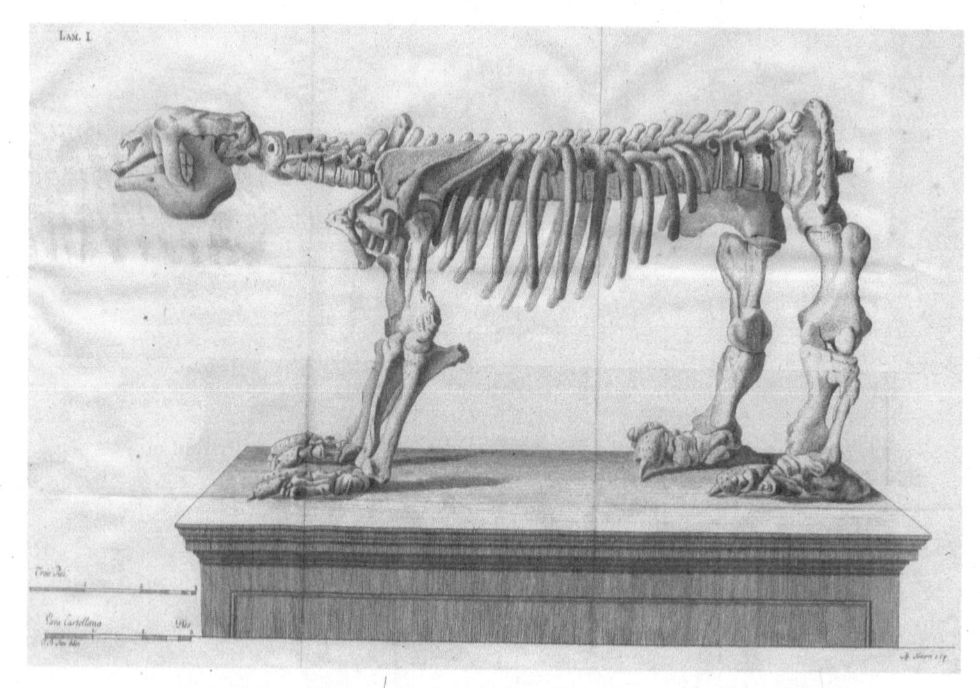

FIGURE 3.1. The Madrid *Megatherium*. Garriga, *Descripción del esqueleto*, plate 1. Courtesy of the Staatsbibliothek zu Berlin-Preussischer Kulturbesitz, Shelfmark: Mi4327.

naturalist Johann Jakob Kaup in the Grand Duchy of Hesse-Darmstadt, "guaranteed to arrive intact." The cast was marketed through controversy and mystery. The advertisement stated how "all zoologists were amazed by its paradoxical shape." Georges Cuvier, rather than being the illustrious namer of this specimen, had misidentified it, describing teeth now known to belong to *Dinotherium* as having belonged to an eighteen-foot-long "gigantic tapir." Kaup himself had presented it as a hippo-like pachyderm, after initially taking it for an edentate; William Buckland thought it was aquatic; and Henri de Blainville imagined it as an enormous dugong. Controversy generated excitement rather than doubt. The "masterfully preserved upper skull" was a scientific sensation and "has given rise to new debates during its exhibition at the Paris Academy, in order to fix the position and lifestyle of the animal."[4]

This chapter uses debates on these two fossil animals, *Megatherium* and *Dinotherium*, to examine how geographies of authority and reconstructions of fossil mammals were contested in the early to mid-nineteenth century. As discussed in the last two chapters, ideas of lost worlds of extinct animals were cemented during this time, and so were methods of rebuilding animals from fragmentary fossils. *Megatherium* and *Dinotherium* held a central importance, simultaneously going against the more confident pronouncements of earlier anatomists, but also bolstering new techniques and valuations within paleontology. Deciphering these animals and placing them within the new chronologies of life required the accumulation of fossils and control over specimens and sites. Scholars could use these creatures to cement their interpretations of life and position their own work and collections as central to understanding important organisms. This could work in various ways, through the gathering of specimens (often through imperial connections), as was the case with *Megatherium*, or through control over productive sites, as occurred with *Dinotherium*. These tensions—paleontology being a wide-ranging science conducted across imperial and international networks, but also thoroughly dependent on specific localities—were highlighted by these creatures.

Accumulation and control simultaneously made these animals unique and valuable, but also wide-ranging. Observations of original fossils were always important, as fossils are by definition individual objects, as well as being used to stand in for entire species or even faunas. Control over original fossils therefore built authority for collections. However, accumulation and control also operated through reproduction and representation. Fossils and reconstructed mounts were reproduced as high-quality engravings, line-drawn illustrations, or tables of measurements. Plaster reproductions were also widespread, to the extent that Lukas Rieppel has discussed fossil casting as an important mode of "publication."[5] Casts could also be sold, generating income for collections that otherwise needed to rely on elite patronage. Casts could also be given as gifts or in exchange; swapping casts of prominent specimens was important for building links between collections. The messy materiality of fossils, frequently fragmentary, damaged, or distorted through long ages in the ground, meant they often needed to be repaired or modeled through casting and manufacturing techniques.[6] Interacting with and debating material became essential to fossil work.

Dinotherium and *Megatherium* were widely debated in these years, across institutional, national, and imperial contexts. They had their imagery partially fixed, through full skeletal mounts and reconstructive illustrations, while still being mysterious creatures that could be interpreted in multiple ways (all of which fed into distinct visions of the paleontological past). The multiplicity of images and potential for various interpretations were important for paleontological research. Differing interpretations, while sometimes presented in acrimonious or belittling terms, did not destabilize the field, but drew in different voices. Uncertainty could raise scholarly profiles and generate excitement—as mysteries around the large bones in Madrid or the character of *Dinotherium* should indicate. These two great beasts brought together claims to authority and imagination through varied means of accumulating, controlling, and working fossils.

Constructing the Great Beast of the Americas

Megatherium, one of the great scientific sensations of the early nineteenth century, was a core example where differences in styles of accumulation and variable interpretations of a specimen were worked out. Debates on *Megatherium* simultaneously drew links between colonial territories and European scientific communities, and varied depending on styles of interpretation and traditions in valuing nature and the life of the past. The animal, often simply named "the great quadruped," is a rare example in the history of mammal paleontology that has been significantly studied, with important works by Juan Pimentel, Irina Podgorny, and Alan Rauch.[7] It is worth tracing the story again here because of the central importance of *Megatherium* to nineteenth-century ideas about the deep past, as it became an icon of fossil mammalian life in its strangeness and diversity.

The initial acquisition of *Megatherium* shows an evident example of paleontology as a colonial science. In 1787 Francisco Aparicio, the mayor of Luján in the Spanish colony of Río de la Plata, was alerted that local workers had found large bones, exposed during the dry season. Similar bones were well known to workers and herders in the area, who used them for temporary stands to build fires or for resting. Aparicio nevertheless reported the finds to the colonial capital of Buenos Aires, and a Dominican friar with naturalistic interests, Manuel de Torres, was called in to supervise an excavation. The extracted skeleton was briefly exhibited in Buenos Aires. The colony's notables marveled at its size, and Indigenous leaders said they were unfamiliar with the animal, and it must have been wiped out by their ancestors before the Europeans came.[8] The bones were then shipped to Madrid, following a decree that all important natural history specimens should be sent to the imperial capital. They were deposited in the relatively recently established Royal Cabinet of Natural History, joining stuffed and skeletal monkeys, rodents, marine mammals, big cats, elephants, and various "monsters," including a two-tailed lizard and a cyclopean cow.[9] Rather than be assembled in the disarticulated manner in which they were found, the bones were mounted by the cabinet's "artist and first dissector," Juan Bautista Bru de Ramón, who modeled the animal's missing tail bones and lower vertebrae in wood.[10]

The Madrid beast is often cited as the first time fossil bones were mounted in a lifelike pose.[11] This was certainly a difficult project, although its novelty may be overstated. Taxidermists frequently reconstructed the bodies of animals they had never seen in life, and "monstrous" mounts were fairly widespread in collections.[12] Bru was skilled in taxidermic and osteological preparation, having reconstructed elephants, sea lions, and other large animals. Bringing back a beast whose bones had been dug up from the ground was therefore not as great a step as the statement "first mounted fossil specimen" might imply. Nevertheless, the creature remained a mystery. Bru wrote a minute description of the bones, but noted how "the lights offered to me by Comparative Anatomy or Zootomy . . . disappear completely when considering this unique skeleton." All possible analogues—"elephant, rhinoceros, horse &c."[13]—were wanting. Comparative collections could give some context, but the beast itself could not be placed.

The mounted skeleton rested in the Royal Cabinet, but news of it spread slowly. Bru commissioned engravings of the bones and skeleton but did not publish his report (various reasons have been suggested, including lack of funds[14] or a falling out with the museum's administration[15]). Diplomatic networks were more important in spreading information: the US chargé d'affaires sent a sketch to Thomas Jefferson, and Philippe-Rose Roume de Saint-Laurent, a French diplomat of Grenadian origin, acquired copies of Bru's engravings while passing through Spain on a diplomatic mission and took them to Paris. The Institut de France then tasked Cuvier to describe the animal, publishing an account in the *Magasin encyclopédique* in 1796.[16] Cuvier followed his techniques of reconstructing the life of the past on what he termed "the animal of Paraguay." He argued the animal must be extinct, as something so huge and ungainly would be known from stories told by the Indigenous population. It was therefore another denizen of the former worlds.

Yet for Cuvier, this could not be a lost monster. It needed to fit somewhere in the systematic arrangement of life, according to "the invariable laws of the subordination of characteristics."[17] Cuvier and his assistants compared the illustrations with other drawings, skeletons, and live animals in the menagerie at the Jardin des Plantes. The vast collections and existing methods for comparing specimens (known from a variety of media) across time and space were important for giving authority over the animal, even though Cuvier never saw the actual bones. The most obvious comparators were large quadrupeds—elephants, rhinos, and horses, as cited by Bru. But comparative anatomy led Cuvier to structural analogues, especially in the key characteristics of teeth and digits (and it is of course possible that not seeing the massive materiality of the bones made links with much smaller animals more apparent). The closest similarities seemed to be edentates, particularly sloths and anteaters. The animal was therefore positioned among the "toothless" mammals, often understood as the most peculiar and (for writers arguing for scales of life) lowest placental mammals. It was also a decidedly South American creature. Giant edentates indicated the past life of the ancient New World mirrored its modern fauna, in large and dramatic ways. The animal also needed a fitting Linnaean name, and Cuvier posited *Megatherium americanum*: "the Great Beast of the Americas."

Cuvier's work was widely distributed, with summaries and translations appearing in Britain, Germany, and the United States. Cuvier's "scooping" of Bru could be taken as an example of an arrogant metropolitan scientist overriding a more peripheral figure. However, statements of the "peripheral" nature of Bru's research may misrepresent the status of Spanish science at this time. Much recent work has argued that Spanish natural history, and particularly the natural history of the Americas, was hugely significant internationally.[18] While the growing status of the Jardin des Plantes and the military and political power of Revolutionary and Napoleonic France (which were to be brutally directed toward Spain) were important issues, the centrality of Paris should not be overstated. It remained one point among many.

Nevertheless, Cuvier's report was received with some anger in Spain. A military engineer, José Garriga, heard of the great skeleton when a friend passed him a copy of Cuvier's account. He quickly visited the cabinet, and tracked down Bru for permission to publish his report on the skeleton. The eventual text—*Descripción des esqueleto de un quadrupedo muy corpulento y raro*—contained an aggrieved prologue by Garriga that Bru's report should be published "not only to do due justice to Don Juan Bautista Bru, but also to our nation, showing the naturalists of Spain have not been so careless as to not describe this skeleton in detail."[19] Garriga called the skeleton "one of the most dramatic, alluring and pleasant spectacles" available to naturalists. Its "corpulence and enormous size" were astonishing, and while it might be regarded as an "unspeakable monstrosity . . . this is simultaneously tempered by how well and exactly it is proportioned."[20] Garriga continued describing all aspects of the skeleton as simultaneously bizarre, but well suited for its life, imagining how its claws when "set in motion would be able to tear and rend the hardest oaks."[21] The creature illustrated the contradictions of the great beasts of the past: wondrous monsters unlike anything alive today, but needing to be understood as functional animals suited for their lifestyles.

The contrast between Cuvier's and Bru's accounts has been interpreted as reflecting a change from the descriptive "classic naturalism" to more interpretative comparative anatomy[22] (notably, Garriga's account is rarely engaged with, other than being called "a pompous description"[23]). However, rather than represent a clear shift, the differing qualities of Bru's, Cuvier's, and Garriga's accounts were to persist through interpretations of *Megatherium* and later discussions of other fossil beasts. Whether these animals should be understood in sober descriptive terms without speculation, whether wider laws could be applied to them, or whether they were objects of pride and artistic wonder were persistent features in paleontological discourse. The drives of the field, aiming simultaneously to build awe at the past while explaining a knowable ancient natural world, with strange but functional animals, gave space for varied modes of description.

Once peace had returned to Europe in 1815, other scholars attempted to interpret the bones, and align them with other conventions in natural history. Two German naturalists, Christian Heinrich Pander and Eduard Joseph d'Alton, traveled to Madrid to study the skeleton as part of a wider work on comparative osteology. The two formed a collaboration while working in Würzburg under the physiologist Ignaz Döllinger on the development of chicken embryos. Pander, from a family of German landowners from the Russian Empire's Baltic provinces, supplied funds and scholarly knowledge, while d'Alton, an engraver, produced the illustrations. Their collaboration continued afterward, although rather than studying the development of individual organisms during their lifetimes, they aimed to trace development within whole branches of the natural world.[24]

Pander and d'Alton's work drew from Romantic and organicist ways of thinking that permeated German cultural and political life in this period—often self-consciously opposed to "French" focuses on mechanics and typology. One important manifestation of this was the growth of *Naturphilosophie*.[25] While literally translated as "natural philosophy" (a fairly uncontentious term in Britain and France, referring to various studies of nature and the world), German *Naturphilosophie* was a particular research school. Promoted by figures like Lorenz Oken and Johann Wolfgang von Goethe, this was a way of thinking about nature that relied as much on intuition and empathy as meticulous study and measurement. Conceiving of the whole of nature as permeated by vital energy, *Naturphilosophie* promoted a holistic view of a world suffused with change and development. Every organism was conditioned by its strivings and life-energy during its lifetime. And each species was determined by these forces, with transmutational change being possible as organisms gained and lost developmental drive over time, either fulfilling their potentials or degenerating.

In the Madrid Cabinet, Pander and d'Alton were permitted to sketch and measure the *Megatherium* bones, eventually publishing these as the initial volume of their *Vergleichende Osteologie*, which also dealt with large pachyderms, rodents, bats, and other creatures. The Madrid beast was placed alongside extant animals and depicted in almost the same manner—its skeleton arranged in a lifelike posture, although with the important exception that living animals like the elephant and hippo had their skin and flesh marked in silhouette around the bones, whereas *Megatherium* did not. Working out the body shape of the animal was too speculative.

Additionally, Pander and d'Alton aimed to revise prior accounts. They opened their work stating that "all images of the skeleton known until now are only downsized copies of the already crude and uncharacteristic drawings of Bru," which were "unsuitable for comparison, whatever value the discoveries have for general views of the lost animal world."[26] They claimed their depictions were artistic, scientific, and accurate, asserting their own authority over the animal. As well as rejecting Bru's measurements, Pander and d'Alton also rejected Cuvier's classification, naming the animal

Bradypus giganteus, citing it was not different enough from the modern three-toed sloth (*Bradypus tridactylus*) to require its own genus. The connection with the sloth, which Cuvier presented fairly neutrally, was here laden with associations. The creature was not a functional and effective organism, but a useless monstrosity. Pander and d'Alton wrote that if "one would choose in one word how to characterize such a monstrous animal (in which nature seems to have taken on the task of surpassing all other animals in clumsiness and awkwardness), one could not choose a more apt term than '*Faultier*'" (using the German term for sloth, literally meaning "lazy beast").[27] These valuations continued: compared with the *Bradypus giganteus*, "the rhinoceros appears delicate, the elephant light and slenderly built; even the hippopotamus may be called well-constituted."[28]

Ascribing uselessness to the creature drew on *Naturphilosophie*. Pander and d'Alton argued all nature was constantly transforming, and the life of the past could be ancestral to the life of the present. Development could either be progressive, leading to diverse and perfected forms, or degenerative, manifesting as inferiority and decline. The sloth was a clear example of the latter. Pander and d'Alton cited Buffon's account of sloths as malproportioned and degenerated creatures, "tormented by nature" and fading away in the dank climate of the New World.[29] The sloth's gigantic ancestor showed this fading as originating deep in the past. Initially using their huge claws to dig in the ground, rising waters forced them out of their burrows and into the trees. With less food available, the beast atrophied to the size of the modern tree sloth, and its bones and muscle lost strength and complexity. This was not, as for Cuvier, a strange yet functional organism; or a wondrous majestic spectacle, as for Garriga; or even a perplexing specimen that could only be described, as in Bru's account. The great beast showed a transforming nature, marked by decline and decay as well as order and perfection—a notion which formed an additional pole around which the deep animal past and its relevance to the present could be understood.

Expanding the "Giant Brood" of Edentates

Early discussions of *Megatherium* focused on the specimen in Madrid to debate the character of the animal and fix particular views of natural history. Similar remains were soon found across the Americas and debated across the Atlantic. These expanded the range and diversity of the great edentates, showing a variety of forms across a wide area. The beast was not an isolated monster, but the representative of a larger group of animals, important in the history of life but now entirely vanished. The extent of the bones acquired in the Americas and the drive to reconstruct the animal enabled other institutions to assert control over the beast and fix its image through the accumulation (and reconstruction) of fossils. These moved interpretation away from Madrid and diversified and reified the animal.

The glamour of the beasts was stoked by their connections with US president Thomas Jefferson.[30] In 1796, the same year as Cuvier's and Garriga's discussions of *Megatherium*, workers digging for saltpeter in a Virginia cave found giant claws and teeth. These generated excitement in the neighborhood, and one Colonel John Stuart notified Jefferson because of his interests in natural history. Jefferson described "bones of a very large animal of the clawed-kind," likely the remains of a lionlike animal. This creature, christened *Megalonyx* (great claw), was, through comparative analogies, assumed to be three times the size of a lion (based on measurements of a lion by Daubenton in Buffon's *Histoire naturelle*) and would have "stood as pre-eminently at the head of the column of clawed animals as the mammoth stood at that of the elephant . . . [and] may have been as formidable an antagonist to the mammoth as the lion to the elephant."[31] Jefferson supposed the animal might still exist, through aligning the fossils with reports from travelers in the region of terrifying howling, a Native American carving of a lionlike beast on the banks of the Kanawha River, and assumptions of the bounteousness of the unexplored North American interior.

These views were soon revised. Caspar Wistar wrote a report on *Megalonyx* in 1799 highlighting similarities with *Megatherium*.[32] Cuvier aligned *Megalonyx* with the Madrid quadruped and other North American fossils, including a tooth found on the same site by Palisot de Beauvois with clear sloth-like qualities.[33] Yet while Jefferson's description of *Megalonyx* was altered, it remained a central referent in debates over *Megatherium*—partly because of Jefferson's celebrity, but also because it demonstrated the wide extent of these animals, living across the Americas. The majority of North American bones were from the southern states, found through the expansion of infrastructure and plantation agriculture. It was also enslaved people rather than learned scientists who seem to have found most of the specimens. *The Georgian* newspaper of 1823 described remains at Skidaway Island: "The negroes on the plantations in the vicinity, recollect to have seen them, twenty years ago, and described them to have been, at that time, quite entire; so much so as to project sufficiently to form a remarkable object, from a boat passing on the river."[34] The planter and slaver James Hamilton Couper described how more "fossil bones of the terrestrial Mammalia"[35] were excavated in the late 1830s during the construction of the Brunswick-Altahama Canal, during which African and Irish migrant laborers endured terrible conditions.[36] These bones, which seemed to include sloth-like remains, were noticed by Charles Lyell on his visit to the United States and commented on by Richard Owen, again demonstrating the importance of transatlantic connections.[37] The original context of the remains was effaced in the rooms of scholarly associations. In 1824 Samuel Mitchell reported the Skidaway bones to the New York Lyceum, comparing them with drawings from Cuvier's *Ossemens fossils* and of *Megalonyx*. While lamenting the specimen was too broken to lead to a complete skeleton, it could "satisfy us, that the United States, which contains so many relics of huge animals, may add to her 'giant brood,' the *Megatherium*."[38]

South America, where the original *Megatherium* specimen had been found, was also a key area for collecting remains of the animal. While later generations of South American scholars used the abundance of local fossils to build their authority and position within international networks (a key theme in chapter 11 of this book), this was more fraught in the early nineteenth century. Irina Podgorny has noted the case of Dámaso Larrañaga, a priest and naturalist in Montevideo who studied documents on the discovery of the Luján *Megatherium* and maintained a natural history collection in his house, which became an important conduit for scholars interested in the living and fossil fauna of the continent.[39] Likewise, the Argentinian doctor Francisco Javier Muñiz was posted to Luján and collected fossils during his residency.[40] While South American scholars were crucial for gathering fossils, and for guiding travelers from Europe, difficulties in distance, recognition, and transportation generally prevented them from taking an international role.

Europeans also settled in South America to pursue interests in natural history, including the Danish scholar Peter Wilhelm Lund who, with the help of a large inheritance and close connections with Danish commercial networks, moved to Brazil in 1832.[41] Peter Clausen, a Dane with scholarly interests in South American animals and economic interests in the saltpeter trade, informed Lund that miners had found giant bones in the region of

Lagoa Santa. Lund settled there, worked with a Danish artist, and purchased two enslaved people to excavate a large number of caves between 1835 and 1844, finding fragmentary remains of several *Megatherium*; a saber-toothed cat, which was named *Smilodon populator*; and fossils of smaller extant animals, including bats, coati, and capybara. Drawing on William Buckland and Cuvier, Lund argued the remains were deposited in the cave during "the last great cataclysm" and concluded "the animal creation of the last epoch of the earth was (at least in this part of the world, and respecting the class of mammals) more abundant and more varied than that which we can see in our days."[42] An important feature of Lund's arguments was that the ancient fauna mirrored modern South American animals. Lund kept a small menagerie of local creatures in his home, examining their behavior and anatomy to help him understand the fossil beasts. *Megatherium* was just one animal within an ancient community, and gave clues to the continent's nature.

The most high-profile reports and collecting efforts tended to be by British scholars, taking advantage of extensive British commercial networks in South America and the prominence of British scientific institutions. Charles Darwin collected many South American fossils during the *Beagle* voyage between 1831 and 1836. While he had to convince his shipmates that these were not "rubbish," they generated huge interest when brought to London, consolidating his reputation as a naturalist.[43] Equally significant was Woodbine Parish, British chargé d'affaires in Buenos Aires, who was directed to giant bones uncovered during an unusually dry season "by a peon in the service of the Sosa family." Parish wrote how vertebrae "were eagerly seized upon" by local people, who used them as kettle rests and seats. It therefore required some negotiation to acquire them, and Parish initially secured only a pelvis, which was apparently useless as a resting place.[44] Nevertheless, Parish was given permission and a letter of recommendation from the governor of Buenos Aires province, Juan Manuel de Rosas, to excavate fossils at other sites, which were then shipped to Britain, exhibited at the Geological Society and Royal College of Surgeons, and described by William Clift as belonging to a "stupendous quadruped."[45] Summaries in popular media like the *Penny Magazine* presented Parish's account of the discovery and speculated that the creature was most likely a digger: "The processes of the bones imply great muscular strength," and the flexible wrists and giant claws "lead us to believe that the animal turned out the earth like the mole" and "searched for roots, and lived on vegetables."[46] William Buckland also

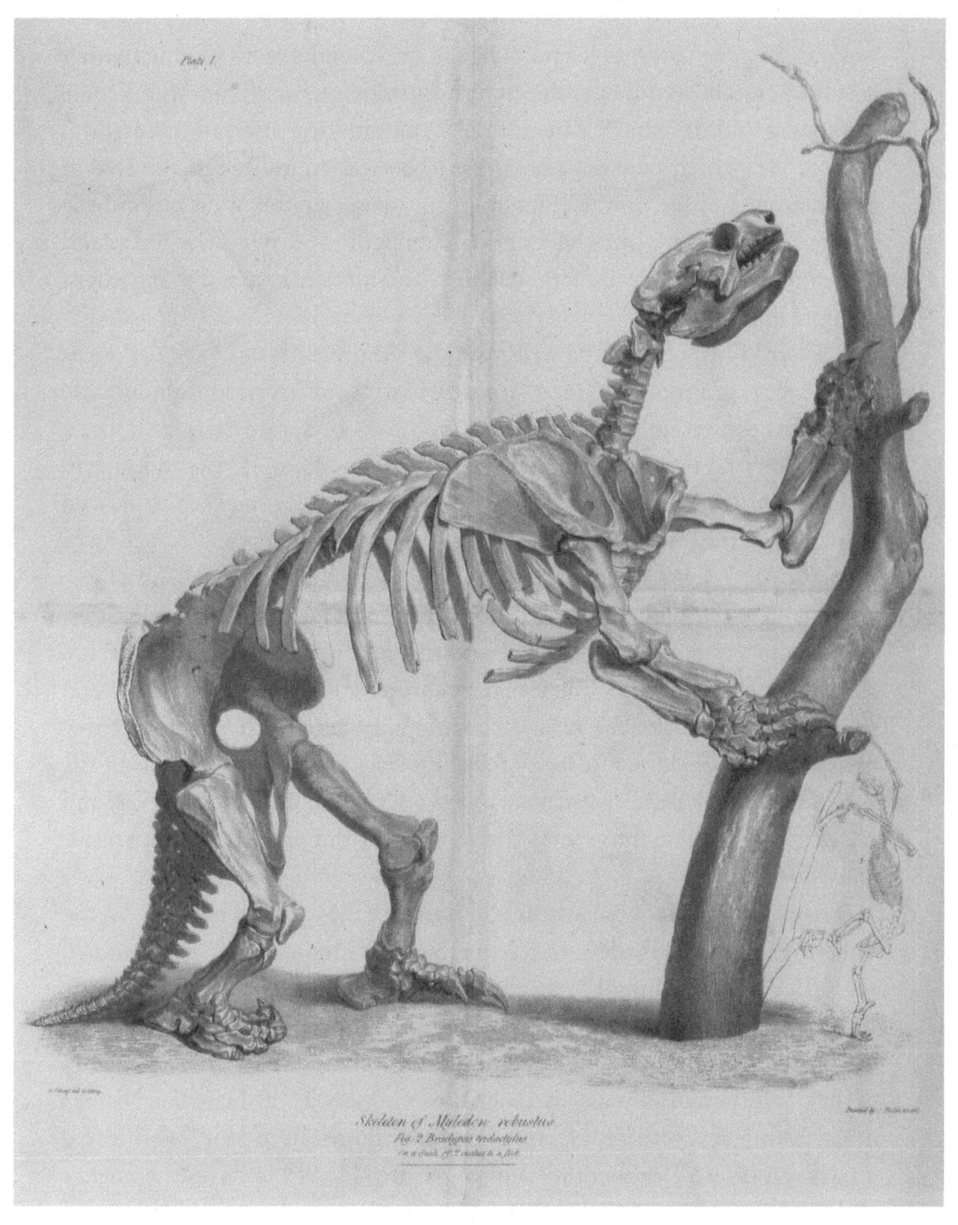

FIGURE 3.2. The skeleton of *Mylodon robustus*, depicted with that of a three-toed sloth. R. Owen, *Memoir*, plate 1. Reproduced by kind permission of the Syndics of Cambridge University Library.

worked *Megatherium* into his publications and popular lectures, christening it "Old Scratch" and describing it as a formidable excavator: "the Prince of sappers and diggers."[47] The animal's anatomy was used to understand its lifestyle, which may have been like the modern mole. But the size of the creature and the power of its limbs meant metaphors went beyond the natural: it was more like the formidable machines driving the industrial economy, and comparisons with machines and infrastructural digging might be more appropriate.

Richard Owen continued writing about the great beasts of the Americas while securing more specimens from British commercial and diplomatic networks. A particularly notable example was in 1842, when he wrote a description of a new specimen purchased by Woodbine Parish and sent to London, which he named *Mylodon robustus*.[48] The *Mylodon* report moved through a series of mysteries, resolved through comparisons and conjectures. Owen started by comparing the fossils with living South American animals, like the sloth and armadillo, as the anatomically closest creatures, and also larger ones like the elephant, rhino, hippo, and ox, which more closely matched its bulk. The visual centerpiece of the work, a full skeletal reconstruction of *Mylodon robustus* next to a modern sloth, showed its tremendous size. Owen wrote how "the singularly massive proportions of the skeleton of the *Mylodon robustus* arrest the attention of every observer, and are not less calculated to excite the surprise of the professed comparative anatomist."[49]

Owen attempted to reconstruct the animal's lifestyle and position "in the organic economy."[50] Its dentition and robust jaws indicated it "was assigned the office of restraining the too luxuriant vegetation of a former world."[51] Its claws and limb bones hinted it was not a burrower or grazer, but used its strong arms to uproot trees and browse the fallen branches. Owen's "theory of the Megatherian animals" assumed they undertook "the Herculean labour of uprooting and prostrating trees, for the acquisition of food," and rested on assumptions of exuberant tropical nature. He wrote how "the energy and rapidity of the growth of trees" in South America meant that huge quadrupeds were required by nature to eat the foliage.[52] The concept of *Mylodon* as a tree-uprooter continued through analyses of the skull, which seemed to show partially healed head wounds. Explaining the injuries was difficult, as Owen conjectured *Mylodon* as having been safe from predators and that it probably had the same "peaceful habits" as modern sloths and anteaters. He therefore argued that a branch had fallen on the creature's head while it was uprooting trees to eat, giving credence to his theory on the animal's lifestyle.[53]

Nearly twenty years later Owen wrote a further monograph on *Megatherium*. This text was an accessory to a larger project, the creation of a complete *Megatherium* mount for the British Museum, where Owen had become superintendent of natural history in 1856. The London mount partly consisted of casts of bones held in the Royal College of Surgeons donated by Woodbine Parish and casts of other specimens in the British Museum. Made as a composite according to principles of comparative anatomy, the new *Megatherium* was simultaneously a museum centerpiece, object of study, and a dramatization of Owen's interpretation of the animal as a gigantic ground sloth. Owen argued the mount was "in a much more complete state, and I believe I may add, more natural attitude, than that of the same extraordinary quadruped, which previously had been unique and the glory of the Royal Museum of Natural History at Madrid."[54]

The London *Megatherium* gained an iconic status. Unlike the unique Madrid specimen, the London mount was an object designed for reproduction. The casting workshop at the British Museum produced numerous copies for exchange and sale, especially to British settler colonies and the United States. In 1864 Henry Ward purchased a copy for the Geological Cabinet at the University of Rochester. However, even with the cast material, assembling the specimen was "a tedious and absorbing task, which occupied the writer and two experienced assistants for a period of nearly two months."[55] By 1866 Ward was selling his own casts of the London *Megatherium* for $250 each (consisting of "124 different casts, representing more than 175 bones"[56]), building a career as a trader in natural history objects for display and teaching. Rather than be regarded as anything inauthentic or conjectural, this cast made the specimen recognizable, exchangeable, and able to be studied around the world.

Juan Pimentel has rightly argued that the image of the *Megatherium* became increasingly fixed in these years, with Owen's reconstruction serving as the model, which "stabilized the anatomy of the *Megatherium* and clarified its habits . . . giving it a skin and a history, an identity and habits—a life."[57] Owen's influence can be seen in fossil mounts, illustrations, and models across the nineteenth century (and beyond). However, the cultural significance attached to the animal still varied considerably, and it remained conceptually flexible. Some commentators likened it to a mythological or biblical giant, drawing on religious traditions around great bones from the earth. The American naturalist William Brown Hodgson combined Owen's interpretation of the great sloths as having "lived by uprooting trees" with a biblical slant, linking it with the Behemoth of Job through quotations from Milton.[58] The account in the *Penny Magazine* conjectured how the bones built a new history of the world and "carry the mind back to the most remote times; not into the contemplation of the ages of mankind, but to the earliest condition of the globe, when it was undergoing a succession of changes, which were, at length, to suit it for the abode of the human race."[59] Contemplating the great beasts led to the deep history of the earth, which had undergone transformations long before humans.

However, the animal's large body, cultural ideas of "sloth" meaning laziness, and concepts of modern sloths as degenerate creatures meant *Megatherium* could also be seen as ridiculous. The German poet Joseph von Scheffel wrote about *Megatherium* in one of a series of poems on geological and paleontological themes that became popular student songs. This developed the theme of the animal as the "lazy-beast" as presented by Pander and d'Alton:

> What hangs there so without movement?
> With the weeds all clumped together
> So monstrously lazy and monstrously big,
> In the deep deep ancient jungles?
> Thrice as heavy as a bull
> Thrice as heavy and dumb,
> A climbing beast, a crawling beast
> The *Megatherium*![60]

After this the reader "to whom such a giant beast does not seem believable," was compelled to "go to Madrid!," but also not to "lose heart, as such monstrous laziness was only good before the Flood."[61] Humans in the postdiluvial epoch needed to exert effort and energy to fulfill their potential. The great sloth could be a moral lesson on the need for a productive life, as the modern world under human dominance was a more active place than the primordial forests of *Megatherium*.

The association of *Megatherium* with student culture was not limited to Germany. In the 1850s a group of researchers in the newly established

Smithsonian Institution in Washington, DC, named their drinking and dining group the Megatherium Club. In Paris the École Normale Supérieure held a partial, mounted *Megatherium* of unclear provenance in the Library of Arts, which played an important role in university hazing rituals. New students would process at night in monkly cassocks toward the *Megatherium* and bow to kiss its tail while being declaimed at by an established scholar in Greek or Roman dress.[62] The beast was partly a symbol of antiquity and partly a comic monster.

While the École Normale Supérieure possessed its partial specimen, the Muséum d'Histoire Naturelle faced difficulties acquiring a complete mountable *Megatherium* (although it did secure several bones). This was regarded as a stain on the national collection, given the role of Cuvier in defining the beast. In May 1837 Michel-Eugène Chevreul and Alexandre Brongniart wrote to the cabinet in Madrid, requesting permission to make a cast of their *Megatherium*, at a cost of 6,000 francs (and an estimated two hundred days of work), while "giving certainty to not damage this precious relic of the ancient world," and offering casts of the Montmartre mammals in exchange.[63] However, this request seems to have come to nothing. Working through French networks in South America also proved difficult. It was only in 1871 that a relatively complete *Megatherium* was mounted at the Muséum. This acquisition had a torturous history, being part of a collection of fossils acquired by François Séguin, a French traveler in Argentina, whose export was initially blocked by the government of Buenos Aires.[64] Séguin's collection was eventually sent to Paris to form part of the Argentinian displays at the 1867 Exposition, and was then purchased by the Muséum for 50,000 francs.[65] *La Liberté* reported on Séguin's *Megatherium* as "a splendid specimen of the colossal species of fossil edentates," and while not the largest of the five known specimens, it was "the most complete."[66] However, it took several years for the skeleton to be fully prepared and mounted and the missing bones cast. *La Nature* explained that this specimen was "the most remarkable of all" the remains in the galleries, "worthy of the collection" of "Cuvier, Geoffroy Saint-Hilaire and their illustrious successors." However, the article ended on a somber note: the galleries of the Jardin des Plantes were "literally full," and so "one of the finest examples of modern paleontology has to be housed in a shed where its conservation seems uncertain."[67]

The comment in *La Nature* raised the flip side of the drive to control specimens—the accumulation of material could be as much a curse as an important mode of scientific work, as space for collections and the need to care for specimens became obstacles. Iconic beasts like *Megatherium* could be gathered from across the world, either as fossils, casts, or (considerably less bulky) illustrations, and securing representatives of the canon of fossil mammals was an essential drive. Accumulation drew on global links and local valuations, and simultaneously expanded and contested understandings of the animal. *Megatherium*'s perplexing character allowed numerous valuations, crossing wonder, humor, puzzlement, and disgust, depending on how its bulk, claws, and conformations were aligned with conceptions of the natural world. Yet the creature also reflected the international expansion of paleontology and changing balances of power. While figures like Owen and Cuvier aimed to fix the animal through control of publications and casts, this was not a steady process, and alternate voices persistently presented other interpretations of the animal. The widespread nature of the great beast made it familiar but variable, usable for different purposes across Europe and the Americas.

Revealing the Terrible Beast of Eppelsheim

Megatherium had a potential rival as the most perplexing ancient beast in *Dinotherium*. Made known and debated across the early nineteenth century, fossils, casts, and illustrations of this perplexing animal were also the center of scholarly debate and drives to acquire material. In doing so, the nature of the ancient world was discussed, as was who was entitled to interpret it (and according to what values). However, *Dinotherium* also shows two significant differences. First was the nature of the specimens. While *Megatherium* and its relatives were known from fairly complete skeletons and casts, *Dinotherium* remains were more fragmentary. First a tooth, then a lower jaw, and then an entire skull slowly revealed the animal. Yet rather than form an increasingly fixed image, *Dinotherium* remained puzzling. If anything, *Dinotherium* indicated the ancient world was far stranger than Cuvier's confident presumptions proposed. Second was geographic spread. While *Megatherium* was known from an increasing number of fossils and went far beyond the initial skeleton in Madrid, the early history of *Dinotherium* was locked around a single site in the central German state of Hesse-Darmstadt, whose controller, Johann Jacob Kaup, built a personal position. This therefore displays a different manner in which authority over the paleontological past could develop—not through the accumulation of vast general collections via imperial channels, but from the control of local sites and specimens valued as key to the past.

Kaup's career shows the strategies which could be followed to develop paleontological work in the German lands. Born in 1803 in Darmstadt, Kaup became interested in natural history from a young age, eventually studying at Göttingen, Heidelberg, and Leiden. The rulers of Hesse-Darmstadt had built natural history collections since the seventeenth century, and the director of the grand ducal Naturalienkabinett, Ernst Schleiermacher, sponsored Kaup to work on the bird and amphibian collections. This was a relatively low position, and Kaup's salary fluctuated depending on the whims of the grand dukes and funding for scholarly activity. However, it nevertheless provided him with an institutional base and the ability to contribute to natural history debates.

The Darmstadt collections contained many fossils. Many were from a town called Eppelsheim, where sand deposits were exploited by workers and peasants, and fragmentary animal bones were dug up along with the valuable sand. Many of these were sent to the Naturalienkabinett, and Kaup began to study them in the 1820s and 1830s. Most seemed to be mammalian remains, and Kaup described a particularly intriguing jawbone in the journal *Isis* in 1829.[68] He argued the molars on the Eppelsheim jaw were almost identical to those Cuvier described as a "gigantic tapir,"[69] but the front teeth were decidedly non-tapir-like. It was therefore a new, unknown animal "perhaps in the vicinity of the Hippopotamus and the Tapir."[70] The three-and-a-half-foot jawbone indicated the animal would have been at least eighteen feet long, if it had the same proportions as a mastodon. Kaup named the jaw *Dinotherium giganteum*, maintaining Cuvier's "gigantic" designation in the species name. But the genus was new. This was the "Terrible Beast," a name reflecting the size of the animal (using the prefix *Dino-* several years before Richard Owen's Dinosauria).

Following his study of the jawbone, Kaup made further trips searching the Eppelsheim Tertiary sands. Over the 1830s he described at least thirty specimens, many apparently of entirely new species, giving the Naturalienkabinett a large collection of local fossils. By 1832 Kaup and the court artist Johann Baptist Scholl published a catalog of casts from the Darmstadt collections, writing, "All naturalists who study fossil osteology feel that plaster casts are essential, since the most exact descriptions and best executed drawings can give us only imperfect notions, and will never be as valuable as solid figures that can be examined in all dimensions." As a stamp of approval, Kaup wrote, "This was at least the opinion of Baron Cuvier, when he wrote to me: 'Your beautiful plaster-casts give anatomists and geologists the means to profit from your treasures as if they owned them themselves.'"[71] The casts were mainly teeth and jaw fragments of various fossil mammals. *Dinotherium* casts were the most expensive, with the palate of a young individual with milk teeth being available for eighteen francs, individual adult molars for two francs each, and a copy of the "lower jaw, which was represented in *Isis*"[72] for seventy francs—by far the most expensive specimen (the nearest were mandibles and palates of rhinoceroses and mastodons, which were twenty francs each).

Kaup was a canny marketer of the Eppelsheim fossils, using casts and publications to connect with other institutions, despite his unsteady position in Darmstadt. His early works were also often written in French in order "to equally communicate to naturalists who do not possess the German language"[73]—an indication of the balance of linguistic authority in the early nineteenth century. The *Dinotherium* also became famous, as foreign scholars attempted to deduce its lifestyle and place in the scale of life. William Buckland argued that it must have been a water-dwelling creature, as the massive tusks would have been "cumbrous and inconvenient to a quadruped living on dry land," but if supported in water, they could have been "employed, as instruments for raking and grubbing up the roots of large aquatic vegetables from the bottom," and would "combine the mechanical powers of the pick-axe with those of the horse-harrow of modern husbandry,"[74] following similar "mechanical" analogies as were used for *Megatherium*.

The fame of *Dinotherium* dramatically expanded in 1835 when the geologist August von Klipstein wrote to Kaup of a new find at Eppelsheim: the entire skull of a huge beast, most likely *Dinotherium*. This specimen formed the basis of a jointly authored monograph. Klipstein wrote a section on Eppelsheim's geology, describing the layering of the landscape through ancient catastrophes and upheavals. Most fossils from Eppelsheim were fragmentary and smashed, showing violent disturbances in the past. The massive skull was unusual, having been laid down in soft marl and "protected against the violent impact of rolling debris."[75]

The specimen was a rare, fragile prize, and extraction required special

FIGURE 3.3. Excavating the *Dinotherium* skull. Kaup and Klipstein, *Atlas* Dinotherii gigantei, n.p. (final page). Courtesy of the Staatsbibliothek zu Berlin-Preussischer Kulturbesitz, Shelfmark: Mi4607.

techniques, mobilizing workers from the surrounding countryside. Kaup described taking the "beautifully preserved yet fragile colossus out from the pit, which was eighteen feet down to the base of the tunnel, where the skull was sunk in its throne of marl." The workers laid gypsum supports around the skull, "anointed with oil and pork fat to prevent any connection between the gypsum and the skull." These were strengthened and underlaid with iron bars. When the gypsum hardened, "twelve strong men" did the initial work of raising the mass from the pit, while twenty-four men at the top "gently" hauled it out. Once raised, the skull was placed on a cushion on "a humble wagon" for a "funeral procession" along bumpy rural tracks to the town of Alzei, before joining the larger road network to Darmstadt.[76] The accompanying image demonstrated the values around paleontological extraction. Securing the skull was a complex technical process, requiring dozens of strong men from nearby villages and a whole system of support. The skull was a precious treasure taken from the earth, the relic of a long-dead creature. The learned scholars descended into the pit—not assisting in the manual labor, but directing the process of bringing the fossil to light. This dramatized the values of paleontology, with scholars supervising systems of work to transform fossil remains into scientific treasures.

A Problematic Animal

Removing the skull from its "marl throne" compounded the mysteries of the animal. The features were judged to be strange and uncertain enough that no one could control the interpretation during those years, and this fluidity of interpretation raised the importance of the beast. The skull and teeth seemed to mix features of different creatures with unique characteristics, notably the huge downwardly projecting tusks. Placing it within the categories of mammalian life was difficult, especially as Kaup possessed little of the rest of the skeleton. So while the head could be deduced, its body and mode of locomotion were entirely mysterious and required comparative inferences. Kaup initially associated *Dinotherium* remains with a large claw found at Eppelsheim, and assumed it must have been a gigantic edentate, like *Megatherium*. Yet this was only provisional. Indeed, Kaup cited *Dinotherium* as a warning against relying on assumed laws of comparative anatomy, as "even the immortal Cuvier thought that the dinotheres were giant tapirs." The strangeness of *Dinotherium* showed "how deceptive some forms are to the human imagination, when it attempts to take fragments of an animal to conceive the whole image of it."[77]

The title page of Kaup and Klipstein's *Atlas* Dinotherii gigantei, depicting *Dinotherium* in its full Tertiary environment, occluded these uncertainties on its character. The creature was surrounded by other Eppelsheim

FIGURE 3.4. Imagining prehistoric Eppelsheim. Kaup and Klipstein, *Atlas* Dinotherii gigantei, title page. Courtesy of the Staatsbibliothek zu Berlin-Preussischer Kulturbesitz, Shelfmark: Mi4607.

animals, including mastodons, rhinos, horses, and crocodiles. *Dinotherium* was seated by the water with limbs folded under itself—depicted so that the ambiguous parts of the specimen, its limbs and the possibility of edentate-like claws, were hidden. The animal's head, known from the skull, was the focal point of the image. The environment itself was lush and verdant, with huge volcanoes erupting in the background, showing a luxuriant watery world where the strange great beasts of ancient Hesse lived their lives, beset by geological catastrophes. *Dinotherium* was surrounded by great changes of the earth and shifts in the animal world.

Aiming to increase his reputation, Kaup took the skull to Paris. A meeting of the Academy of Sciences on March 20, 1836, was almost entirely devoted to *Dinotherium*, with various luminaries pronouncing on its characteristics. First to speak, and in the greatest depth, was Henri de Blainville. He thanked Kaup and Klipstein for sending the skull, "allowing the naturalists of Paris to gain an exact idea of the affinities of this gigantic animal with those which still live on the surface of the earth," nevertheless indicating that Parisian naturalists were best suited to understand the creature, rather than German collectors in Darmstadt. Blainville's interpretation was also different from Kaup's, arguing it was a "gravigrade" animal, close to dugongs and manatees. He drew attention to the importance of the skull, that "the jaws or its appendages, and the teeth which arm them, is indeed in zoology, as in physiology, the dominant part of the organism." In Blainville's system gravigrades included elephants and manatees, "one having to seek its food on plains close to large rivers, and the other in the rivers themselves, but both offering remarkable and common peculiarities in their dental, digital, and mammary systems."[78] *Dinotherium* was therefore slotted into this group, as a link in the chain of life.

Blainville's interpretations dominated the other Parisian scholars. André Duméril attempted to argue that it might have been an edentate due to its lack of a zygomatic arch, but Blainville brushed this aside, stating that his observations had "victoriously overthrown" this hypothesis. Isidore Geoffroy Saint-Hilaire stated that further observations from himself would be "useless," as he agreed entirely with Blainville's placement of the creature among the manatees, though he did have misgivings about Kaup's monograph, backhandedly citing how the "handsome plates supplement what might be desired in the zoological description." The preservation of the skull, presented with such pride by Kaup, was for Geoffrey Saint-Hilaire a problem, as "various agglutinating substances, particularly mastic, hide the sutures and various details of the structure of the ossification of the bones, the consideration of which is far from unimportant." The expectation was that "when the skull of the *Dinotherium* takes its final place in one of the scientific museums of Europe, we can remove the solidified mastic."[79] Such a specimen could not have its final resting place in Darmstadt and ought to be cleaned, restored, and held in a grander institution like the Muséum in Paris.

While these dismissals of Kaup's interpretation (and care of the fossils) could be seen as marginalizing a more peripheral figure, the Parisian scholars, despite their confidence, could not dominate or control the specimen. Kaup took the skull back to Darmstadt and was given a right of reply.

He sent a letter comparing the *Dinotherium* skull with other specimens in the Muséum d'Histoire Naturelle, which he had been permitted to examine while in Paris. He revised his initial opinion that the creature was an edentate, arguing that the associated claws probably belonged to a different animal. However, Kaup did not accept Blainville's idea that it was like a manatee. "The texture of the bones" were hard like a pachyderm's, and not "fibrous" like those of marine mammals. And from "the shape, structure, and method of replacing teeth, *Dinotherium* is obviously a pachyderm, and in this respect has only a slight analogy with manatees, and absolutely nothing with dugongs."[80] Kaup therefore maintained an independent position, even against a centralizing figure like Blainville.

The Paris discussions were not simply metropolitan authorities erasing the work of a "provincial" scholar. The continued uncertainty over the animal meant that debate grew wider. The Munich paleontologist Andreas Wagner wrote, "The formation of the skull being so peculiar is precisely why various interpretations have been made of the overall construction of this problematic animal."[81] The otherwise descriptive *Petrefactenbuch* of Friedrich Albert Schmidt produced reconstructions of the "mighty monster" *Dinotherium* "in life," based on the two major hypotheses—pachyderm-like or manatee-like. The skull and teeth could create two completely different renderings of the animal, one terrestrial and one aquatic. Schmidt noted, "We must wait for the time when we have more information to determine a certain judgment."[82] *Dinotherium* pushed the mode of reasoning cemented by Cuvier, that particular anatomical features could illustrate an animal's affiliations and lifestyle, to the breaking point. But this was not a mark against paleontological authority. It was instead a mystery, which drove debate and the search for new fossils.

In later years, reflecting the growing status of German science, Kaup produced more works in German and corresponded with and visited collaborators in Paris and London. He used casts of *Dinotherium* and other Eppelsheim specimens to build the Darmstadt collections further, even purchasing Peale's mastodon skeleton in 1854 after the Philadelphia Museum closed and its collections were sold and dispersed.[83] However, material did not just flow into Darmstadt, but also moved out. The Eppelsheim *Dinotherium* skull was cast by the Darmstadt workshop, but the specimen itself belonged to Klipstein, who in his later years decided to sell it. After some negotiations, it was eventually purchased by Thomas Oldham, director

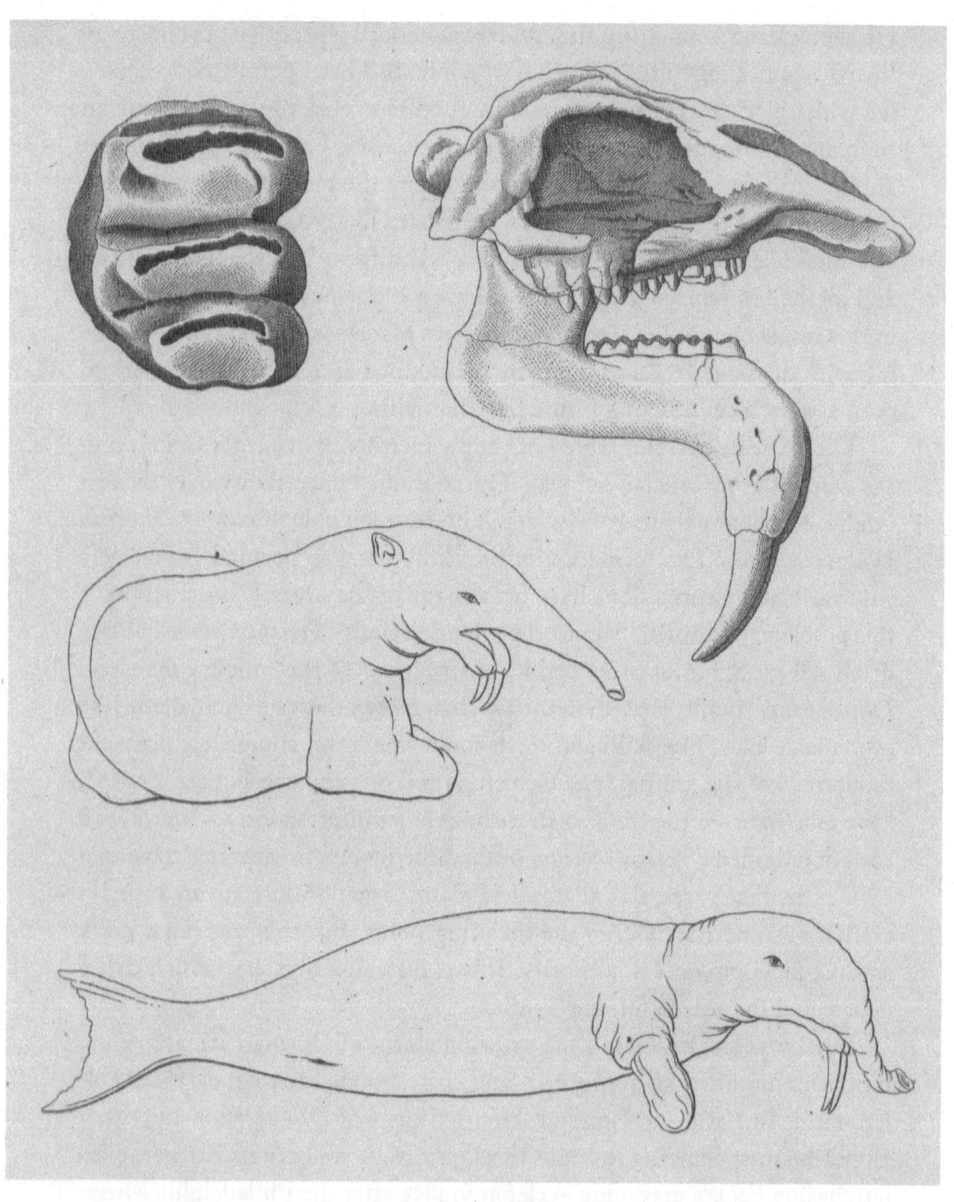

FIGURE 3.5. Debating the *Dinotherium* as pachyderm- or manatee-like. Schmidt, *Petrefactenbuch*, plate 61. Staatsbibliothek zu Berlin-Preussischer Kulturbesitz, Shelfmark: Mi422.

of the Indian Geological Survey, who in turn donated it to the British Museum. Yet rather than be displayed as a prize specimen, it seems to have fallen into obscurity. Rumors abounded that the skull had been destroyed, either during Oldham's possession or during its transit to London—although Charles William Andrews, cataloging the museum's proboscidean collections in the early twentieth century noted this was incorrect. What seemed to be the original skull had been "much broken" at some point, but "has been, on the whole, skilfully mended since then."[84] The British Museum however continued to exhibit a cast rather than the original specimen, possibly because of this damage.

Debates on *Dinotherium* continued into the following decades, with new finds and concepts shifting the animal's status. *Dinotherium* skull and teeth fragments were identified from a range of places, including central and southern Europe and India, indicating the wide spread of the animal and how well known it had become. In 1853 what appeared to be the complete skeleton of a *Dinotherium* was excavated by workers constructing a railway line in Bohemia.[85] However, this was not widely publicized until the late nineteenth century, when interest in the *Dinotherium* rose once again as extensive remains were found in Romania, becoming important scientific currency.[86] These cemented it as a more elephantine creature, undoubtedly huge but primarily terrestrial.

Uncovering and possessing *Dinotherium* required tremendous technical and artistic feats, with excavation, production of casts, and transmission around scholarly networks leading to the debate, while enhancing the position of the Darmstadt collectors who controlled the specimens and localities. Kaup became an early example of a major phenomenon across nineteenth-century paleontology, of scholars using control of sites to build their reputations and positioning themselves as important nodes within paleontological networks. However, this contrasts with scholars in Argentina and the United States, who were much less able to do so despite controlling access to giant sloth material. Partly related to how Kaup managed information about the specimen, it also relates to the position of German science in these years, which was moving from a relatively secondary position to a much more important one. Kaup and his terrible beast would become major points within scientific debate, even against criticism from established scholars in Paris.

More broadly, the fate of *Dinotherium* represents a key theme within this chapter: accumulation and control could work in several ways, potentially bolstering the importance of establishing centers, but also allowing particular sites and localities to play an outsized role through control of novel and well-marketed specimens. The geographies of paleontological authority could be varied and fluid giving way to numerous interpretations and actors. From the conventions of the field, strange huge beasts were discussed as fossil wonders and given diverse values connected with particular places and views of nature.

CHAPTER 4

Uncovering Siwalik and Pikermi

Colonialism and Antiquity in India and Greece

HUMPHRY DAVY, AWARDING THE COPLEY MEDAL OF THE ROYAL SOCIETY OF LONDON to William Buckland in 1822, gave a speech linking the human and animal pasts. While modern observers may "look with wonder upon the great remains of human works, such as the columns of Palmyra, broken in the midst of the desert, the temples of Paestum, beautiful in the decay of twenty centuries, or the mutilated fragments of Greek sculpture in the Acropolis of Athens," the "grand monuments of Nature, which mark the revolutions of the globe," were possibly even greater, showing "new generations rising, and order and harmony established, and a system of life and beauty produced, as it were out of chaos and death; proving the infinite power, wisdom, and goodness, of the GREAT CAUSE OF ALL BEING!"[1] Davy's elaboration illustrates that the natural past was not the only deep past examined and valued in the early nineteenth century. Ancient monuments in Europe, Asia, the Americas, and North Africa absorbed much attention and were worked into new human histories.[2] Objects of the human past and fossil remains of extinct animals were often linked. They could also be understood in the same manner, showing drama and change, but also providence, progress, and the work of a Creator. Moreover, they were both subjected to new scholarly approaches. Antiquarianism, geology, and natural history were often part of the same project (and indeed, naturalists and antiquarians were often the same people, or at least active within the same networks).[3]

Connections between the human and animal past were present in much paleontological work. They were particularly marked in projects undertaken in territories marked as being "ancient" by European scholars. Two stand out, constructing iconic sites linking human and natural antiquity. The first was the Siwalik formations in northern India, where British colonial officials directed a series of excavations and worked these into visions of the ancient Indian landscape, as well as wider patterns of development.[4] The second was in Pikermi, a village outside Athens, where fossils were collected beginning in the 1830s, first as fragments by German soldiers and scholars and then through a system of works under the French geologist Albert Gaudry.[5] These projects form the focus of this chapter, examining how control of these sites constructed new visions of the deep past.

These projects built images of natural antiquity by taking advantage of new structures of northern European authority. The early nineteenth century saw a tremendous increase in informal commercial penetration and political influence in some regions (such as the eastern Mediterranean) and formal colonial control in many others, with India providing a prominent example. British and French science, commerce, and military power interacted with and subordinated local knowledge and social structures. Paleontology became part of this expansion of influence, as paleontologists took advantage of commercial connections, colonial control, and the desire of elite figures for recognition and engagement in regions where fossils were found. The fossils of Greece and India were not important just for their antiquity, but were also significant for their extent, variety, and—in another echo of imperial policies—the quantities that could be extracted and shipped out. Tons of fossils from both the Siwaliks and Pikermi were sent to centralizing collections. Hierarchies of knowledge were key, and fossils were understood in metropolitan collections.

While the fossils themselves were extracted, knowing and interpreting them still depended on encountering the territory where they were found. Fossils had been known in both Greece and India for millennia, and people often collected and used them for a variety of purposes.[6] New paleontological projects in some ways used these local ideas, but subordinated them to new ways of dominating the landscape and its pasts. Assertions of wonder at the landscape and imaginative reveries of bringing the past to life were as significant aspects of paleontological work as the supervision of excavations, the laborious transportation of fossils, and the comparison of bones and teeth. Yet these were forms of imagination that reworked previous interpretations. That fossils were deeply rooted within particular places, but also needed to be understood within central collections, was key to constructing meanings for the finds. Senses of exoticism and connections with landscapes and human cultures presented as antique and ancient gave fossils and conjectures drawn from them a powerful world-building role.

Hierarchies and Geographies of Knowledge in the Siwaliks

The excavations in the region which became known as the Siwaliks provide possibly the most unambiguous early nineteenth-century example of the extraction of fossils within colonial systems. They drew from British scholarly and military interests in India, particularly around two individuals reflecting different parts of the colonial structure. The first was Proby Cautley, a military engineer with the East India company, who worked on infrastructural projects across northern India, most notably the Doab Canal between the Ganges and the Yamuna Rivers. The second was Hugh Falconer, a Scottish naturalist and medical officer appointed superintendent of the East India Company's botanical gardens at Saharanpur—an institution mixing natural history with governmental utility, maintaining collections of Indian plants, especially those with medicinal or commercial potential.[7] Falconer and Cautley of course did not work alone, but instrumentalized Indian expertise and labor. Charles Murchison recounted this as crucial for Falconer's work at Saharanpur, as he utilized Indian workers in this "remote provincial station, with at that time only half-a-dozen European families." Murchison described how Indian workers under Falconer's authority improvised barometers from broken tumblers, distilled mercury from cinnabar, and used "some indigenous oil as a substitute for salad oil, when the European supply had been exhausted." Falconer's "discipline" and "self-reliance," and the "intelligence, docility, and exquisite manual dexterity of the natives," was seen to enable him to further "that great fund of universal information."[8] Of course, this was an ideal view which obscured and marginalized how Falconer and other British scholars depended on Indian knowledge.

Cautley and Falconer took advantage of the early British colonization of

India, which integrated control over land and people, with cultural, historical, and geographic knowledge. Over the eighteenth and early nineteenth centuries, the East India Company annexed large territories in South Asia and subordinated numerous Indian states into client relationships. As well as military expansion, taxation, and resource extraction, the company was also a patron of scholarly work within India. Indeed, a wide literature has developed using the promotion of science in British India as a paradigmatic case of scientific work through colonial administration.[9] Scholars connected with British rule simultaneously aimed to know the territory and its inhabitants, creating knowledge useful for administration, commerce, and military activity. The company sponsored botanical gardens, scholarly associations, and collecting and surveying expeditions—which, as argued by Jessica Ratcliff, were driven as much by local military and governmental needs as by "scholarly curiosity" and demands from the metropole.[10] Additionally, British colonial administration depended on local collaboration. Kapil Raj has stressed how scientific activity in India was often based on exchange and circulation between Indian and British contexts.[11] And this was of course not an equal relationship. As with Falconer's use of Indian expertise in Saharanpur, officials attached to the East India Company worked with Indian political and scholarly elites and deployed a huge amount of labor, in attempts to control and know the territory.

British officials often reported on Indian knowledge, including understandings of fossils such as the "Bijli ki har" (lightning bones) used in medicine and fossil shells in Hindu temples.[12] Fossils were also status symbols. The raja of the Kingdom of Nahan in the Himalayan foothills amassed a collection of teeth and bones of what seemed to be elephants. These were given to him by farmers and traders and were in turn given as gifts to notable visitors. British officials visiting Nahan would frequently describe it as a model client state, with one traveler noting (in a patronizing manner) that the raja himself "had been educated almost entirely under the kind and fatherfuly superintendence of captain Murray, (the superintendent of the Seik states), with whom he was esteemed as a pet child."[13] The gathering and exchange of fossils therefore fostered relations between Indian elites and British rulers, although with the former being subordinated (and in this case, infantilized).

British projects also drew on longer traditions of Indian development. Transportation and mapping were key concerns of the colonial government, to facilitate movements of goods and troops and expand agricultural and mineral exploitation.[14] The Doab Canal on which Cautley worked, discussed in depth by Pratik Chakrabarti, was based on older canal systems constructed by previous Indian rulers.[15] Digging the canals had excavated human and animal bones. British scholars highlighted references to the water-channel building under Firoz Shah Tughlaq in the fourteenth century, during which the "bones of elephants and men were discovered."[16]

Indian fossils were often reported in the journals of the Asiatic Society of Bengal, an institution binding knowledge of modern Indian nature with antiquarianism. The Asiatic Society was founded in Calcutta in 1784, and its members ranged across Indian philosophy, religion, literature, natural history, medicine, and geography. There was consistent emphasis on languages, literature, and law (subjects of governmental importance), and in this period Indian scholarship was still framed in an idiom of wonder and romanticism, rather than the disparagement of Indian culture that would take hold in the late nineteenth century.[17] British orientalists often saw the Hindu Puranas as representing ancient history (albeit in a refracted form), while comparative studies of Indian religion and languages reconstructed an ancient past for the subcontinent.[18]

These structures provided the basis for Falconer and Cautley's collaboration, which represented an almost ideal division of labor within colonial paleontology: Cautley the engineer, mapper, and geographer experienced in supervising work teams; Falconer trained in anatomy and natural history. After meeting in the circles of colonial sociability, the two examined fossil collections in the East India Company's and Asiatic Society's collections, then traveled to Nahan in 1831, where the raja "presented them with a fossil tooth found at Sumroti, near the valley of Pinjore."[19] Mobilizing Indian knowledge, Cautley wrote, "There is a tradition existing, of the remains of giants having been discovered in the neighbourhood of the Pinjore valley, near a village named Samrota, the said giants having been those destroyed by the redoubtable Ramchandra."[20] While presented as picturesque local understandings, Indian interpretations were nevertheless crucial for finding specimens.

In 1834 Falconer and Cautley traveled to the Himalayan foothills with a team of workers, including two British officers and numerous (yet undescribed) Indian excavators. The excavation is presented by Savithri Preetha Nair as a key example of "boundary work," linking diverse interests in the project of interpreting fossils.[21] The scholars documented their work in a series of letters to the Asiatic Society. Rather than dry catalogs of finds, these expressed exclamatory wonder at the landscape and extent of the material. Falconer wrote how he "reaped a splendid harvest. Conceive only my good fortune: within six hours, I got upwards of 300 specimens of fossil bones!"[22] Cautley likewise exclaimed, "The hills were covered with fossils . . . (how they could have escaped observation before, must remain a source of wonder). Mastodons and hippopotamus's remains looking one in the face at every step!," and asked, "What shall we have next?"[23] Indian fossil fields mirrored but surpassed those known from Europe and the Americas and inspired amazement at their potential for extraction.

The letters contain much information about the emotional reactions of the colonizers to the fossil fields, but lack details on how the work was managed, only bringing this up when there were difficulties (a common point mentioned by Stephen Shapin in his discussion of "invisible technicians" in scientific work).[24] These difficulties included many problems with transportation of the material. Cautley wrote that the roads out toward the Yamuna River consisted mainly of "bad footpaths" along rivers and streams through "an utter confusion of mountains."[25] However, they also included Indian interest in the fossils. Cautley wrote that a "splendid specimen" of a fossil skull was interpreted as a "Deo ka sir" (literally, the head of a god) by one of the workmen, who subsequently "bore off the head in triumph to the Nahan Raja." Cautley was pessimistic about getting the skull back, but this was not a major upset: given the richness of the site, "there is no doubt of our finding many more, as the fossils are in abundance."[26] The unnamed workman's valuation of the skull demonstrates how fossils were of symbolic, cultural, and religious importance. The raja had a large collection of fossils which he would present to honored visitors. The case of the skull potentially indicates wider practices, with local people bringing fossils to Nahan and participating in the circulation of this material. Unlike in Western contexts, where fossils would move but then be fixed in large collections, the use of fossils in India seems more fluid, connected with negotiated social and political relationships.

The privileging of comparison and collection for natural history made studying the fossils in India difficult. Cuvier was an important reference for Falconer, but attempts to secure a copy of his *Recherches sur les ossements fossiles de quadrupèdes* were initially fruitless. Instead, Falconer and Cautley built up their own comparative collection, slaughtering a large number of Indian animals to compare with fossil specimens. Charles Lyell cited "their extraordinary energy and perseverance" as they "found a Museum of Comparative Anatomy in the surrounding plains, hills, and jungles, where they slew the wild tigers, buffaloes, antelopes, and other Indian quadrupeds, of which they preserved the skeletons."[27] Building a collection was a critical part of comparative anatomy, and the "ingenuity" of Falconer and Cautley connected modern and ancient Indian nature. Knowing these animal worlds depended on their placement in the territory—and tremendous violence against its inhabitants.

Falconer and Cautley emphasized that fossils seemed to represent the whole of creation, being "a mixture of the new and of the old, of the past and of the present, of familiar with surprising forms, together with a numerical richness, such as no other explored region has exhibited within so comparatively limited a space."[28] The wide-ranging fossil fauna mirrored various places. Some of the animals were known from Africa, some from modern India, others from European fossils, and some were completely new. Falconer also linked the Siwalik fossils with Indian culture and religion. Pratik Chakrabarti has discussed how one specimen—the giant tortoise *Colossochelys atlas*—became an icon of the excavations, bolstering theories that Indian religions recalled the past environment, with the Hindu tradition of the world being carried on the back of a turtle possibly reflecting encounters with this immense creature.[29] Indian knowledge also contributed to geography. Cautley noted that the region did not have a specific name in British maps prior to their work, so he presented the term *Sevàlik* for the territory (which was later conventionalized to Siwalik). This term, which indicated that the land was a residence of Lord Shiva, had been reported by some British travelers as a local name for the region, and was also presented in "some traditional writings in the possession of the high priest of Mahant residing at Deyra" who was apparently in contact with Cautley,[30] again showing the use, but also subordination, of Indian knowledge within these processes.

Understandings of fossils were bolstered by observations of modern Indian animals. Some of the prize Siwalik specimens were small primate fossils, which were incredibly rare. Their rarity was explained through modern Indian monkeys, which would die, fall from trees, and "immediately [became] the prey of the numerous predaceous scavengers of torrid regions, the

Hyaena, the Chacal, and the Wolf," which, according to Falconer, happened so frequently "that the simple Hindoo believes that they bury their dead by night."[31] In this case, Indian culture was highlighted for misunderstandings rather than deep knowledge, but still explained why monkey fossils would rarely survive. The Siwalik fossil camels also compared unfavorably with those of modern India. They were larger and more robust than the domesticated "common Camel in use in the Bengal provinces,"[32] showing "domestication may have caused the deterioration or otherwise of the *Camelus dromedarius*," especially "in a country and amongst a race of people who pay little attention to their improvement," following the common colonial trope of regarding Indian animal husbandry as deficient. Falconer hoped for camel specimens from central Asia, either wild or owned by nomads, so that the Siwalik fossils could be compared with "the camel in all its perfection."[33] Fossils indicated the power and strength the animal could attain, were it not stymied through lack of care.

While most Siwalik fossils corresponded to known creatures, some were stranger—and here linkages between myth, comparative anatomy, and modern environments were especially pronounced. One mammal rivalled the *Colossochelys* as the highlight, named *Sivatherium giganteum* (the beast of Shiva) and reported as "a new accession to extinct Zoology." The principal fossil was "a remarkably perfect head," which required considerable work to render into a usable object. It was originally "completely enveloped by a mass of stone" and "might have been passed over, but for an edging of the teeth in relief from it, which gave promise of something additional concealed." The extraction of the fossil "after much labour" from "the hard crystalline covering of stone"[34] (most likely by an unnamed Indian mason) was a process of discernment and revelation. Once uncovered, the *Sivatherium* was a fossil wonder, connecting several animals: "The form of the head is so singular and grotesque, that the first glance at it strikes one with surprise," with great horns and with a nasal opening indicating "the *Sivatherium* was invested with a trunk like the Tapir."[35] It was of "large size, surpassing the rhinoceros"[36] and mixed characteristics of camels, giraffes, and tapirs. This was "a very remarkable animal, and it fills an important blank in the interval between the Ruminantia and Pachydermata,"[37] taking a vacant position in the scale of life.

Sivatherium was impressive in itself, but also significant for its position within the assemblage of ancient Indian animals. Falconer and Cautley

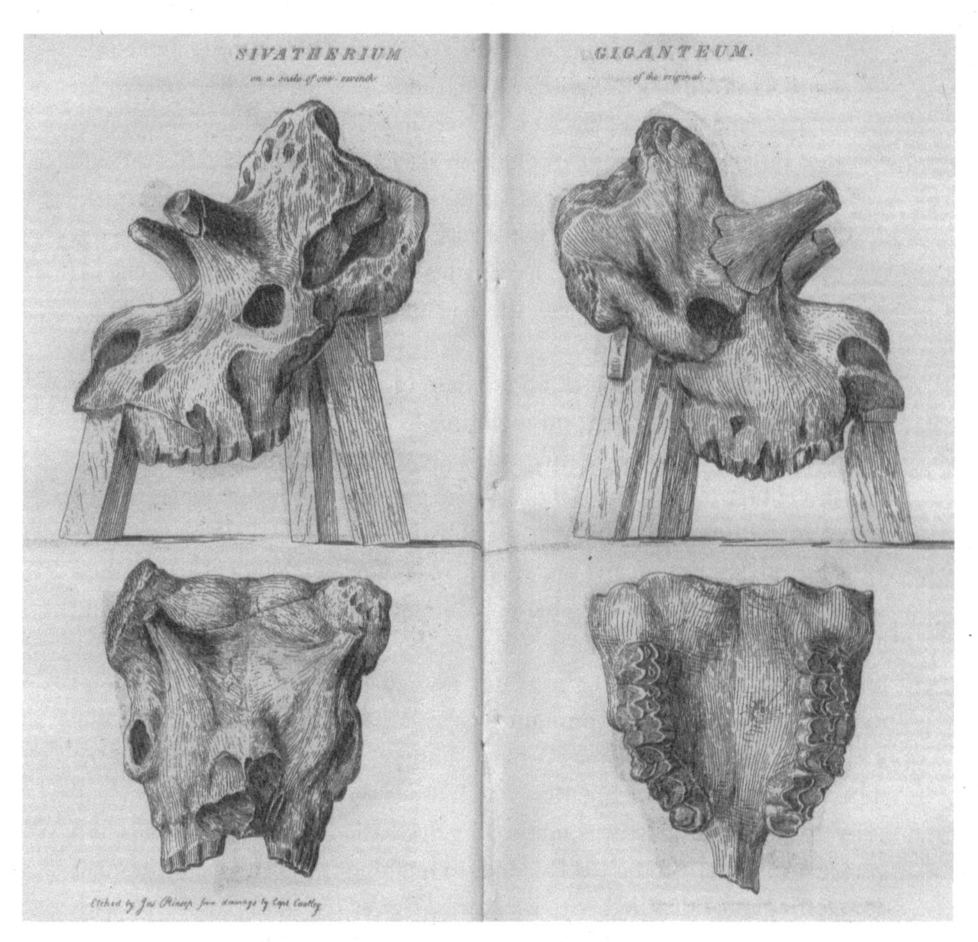

FIGURE 4.1. The skull of *Sivatherium*. Falconer and Cautley, "*Sivatherium giganteum*," plate 1. Courtesy of the Staatsbibliothek zu Berlin-Preussischer Kulturbesitz, Shelfmark: Uk139–5=49/60.1836.

engaged in several imaginative reveries of what the land had been like when these animals had lived, with one particularly florid example drafted by Falconer in 1840:

> What a glorious privilege it would be, could we live back—were it but for an instant—into those ancient times when these extinct animals peopled the earth! To see them all congregated together in one grand natural menagerie—these Mastodons and Elephants, so numerous in species, toiling their ponderous forms and trumpeting their march in countless herds through the swamps and

> reedy forests: to view the giant Sivatherium, armed in front with four horns, spurning the timidity of his race, and, ruminant though he be, proud in his strength and bellowing his sturdy career in defiance of all aggression. And then the graceful Giraffes, flitting their shadowy forms like spectres through the trees, mixed with troops of large as well as pigmy horses, and camels, antelopes, and deer. And then last of all, by way of contrast, to contemplate this colossus of the Tortoise race, heaving his unwieldly frame and stamping his toilsome march along the plains which hardly look over strong to sustain him.[38]

Ralph O'Connor has discussed how these imaginative reveries, where the learned scholar traveled back through time to the ancient landscape through the powers of his own imagination, were an important feature of nineteenth-century paleontological discourse (numerous other examples will be seen across this book).[39] They were usually the culmination of long processes of technical description and saw the balance of paleontological articulation swing decisively toward imagination. Extracting fossils from the earth and comparing them with modern forms enabled some knowledge. But true understanding of the "living pageant" needed the creativity of the scholar to position the bounteous community of organisms in its ancient landscape. This sight had never been available to humans, but paleontology, and the mix of typology and imagination which drove the field, enabled its revival.

Yet the ancient past was marked by dramatic change. The fossils and geology of modern India led Falconer to imagine the ancient subcontinent as a verdant island, filled with creatures. But as the land connected with the rest of Asia and the Himalayas rose, the climate shifted and dried, and the animals could not survive. The diversity of ancient beasts was tinged with a melancholic reflection on modern life. Cautley wrote, "We appear to be gradually losing all the larger forms of creation. The elephant and giraffe of the present period will in all probability share the same fate as the Mastodon and Sivatherium of former eras, and be only recognized in the proofs exhibited by the researches of the geologists."[40]

While the Siwalik fossils were initially studied in India, Cautley and Falconer felt the specimens needed to be sent to Europe to be truly understood. They sent reports to British and French institutions, where their finds were valued as the first fossils from "the tropics." Drawings of the *Sivatherium* skull sparked a debate at the Academy of Sciences in Paris. Étienne Geoffroy Saint-Hilaire believed *Sivatherium*'s anatomy, and especially its hornlike structures, meant the creature was "nothing other than an antediluvian giraffe."[41] This was a creature he claimed to know well, having been responsible for transporting the giraffe gifted to France by the Pasha of Egypt in 1826, which famously walked from Marseilles to Paris after arriving by ship. Falconer and Cautley's interpretation of it as a mixed animal between pachyderms and ruminants showed how they "had the misfortune of too keenly adopting . . . the theory of the scale of beings."[42] If the "colossal quadruped" was a great giraffe, it showed (in a similar way to *Megatherium* and fossil elephants) that the life of the past was on a grander scale than that of the present and bolstered transformist ideas that animal species could change into other types over time. Blainville meanwhile rejected these designations, calling the animal "truly something very extraordinary, a large species of antelope, even more hideous than the gnu."[43] It was a misshapen monster and could not change into another creature—simply being another lost link on the chain of life.

Finding a final (metropolitan) home for the collections was also a concern. While Cautley initially intended to donate his collection to the Geological Society of London, it was too large for their stores, and so the 214 large chests were transported to the British Museum at a cost of £602. Falconer returned to Britain in 1842, bringing "seventy large chests of dried plants" (which suffered damp conditions and mostly rotted on the journey) and "forty-eight cases containing five tonnes of fossil bones."[44] The fossils were divided between India House and the British Museum, and Falconer spoke on the Siwalik fossils to learned societies, with life-size paper drawings of the *Sivatherium* skull and the entirety of *Colossochelys.*

After the initial excitement, matters died down to the grind of comparative work. Following lobbying from the British Association, Falconer was awarded a government grant of £1,000 to work on a monograph, *Fauna antiqua sivalensis.* The first volume was published in 1846 and attempted to decipher relations between the Siwalik elephants. Links between fossils and myth remained. Falconer argued that elephant fossils "have in all ages attracted more attention, both from the learned and from the unlearned, than perhaps those of any other family of extinct animals." In Europe they were "regarded as demonstrative evidence of the former existence of Titans, Giants, and other fabulous beings handed down to us in the records of superstition and mythology," while "the people of India even now, usually refer such remains to the Rakshas or Titans, who hold so prominent a place in the ancient writings of that country." While Falconer was interested in mythology, he nevertheless praised "the severe investigations of modern science, [which] have expelled these fictions from the belief of civilized mankind; and reconstructed the true forms of the animals which appear in many instances to have given rise to them."[45] Paleontology had an ambiguous relationship with myth. Scientific rationality dispelled ancient monsters, but paleontology created new lineages of wondrous beings retaining traces of older myths. The ancient world of the Siwaliks, with its large beasts like *Sivatherium* and *Colossochelys*, were knowable through fossils and imaginative reveries. The creatures were rooted within the landscapes and institutions of colonized India, while creating a modern mythic system from the deep past.

Glittering and Crumbling Bones between Greece and Bavaria

Formal colonial control was not the only way political and economic power influenced fossil work. More informal links were also important. Across the nineteenth century direct colonial control was just one way that economic, commercial, and military powers expanded their influence. Diplomatic intervention, unequal commercial treaties, and dominance of economic sectors were also important strategies, which often did not necessitate formal control (and were particularly marked in South America and eastern Asia). These informal links could also be important channels for extractive sciences like paleontology. Indeed, British influence over *Megatherium* fossils from South America in the mid-nineteenth century, outlined in the last chapter, illustrates how informal power could be a driver of paleontological collecting.

An important example of these trends was excavations in Greece at the site of Pikermi near Athens. They show similar issues to the Siwaliks, of human and natural pasts being linked, leading to the extraction of fossils and imagining of ancient faunas. Yet Pikermi shows the heterogeneity of networks around new deep pasts. Of course Greece's position in the early nineteenth century was very different to India's. The Kingdom of Greece secured independence from the Ottoman Empire during the 1820s, with support from France, Britain, and Russia. Following this, a monarchy—under a king of the Bavarian Wittelsbach dynasty—was established. Britain and France maintained significant economic power, securing concessions over Greek commerce, while a strong Russian faction exerted political influence.

Greek scholarly institutions also developed, largely based in Athens and devoted to antiquities and natural history, keen to build a national scholarly community.[46]

Like India, Greece was privileged by northern European scholars as an ancient land. Greek classics formed much of the educational background of Western scholars, with knowledge of ancient Greek being a mark of status. Classical languages were also essential for taxonomy: in the Linnean system, species were usually given Greek and Latin names (which emphasizes how radical the name *Sivatherium* was). A working knowledge of Greek was expected of an educated naturalist. Yet the importance of ancient Greek meant that Greece's modern status was often problematic. That Greece was now a small country subjected to great-power rivalries seemed to contradict its exalted condition in the classical era. Many foreign travelers wrote of modern Greeks in terms of decline, with the modern land still holding echoes of the old culture, but having lost a great deal (to the extent that the British, French, and Germans often saw themselves as the true inheritors of ancient Greek culture).[47] Among Greek scholars, this classical heritage was looked on with some ambiguity, and there were persistent tensions over whether to emphasize classical Hellenic heritage or the Orthodox Byzantine past.[48] While the former was useful for interacting with external partners, the latter more often resonated within Greece itself.

Early fossil finds in Greece were often by-products of studies of the classical past by expatriate scholars. Anton von Lindermayer, personal doctor of King Otto, recalled how in 1836 the British antiquarian George Finlay hosted a salon where he "showed me a small fragment of bone, which he could not suggest as a natural or artistic product," and "after a long period of consideration I determined it to be a fossil bone." Lindermayer and Finlay visited the village of Pikermi, where the bone had been extracted. A local worker "showed us a place where already on the first glance a few broken bones were visible." The two began to dig, to the mystification of "the curious people who watched our activity with great attention, and gave us the impression, that they were sorry for our vain efforts on these insignificant worthless things." After finding a few more bones, the two returned with four German soldiers "with the necessary tools and organized a rich excavation."[49] References to local people finding fossil digging incomprehensible, or assuming that scholars were digging for gold rather than antiquities, is a common trope in nineteenth-century antiquarian and paleontological discourse. This can be read in different ways. For the scholars, the meaning was fairly straightforward—the ignorant people were unaware fossils were themselves "treasures." However, from the viewpoint of the villagers, these were strange outsiders engaging in puzzling activities—not domineering colonial figures, but eccentric travelers to be observed and managed.

Finlay donated the fossils to the Natural History Society of Athens, along with some worked flints interpreted as "Persian arrowheads" (as the site was reasonably close to the battlefield of Marathon). Beyond this, he felt he did not have the expertise to interpret the fossils and was mainly interested in archaeological artifacts and whether the stream around Pikermi corresponded to geographic features mentioned by Strabo. While he sent a box of fossils to the London Geological Society, the metropolitan institution did not follow up on these researches.[50] Lindermayer meanwhile organized a separate excavation in 1837, employing German soldiers for digging, many of whom picked up fossils themselves and "especially found pleasure in the small holes of the long bones, since the marrow cavities were filled with beautiful barite crystals."[51]

The story of the fate of these remains became increasingly embellished over the years, not least owing to a dispute between the Munich natural historian Andreas Wagner and Lindermayer on the history of the site, with Lindermayer claiming that his initial involvement in the project had been effaced and especially that "a very small head" of a fossil monkey had been discovered in one of his excavations.[52] Later writers talked about how the soldier found a glittering skull covered in crystals he mistook for diamonds, and one account stated that the fossils were discovered only when the soldier was arrested in Germany on suspicion of grave robbing.[53] Wagner claimed that this story had "been distorted beyond recognition by poetic legend," and presented a more prosaic version (which gave him ownership of the fossils). Disentangling myth from authenticity was not just necessary for extinct beasts, but also excavation accounts.

"The real story" according to Wagner was a Bavarian soldier returning home through Munich brought him two long bones, whose "marrow cavities were covered with beautiful shiny white calcareous crystals," which he thought might be diamonds. Wagner informed him the crystals were calcite and might be worth a small amount to a jeweler, but inquired whether he had any other bones. The soldier said he did but "it was such bad stuff, that he did not dare to bring it." The next day the soldier returned with a box containing "all kinds of small pieces of rock," along with the teeth of horses, ruminant foot bones, but most importantly "a fragment still partially encased in rock," which Wagner identified as "the snout of a small primeval monkey." While not an entire crystal-encrusted skull, this was still a major find, as fossil primates were known only from "insignificant pieces" from France and "somewhat better" specimens from Siwalik. Not wanting "to abuse the simplicity of the poor fellow," Wagner paid the soldier four Kronenthalers for the box (something the soldier found surprising, assuming it would only "be worth a few flagons of beer"). So according to Wagner "both parties separated from one another in the greatest spirits: the soldier had ample travel money to return to his home (not far away), and I had in my possession some scientifically very important, if rather unassuming, pieces."[54] Presented in terms of mutual benefit, this story illustrates how the confluence of overseas military influence and scholarly collections could expand paleontological work.

Wagner classed the primate skull as somewhere between a gibbon and a langur monkey, and placed the specimen within a new genus: *Mesopithecus pentelicus*. The fossils raised tantalizing clues for future research, especially when Lindermayer sent him more material. While "very much shattered, making determining them very difficult,"[55] they seemed to include horse, rhinoceros, antelope, and possibly mastodon and *Dinotherium* remains. The fossils were sufficiently interesting that Wagner organized a small expedition. Johannes Roth, a geologist and geographer, directed a series of excavations in Pikermi in 1852–1853 as part of a general expedition to "the Orient," with support from the Bavarian State Collections. However, excavations were difficult. While fossils were abundant, they were also fragmentary; "skeletal pieces of the most diverse mammals were mixed up, almost all broken, bent, crushed." The bones were filled with calcite, sodden with water, often broken by plant and tree roots, and "crumble into innumerable pieces, for they are penetrated in every direction by cracks."[56]

Roth and Wanger cowrote two articles after the excavations, describing the "rich bone-breccia" of primate, carnivore, rodent, edentate, pachyderm, equid, and ruminant fossils.[57] Overall, they described eighteen species, although the proboscideans and ungulates were so numerous and fragmentary that categorizing them was impossible. The fossils gave clues to an abundant former world, yet their materiality—fragmentary and broken—caused problems. And once removed from their locality, the specimens became difficult to interpret. They could be typologically arranged onto a general grid of life, but fuller understanding required placement in the landscape. The site was rich, with diverse animals. However, determining the "value" of the finds was difficult.

Albert Gaudry and the Geological Missions to Pikermi

In the early 1850s Greek scholars like Heracles Mitzopoulos of the University of Athens and the physician Aristeides Chairetis excavated at Pikermi. Mitzopoulos sent a box of Pikermi fossils to France in 1854, "taken myself with hammer and scalpel," including the jaw of a fossil primate and equid and antelope remains. He felt these showed the "modifications of the revolutions of the centuries, from which no part of Creation can escape, for nature itself is nothing more than perpetual birth and perishing," and declared the fossils a gift "in the name of the fatherland of Aristotle to the illustrious fatherland of Cuvier."[58] The reference to the respective fatherlands of Aristotle and Cuvier emphasized the conceptual importance of ancient Greece (still valued by French scholars, even as they moved away from Aristotelian concepts), while acknowledging the authority of Cuvier and other Parisian scholars. Mitzopolous's gift sought to enmesh the Greek scholarly community with the French.

The prestige and vast collections of the Jardin des Plantes and the use of French as an international language of science meant that French scholars maintained wide influence. Museums and natural history associations in Europe and the Americas often turned to French institutions to recognize their researches. The scholars at the Jardin des Plantes could take advantage of this, expanding their holdings through gifts and working with intermediaries. This status also linked with France's colonial and commercial power, which was expanding, although somewhat unsteadily. France invaded Algeria beginning in the 1830s and Indochina and Mexico from the 1850s and 1860s, but this formal empire building occurred in fits and starts. French informal influence also expanded throughout the Mediterranean, South America, and Asia through commerce, diplomatic pressure, and sometimes military threats—a model of "informal empire" which has traditionally been studied in British imperial history, but recently highlighted in the French Empire by David Todd.[59] For French paleontologists, these informal links were often more fruitful.

The connection between French scholarship and informal power can be seen in the career of Albert Gaudry, who not only became the prime investigator of the Pikermi fossils, but the highest authority in French paleontology in the late nineteenth century. Gaudry was initially educated in the geological laboratories at the Muséum d'Histoire Naturelle, and took part in a governmental "mission to the Orient" in 1853–1854 to examine agriculture in Syria, Egypt, Greece, and Cyprus. The mission had colonial imperatives. Levantine products, particularly textiles, silk, and wine, were important imports and could potentially be produced in France. Agricultural techniques practiced in the eastern Mediterranean could also be models for France's North African empire. Gaudry assessed the geology, climate, and agriculture in Cyprus as matching that in Algeria, possibly illustrating "the probabilities of success of cultures and educations which we are undertaking in our African colony."[60]

Gaudry observed landscapes and economies with a commercial and colonial eye, and was aided by French diplomats and traders. The French minister in Greece, Alexandre de Forth-Rouen, put him in touch with the Greek government and scholarly establishment and took him to Pikermi. Gaudry regarded the limestone geology with an acquisitive gaze, describing an "inexhaustible" mountain spring "which forms a charming oasis of vegetation and freshness" and if exploited could "provide Athens with the water necessary for the city and surrounding countryside." Evidence of "the raising of blocks of marble which have seduced artists" were seen in large cuttings, and Forth-Rouen was instrumental in reclassifying this Greek marble as "first-quality marble," allowing it to be imported into France. Pikermi itself was described as a large ravine with a stream running through it "near a group of hovels." Gaudry's eye for commercial exploitation turned to paleontological prospecting. The fossil site also gave opportunities, as "the deposit of bones is far from exhausted" and "the abundance of fossil debris seems immense."[61] Marble and fossils were both resources to be extracted and claimed for "progress" and "civilization"—one for the construction of huge edifices (including new museum buildings) and the other for building new ways of knowing the natural world. At Forth-Rouen's instigation, three boxes of fossils were sent to Paris. Georges Louis Duvernoy reported on these, noting the site was "an ossuary similar to that of Sansan"[62] and describing a similar variety of fossils as Wagner and Roth.

Duvernoy's report raised French interest in the site. However, rather than rely on Greek institutions to supply more fossils, the academy voted to support French research, selecting Gaudry to direct excavations in the winter of 1855–1856. As was typical in mid-nineteenth-century paleontological excavations, Gaudry supervised a team of local workers who dug up and packed the material, while villagers provided supplies and assistance in locating fossils (he noted on one occasion how "a local child" led him to "huge bones at the very bottom of the stream").[63] He mentioned his foreman, Giacinto Guicciardi, on several occasions, who seems to have been essential for organizing the work. In Gaudry's letters to the academy, he wrote how Guicciardi "perfectly assisted" him in "the most meticulous precautions," removing entire blocks of rock containing friable bones and hardening them in gum."[64] In Gaudry's popular articles for the *Revue des deux mondes*, he described his interactions with local people in more detail—arguing for the "improvability" of the Greeks. He wrote how his workers quickly learned to "manage their picks so as to preserve the fossils they discovered, and as soon as an unusual-looking piece appeared in the rock, they redoubled their attention," and "the most skillful of them could even say the names of our most common species." He contrasted the Greeks with the "indolence" of the "Orientals" he had observed on his trade mission, as "the modern Greeks [are] an eminently spiritual people and [had] a singular aptitude for all that they would like to undertake."[65]

Gaudry's accounts did not just use the language of paternalism and improvability, but also discussed the work in romantic and the picturesque terms. In several publications, he repeated the same anecdote of feasts after a day of excavation:

> After hours of suffering, we had moments of pleasure: finding an unknown fossil often redoubled our courage. In the evening of each day marked by an important discovery, we had a small party—a skin of resinated wine and honey was brought from the Hymettus; sometimes we felled the branches of an old tree and roasted a sheep *à la pallicare*, that is a whole sheep, as in the time of Homer. When the wine spread gaiety, workers, shepherds and gendarmes surrounded the remains of the hearth; they sang old Albanian songs, then some began to dance, while others clapped their hands to mark the rhythm. If a traveler lost at the foot of the Pentelic had seen our encampment then, he would have thought he saw a ring of fauns from the time of Greek mythology.[66]

This was another reverie by a paleontological scholar linking the modern world and antiquity. However, rather than communing with the animal past, Gaudry linked the excavators and military escort with ancient Greece. He also acknowledged the local people more than Falconer's and Cautley's accounts, indicating these Europeans were considered more worthy of appreciation than the Indian workers in the Siwaliks.

Behind the scenes Gaudry's excavations were entangled with French interests in the eastern Mediterranean. The expedition coincided with the Crimean War, and conflict erupted in Greece between pro-Russian and pro-Anglo-French factions (exacerbated as the Ottoman Empire, Greece's former overlord, fought on the Anglo-French side). Gaudry told stories of the countryside being infested with bandits, and how he was advised not to travel lest he be abducted—and if no ransom were paid, the bandits "will cut off your ears or kill you."[67] Despite evocations of wildness and danger, Gaudry was supported by Greek and French state power. The Greek government supplied letters of recommendation and an armed escort, and Lindermayer recalled how Gaudry was watched over by French naval and army detachments based in nearby Piraeus.[68]

At the end of the season a large quantity of fossils was shipped to Paris. Gaudry stated that Pikermi was "one of the richest in the world for the remains of fossil mammals," and "it is impossible for me to precisely number the specimens I have received—it is immense." While no complete skeletons were found, there was a large variety of material: the skull and hand bones of fossil primates, remains of unknown carnivores similar to hyenas, pachyderm bones, almost innumerable fragments of ruminants (including three or four hundred pieces of antelope jaw), and some prize specimens "of great beauty," including two entire fossil horse skulls which "seem to me destined to feature among the most remarkable specimens of the Muséum of Paris."[69]

The academy supported a second expedition under Gaudry in 1860, with more funds "to employ a greater number of workers." The excavations took place in summer rather than winter, and "the intensity of the heat made work very difficult, and most of my workers were struck by intermittent fevers."[70] While the political situation in Greece was more stable, there was still difficulties. Gaudry packed forty-three boxes of fossils, but the Greek government resisted their export, following new laws on the exportation of antiquities. Delicate negotiations were required. Gaudry wrote how the stipulations of the Academy of Sciences' grants meant he could not surrender any material, but he promised that "when the bones of Pikermi have been classified and studied in Paris, I will choose duplicates which will be sent to the Museum of Athens."[71] This was apparently acceptable, demonstrating an early instance of political and scholarly communities attempting to maintain fossils in the territory where they were excavated.[72] Greece's position as an independent European state was an important factor here, although allowances still needed to be made for stronger economic and political powers like France.

Gaudry spent much of the 1860s working on the Pikermi fossils, publishing several reports and eventually a two-volume monograph. Dialogue with the Siwalik fossils was particularly significant, with Gaudry writing, "No European deposit surpasses Pikermi for the number of gigantic species," but nevertheless he noted how the even greater extent of the Siwalik fossils meant that "India has surpassed Greece." The Siwalik fauna "presents the most beautiful association that can be imagined; their multitude equals their grandeur."[73] Analogies with Siwalik stretched to the animals themselves. Gaudry named one large ruminant *Helladotherium duvernoyi* (the beast of Hellas). It was "remarkable not only for its gigantic size, but also for the shape of its skull and the bones of its limbs." Gaudry compared *Helladotherium* with the cast of a female *Sivatherium* skull held in Paris, which was roughly the same size and bore a "striking" resemblance.[74] *Helladotherium* also mixed features of different animals. The skull was "heavy like that of a cow, but more elongated and without horns"; "its enormous teeth" were like an antelope's but much larger; its neck had "roughly the same proportions as in the *Megaloceros*"; its limbs partly like those of an ox and partly like those of a giraffe. While it was not as large as *Sivatherium*, that Pikermi contained a parallel huge herbivore, linking giraffes, antelopes, and pachyderms, was another similarity.

Rather than describe the ancient animals as strange, Gaudry emphasized how they connected to modern creatures: "When we compare the mammals of geological times and those of the current era, we are struck above all not by their external diversity, but rather their unity of organization." Few of the animals were completely new, and those that were seemed to "establish new links" between already known creatures.[75] Gaudry engaged in a similar reverie to Falconer, visualizing the diversity and harmony of ancient Pikermi:

> The landscapes were animated by the most varied mammals; here, the two-horned rhinoceros and huge boar; there monkeys frolicking among the rocks, or carnivores of the civet family, martens and cats watching their prey; the Pentelic marble caves served as a dwelling place for the hyenas; the hipparions ran in immense troops on the plains, like the quagga and zebra of Africa. No

> less fast and even more elegant, the antelopes also composed large bands. The herds of each different species could be recognized by the form of their horns. . . . The *Helladotherium* and a giraffe close to the current species dominated the ruminants. . . . [T]he most majestic of all animals was the *Dinotherium.* How beautiful he must have been to see, when he stepped forward, escorted by the mastodon with mammoth teeth and the mastodont with tapiroid teeth! We hear the roar of the terrible *Machairodus* with dagger-like canines. Many other species accompanied those which I have indicated, their cries mixed with bird song. In the concert of all these beings, only the voice of man was missing.
>
> No region of earth offers such a spectacle.[76]

Once again, the variety of animal life was wondrous, and required empathy as well as technical skill to understand. The elephants and their relatives (including the now fully enshrined *Dinotherium*) and herds of ungulates were revitalized from the fragmentary and confused mass of bones into a majestic scene surpassing that of any country now known—although expressed in similar terms employed by Western travelers in Africa reporting on the continent's abundance. Ancient Europe showed that even these modern spectacles were shadows of past life. Potentially referring to contemporary debates over Darwinian evolution, Gaudry wrote how "there was no competition for life, all was harmony."[77] Nature was not violent and disruptive, but was a world of fellowship and order.

The Pikermi fossils also gave lessons in ancient geography. Such large numbers of animals could not have been confined to Greece, but must have inhabited a greater landmass linking Africa, Europe, and Asia. The animals ranged across "a vast continent" where "the giraffes, rhinoceroses, mastodons, and troops of Hipparion and Antelopes could develop freely." Yet this state of plenty was destroyed by geological forces. "When the ground was dislodged so that one part sank into the sea, and the other rose to form the Tertiary hills which cover the secondary terrains of Attica, many animals, fleeing the flood, climbed the Pentelic. There they died, for lack of space and food, and little by little their bones were carried away by the waters." The richness of the site relayed a story of geological tragedy, as the starving animal exiles of the formerly bounteous land died, and their bones "were buried in the silt which was deposited at the bottom of the mountain."[78]

Another question was how fossils related to ancient knowledge and whether they fed into ancient Greek and Roman culture. The abundance

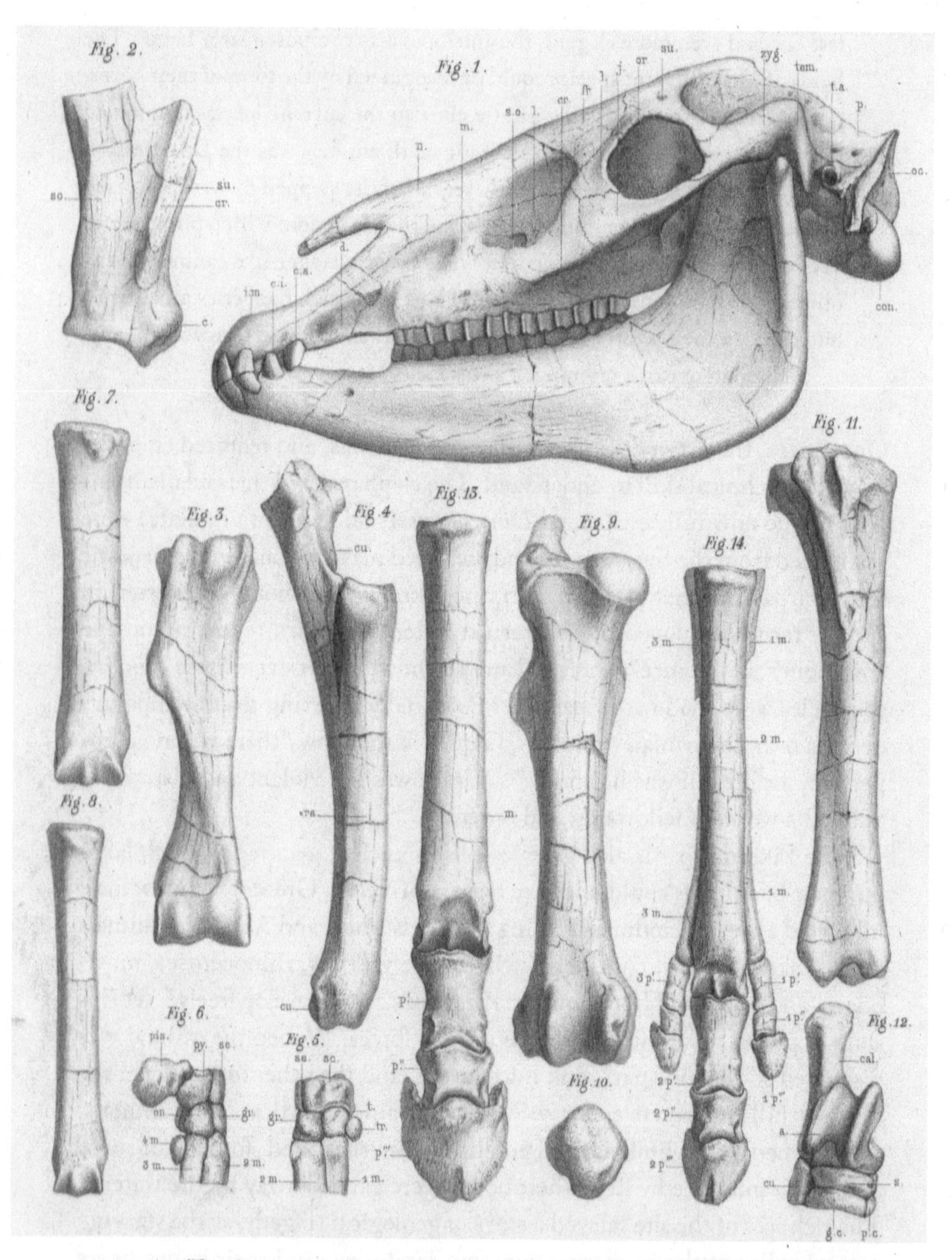

FIGURE 4.2. The three-toed *Hipparion gracile*, one of the most abundant fossil animals from Pikermi, depicted as bones. From Gaudry, *Animaux fossiles et géologie*, vol. 2, plate 35. Courtesy of the Staatsbibliothek zu Berlin-Preussischer Kulturbesitz, Shelfmark: 4"Mn16436-Atlas.

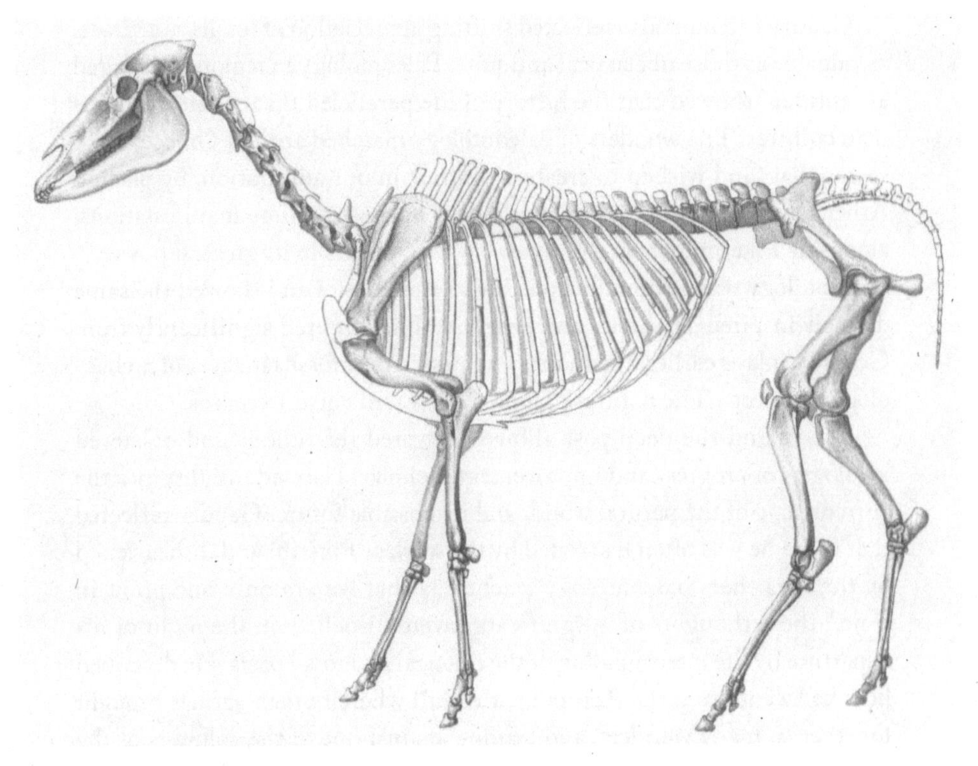

FIGURE 4.3. *Hipparion gracile* as a complete skeleton. Gaudry, Ani*maux fossiles et géologie*, vol. 2, plate 36. Courtesy of the Staatsbibliothek zu Berlin-Preussischer Kulturbesitz, Shelfmark: 4"Mn16436-Atlas.

of the bones and references to fossil ivory in Pliny meant classical observers must have been aware of fossils, and Gaudry thought that observations of fossils in the landscape may have influenced cultural and religious traditions. Considering Ovid's *Metamorphoses*, Gaudry imagined "the sight of the fossils may have inspired the Greeks to think that organized bodies had turned to stone."[79] Likewise, large fossil remains could have been interpreted as those of mythological creatures; "perhaps the assemblage of quadrupedal bones, apparently little different from those of humans and ruminant animal bones, lent credence to the fable of satyrs and fauns. Perhaps the gigantic fossils at the base of the Pentelic passed for the remains of the Titans struck down by Jupiter." While Gaudry remained skeptical that fossils were the "primary source" of the myths, he nevertheless asked, "Is it unreasonable to think that it has facilitated these beliefs?'[80]

Gaudry's account also reflected shifting appreciation of fossils as artifacts as valuable as those of classical antiquity. Paleontology in regions privileged as "antique" showed that the history of life paralleled that of esteemed human cultures. The wonders of paleontology matched ancient Greece, as "it seems that God wished to create a contrast in our admiration, by placing Athens, where the intellectual world gave its most sublime manifestations, alongside Pikermi, where the organic world appears in its greatest power."[81] Paleontology was as morally uplifting as the classics and showed the same divinely inspired aesthetics and progress. This differed significantly from George Finlay's earlier valuations, who passed over fossils in favor of archaeological objects. The natural past now possessed equal treasures.

Fossils and the deep past allowed elevated reflections and bolstered ideologies of progress and improvement. Scholars claimed insights into the current state of the natural world, and its possible future. Gaudry reflected that while he was often frustrated by the work at Pikermi and disheartened by the idea that "paleontology teaches us that we are only one point in time," these thoughts of insignificance were dispelled on the night of his departure by the intermingling of the classical and fossil pasts. He described how he "went up to the Acropolis, this hill where human art has brought together so many wonders, and leaning against one of the columns of the Parthenon, I said to myself: 'It does not matter that man has a very small body, for God has endowed his soul with genius; it does not matter that we were born yesterday, that the past was for beings without reason, if the present is ours, and the future is reserved for us!'"[82] All forms of antiquity could be brought together, highlighting the importance of humans within creation—and human mastery over culture and nature.

PART II

CONSOLIDATING THE AGE OF MAMMALS, 1850s–1880s

FIGURE 5.1. The 1885 gallery of paleontology in Paris. Gaudry, *Ancêtres de nos animaux*, frontispiece. Reproduced by kind permission of the Syndics of Cambridge University Library.

CHAPTER 5

A Tale of Two Elephants

Fossil Mammals in Paris and London

On March 17, 1885, the Muséum d'Histoire Naturelle in Paris finally opened a public gallery specifically devoted to paleontology. Located in a newly built wooden annex of the Cour de la Baleine (a large room belonging to the comparative anatomy collections previously used to display a whale skeleton), this was hardly in a prime position. Yet it was arranged to give a dramatic vision of the life of the past. Albert Gaudry, who had been the Muséum's professor of paleontology for thirteen years, wrote how, "despite the simplicity of the premises, the gathering together of these specimens of the ancient world . . . seems worthy of interesting naturalists and philosophers." On entering, visitors were confronted with the most impressive specimens from the Muséum's fossil collections. They would see the skeleton of *Megatherium cuvieri* acquired by François Séguin, "strange, with its descending jowls, prismatic teeth, crooked digits, its massive hindquarters."[1] Alongside the *Megatherium* were carapaces of glyptodons from the Pampas of Argentina, mammoth tusks, the shells of fossil giant tortoises, a complete skeleton of *Mastodon augustidens* from Sansan, a giant deer, wall panels of ichthyosaurs from Germany and England, and fossils of *Palaeotherium* and *Anoplotherium* from the Paris basin. The collections showed the past diversity of life, with specimens from France and across the world.

The centerpiece of the gallery was the complete skeleton of "the Elephant of Durfort" from southern France. At over four meters tall and nearly seven meters long including the immense tusks, it towered over even the *Megatherium*. Gaudry described how "the elephant of Durfort is the most majestic of the animals in our provisional gallery. No foreign museum contains such an imposing skeleton of a mammal."[2] Appreciative statements were repeated in the press. Maurice Daubin in *Le journal de la jeunesse* drew attention to the specimen's great size, and "if one were to place a modern elephant next to it, it would give the impression of a young foal by the side of its mother,"[3] and Stanislaus Meunier in the *Nouvelle Revue* wrote how "its skeleton is one of the marvels of the Gallery of Paleontology of the Muséum; it greatly reflects the work and ability of the naturalists who have extracted it from the soil and mounted it."[4] This elephant was not just a prize specimen, but gave lessons about ancient France. Gaudry compared it with the better-known (and later) woolly mammoth, noting it differed "not only by its great size, but also by its more protruding chin, less curved tusks, and larger molars, covered by a thicker enamel. It is obvious that it lived in a warm climate and did not have thick fur, like the mammoth of Quaternary times."[5] This elephant lived in a lush and warm prehistoric France, which illustrated dramatic changes over earth's history.

The elephant of Durfort however had a difficult past, as it moved between a range of contexts to become a centerpiece of the natural history galleries in Paris. Its coming to the capital was the work of provincial scholars, local conflicts and connections, and contestation within the Muséum itself. These features defined paleontological work more generally, ensuring the field was never under the sole authority of great collections. The remains were initially retrieved far from Paris, in the village of Durfort in the Department of Gers in 1869.[6] According to a likely embellished anecdote presented by Gaudry, two aristocratic antiquaries, Paul Cazalis de Foundouce and Jules Ollier de Marichard, came across a road-building site while searching for prehistoric human remains. They saw a pile of fossil bones dug up by the road builders, one of whom led the antiquaries to what he described as the still-buried pipes of an old fountain. The antiquaries identified these "pipes" as the tusks of a fossil elephant. While the story entrenched the idea that the aristocratic scholars were central to identifying the fossils, they nevertheless relied on the labor of local people to find and excavate them.

The skull and tusks were extracted and sent to the nearby University of Montpellier. Cazalis wrote a report for the Société Géologique de France that the rest of the specimen was remarkably complete.[7] However, this appeal to central authority did not prevent the work from being affected by local considerations. The landowner forbade further excavations until he was reimbursed for the use of his land. Long negotiations took place (delayed even further by the Franco-Prussian War), and the site was almost taken over by the Swiss paleontologist François-Jules Pictet de la Rive, who only relented when he learned he would not be able to acquire the skull from Montpellier. It was only in 1872 that an accord was struck between Cazalis and the landowner, with the backing of the Muséum, to allow three years to excavate the rest of the specimen.

The materiality of the fossils also caused problems. While the skeleton was relatively complete, it was also fragile and broken. The skull in particular was partially smashed and needed considerable restoration. Jean-Benjamin Stahl, head of the Muséum's casting workshop, went to Montpellier to deploy a new technique of hardening fossil bones using resin and spermaceti.[8] The bones of the fossil elephant were then taken to the Department of Comparative Anatomy at the Muséum and mounted on a metal armature in its main laboratory (which quickly became known as "the elephant room" among the workers). The elephant of Durfort was publicly displayed only when the new Gallery of Paleontology opened, which was itself the result of conflict over the remit of paleontology and its demarcation with other fields.

The story of the elephant of Durfort is emblematic of a central tension examined in this chapter. In the mid-nineteenth century paleontology developed as a science based on presenting the wonders of the fossil past to urban audiences and accumulating fossils in large collections. Fossil mammals played important roles in these processes. "Exotic" and foreign specimens, like *Megatherium* and *Sivatherium*, became important display objects, but attention focused especially on fossil elephants and other proboscideans. These could be great centerpieces, linked with the cultural valuation of elephants, and the specimens were often decidedly local or national (much like Peale's mastodon had been several decades earlier). Paleontology developed as a museum-based science, but also as an urban one, tied to large scholarly institutions in developing cities.

Yet, as the provisional gallery of paleontology in France's national museum of natural history demonstrates, paleontology and fossils often had problematic positions within large museums. An earlier literature, epitomized by Tony Bennett's *Pasts beyond Memory*, saw late nineteenth-century natural history and ethnographic museums as crucial for spreading didactic ideological messages around evolution and development according to powerful curatorial visions.[9] Yet the growth of museums was always complex, and messy resolutions of long-running disputes were as important as grand unifying visions.[10] This was especially the case for fossils, which were frequently dispersed between different remits. Arguments over who had authority over fossils affected not just the placement of the elephant of Durfort, but the role of the deep past itself.

The gathering of fossils also went beyond the museum, as the metropolis itself became a place where the deep past was encountered and reworked. While paleontologists would emphasize fossils being taken from distant regions, the acquisition of fossils was also a decidedly urban phenomenon. This partly operated in the expected manner of urban museums accumulating specimens from diverse places to display life's history, but the city also became an important site for the paleontological work itself. The trends seen in the fossils of the Paris basin continued across the nineteenth century, as cities became important excavation sites in themselves. As the city became rooted in the deep past, the worlds of prehistory and the Age of Mammals became part of the urban fabric.

Competition over Fossils at the Muséum

The lack of a fossil gallery in Paris before the 1880s is possibly surprising. For if there was a global center for research on extinct life in the nineteenth century, it was the Muséum.[11] The institution maintained the collections (and inherited prestige) of Buffon, Cuvier, and Blainville, and was central for studying many important specimens, such as *Palaeotherium*, *Anoplotherium*, *Megatherium*, and the animals of Pikermi. The Muséum was also a huge public attraction. An 1884 guide presented the Muséum as it wanted to be seen: "the Louvre of the natural sciences" and a manifestation of French prestige. "Its collections of natural history are the richest in the world, and represent enormous value; almost all learned French naturalists have been educated in our galleries and laboratories; many new sciences, including comparative anatomy and paleontology, were born in the Muséum."[12]

However, behind this facade of universal science and harmonious order, the Muséum was a complex institution, riven by internal tensions. A series of departments, corresponding to different branches of knowledge of the natural world, were each headed by a professor, who taught courses, directed research, and maintained collections, staff, and buildings. Each professor was "the veritable director, administrator and conservator of the collections which he knows better than any other,"[13] and aimed to control his discipline as a scientific fiefdom. While the Muséum was headed by an elected director, authority rested with the Assembly of Professors. They regularly met to debate administrative issues, salaries, budgets, extraordinary purchases, and new appointments. While this arrangement was intended to promote collaborative ways of knowing the natural world, it frequently led to conflict. Budgets, appointments, and specimens were fought over. Chairs and departments were not static, but were created and abolished depending on discussions among the professors. If there were deadlocks, the professors appealed to the Ministry of Public Instruction (the government department devoted to education and cultural affairs) to resolve them.

While a great deal has been written on the "golden age" of the Muséum in the early nineteenth century, its later development is much less studied. This is a major gap because it underwent continued expansion and reform across the century.[14] The Department of Comparative Anatomy continued to build on Cuvier's collections, and fossils were accumulated by the professors of mineralogy, botany, and zoology. From the 1840s inquiries were made by the Ministry of Public Instruction on whether a new chair of paleontology should be created—taking the term coined by Blainville in 1822 referring to the study of the life of the past. However, Michel-Eugène Chevreul, the Muséum's director, argued that a separate chair of paleontology was not required, "because paleontology is here taught by the different professors, and each has authority over the branch of paleontology relating to the field which he is in charge of . . . the history of fossil plants is held by the Professor of Botany . . . the history of fossil vertebrate animals is held by the Professor of Comparative Anatomy . . . and finally the two professors of Zoology are charged with the history of mammals, birds, reptiles, and fish."[15]

The resistance to a chair of paleontology was partly due to factional infighting over collections and authority within the Jardin des Plantes. However, the neglect of an independent chair did not reflect a neglect of fossils. Cuvier, Blainville, and Paul Gervais (all holders of the prestigious chair of comparative anatomy) had been central pronouncers on paleontological specimens. The study of fossils, utilizing methods, techniques, and concepts from different fields, was always "boundary work," subject to institutional debate. The blocking of a chair in fact demonstrates how deeply the study of fossils was entangled with the overall study of life.

Despite the misgivings of the professors, a chair of paleontology was created in 1853 (taking the place of the defunct chair of "Plants of the Countryside"). Its establishment was primarily due to lobbying by Alcide d'Orbigny, a naturalist and expert in fossil invertebrates who made a reputation through expeditions in South America, and gained support from the Ministry of Public Instruction. D'Orbigny partly justified the need for such a chair by comparing the situation in France with other countries. He wrote to the ministry that while "special chairs of Paleontology have been created everywhere: in England, Berlin, Vienna, St. Petersburg, Liège, Geneva, Philadelphia, and New York. . . . France—where the first serious elements of paleontology were developed, where this science has been most cultivated, France—alone among the civilized nations—still lacks a public chair of paleontology."[16] The lack of a chair was a stain on national honor and an indication that French institutions were not living up to the status of French science. Other factors in d'Orbigny's long report were the expansion of paleontology over the past decades—highlighting how thirty thousand fossil species were now known—and that the subject was useful for industry and in finding mineral resources. This highlighting of national pride, the extent of the field, and its economic importance built a successful case for a paleontological chair.

However, the chair of paleontology was far from prestigious or well resourced. D'Orbigny requested that all fossils in the Muséum be transferred to his ownership, but this was rejected by the other professors. Subsequent developments underline this problematic position. When d'Orbigny died in 1857, the chair was left vacant until 1861, when another geologically focused invertebrate specialist, Adolphe d'Archiac, was appointed. Further attempts by this chair to acquire fossils were unsuccessful, apart from a small victory in 1861 when the Departments of Paleontology and Comparative Anatomy divided the fossils in the mineralogy collections between them. After d'Archiac's death in 1868, the chair was granted to Edouard Lartet—an important shift, giving the chair to an expert in fossil vertebrates. However, illness and the Franco-Prussian War meant Lartet was never active in the role, and he died in 1871. Following this, there was a vote on whether the chair of paleontology should be abolished. While the majority of professors voted in favor of retaining the chair, this was only on the proviso that paleontology at the Muséum "should be from the point of view of stratigraphic geology."[17] Though fossils were significant, paleontology as a field remained precarious.

The Epoch of Gaudry

Paleontology at the Muséum was wracked by conflict, as well as a rapid turnover of professors. These obstacles were overcome by the next chair of paleontology, Albert Gaudry. While his career at the Muséum was blocked for a time owing to his strongly transformist views, by 1872 Gaudry had gained enough support to secure the chair of paleontology, which reinforced the emphasis on fossil vertebrates suggested by Lartet's appointment. Gaudry held the chair for thirty years, offsetting a major problem with previous professors of paleontology—their relatively short tenure and often outsider status within the Muséum.[18] During this time Gaudry built up a powerful Department of Paleontology, demarcating the subject as a distinctive presence, with its own collections and displays. Yet this was not easy, requiring protracted struggles within the Jardin des Plantes.

Gaudry was an adept player in museum politics, a proactive gatherer of specimens, and a committed promoter of paleontological science. In his popular works and public lectures, he discussed paleontology as having practical and theoretical importance. Stratigraphic paleontology was crucial for knowing the history of the earth and for studying it through layers of rock observable in the landscape. Yet there was a higher element, a more philosophical paleontology, which investigated the history of life and connections between life-forms in the past and the present, to show the principles organizing the whole of creation. In Gaudry's conception these two areas needed to be brought together, narrating a history of progress and harmony, with lessons for understanding nature and humanity.

It took seven years for Gaudry's position to seem secure enough for him to move on the core prize—the fossil collections held by the Department of Comparative Anatomy. In 1879, following the death of Paul Gervais, there was a short vacancy in that chair, and Gaudry proposed bringing the fossil vertebrates into the paleontology collections. This was unanimously agreed on by the Assembly of Professors. However, Georges Pouchet, appointed chair of comparative anatomy a few months later, was aggrieved. Conflict erupted over the ownership and display of the fossils, not just playing out in Muséum meetings, but exploding into the press. The radical left daily *Le rappel* rather mockingly described the "Fight at the Jardin des Plantes": "The Director of the Muséum agreed with M. Gaudry, who, aided by his laboratory assistant, started to remove the collections of his rival and replace the fetuses and skeletons with mastodons and cetaceans. M. Pouchet, informed of this sacrilege, demanded the despoilers leave. They stayed, and an altercation arose. Hearing the call, the old gentlemen, soldiers and nurses of the Jardin des Plantes, rose up. Some of them took up the cause for paleontology, others for comparative anatomy; they came to blows, tibias flew through the air, and spirit-jars were thrown."[19] Stories abounded of Pouchet obstructing the removal of specimens by placing an elephant skeleton near the entrance of the galleries, blocking Gaudry's workmen. Eventually, Pouchet's intransigence meant the minister of public instruction had to intervene. The fossils were awarded to Gaudry—but even in 1894 he still claimed Pouchet was hiding fossils in the comparative anatomy stores.[20]

The end result was that by 1885 paleontology gained most of the fossils from comparative anatomy and was able to open its public exhibition space. Gaudry was an effective media operator, and reports of the galleries in the Parisian press often paraphrased his descriptions. *L'année scientifique* noted that "the Muséum d'Histoire Naturelle has had a collection of fossil animals (both complete and fragmentary) for a long time, which the public did not know existed, because these precious bones were relegated to the most isolated and obscure corners of the establishment."[21] And *Le petit moniteur universel* wrote that this was "no more than the first stone of a great scientific edifice which they wish to devote to the study of the development of life on our globe in the times which precede the coming of man."[22]

Not all press reports were so supportive, and many illustrated problems with the gallery. Practical issues of access were often as important as grand theoretical debates. The popular periodical *La semaine des familles* was concerned that visitors needed to obtain a special ticket to see the gallery of paleontology, asking, "Why this semi-restriction to free entry? Alas! It is necessary to make a very sad confession for our country: the Muséum does not have the funds to pay a sufficient number of guards for the service of the collections." When the correspondent arrived in the exhibition room itself, he saw that "the new museum is installed in nothing more than a wooden gallery; I would not want to use the irreverent word of 'shack' [*baraque*]."[23]

The paleontological collections remained hampered by lack of funds and a secondary position. Yet this also implies the important expected features of museums: they were supposed to be grand and imposing, where the locale as much as the objects inspired wonder.

Despite these misgivings, reports on the paleontological gallery shifted in tone when describing the specimens, moving into awe and sensation. The report in *La semaine des familles* continued that "as soon as you cross the threshold of the paleontological museum, you feel a sort of stupor, a sensation of astonishment and being overwhelmed similar to that which you feel in the rooms of the colossi of the Assyrian museum in the Louvre." Following this "the eye is immediately seized, dominated, by the giant of this place—a place which does not lack for giants. But all the others seem small next to their powerful lord,—His Enormity, the *Elephas meridionalis*, or to call him simply in French, the elephant of the Midi."[24] Despite the lackluster setting, the fossils brought urban audiences into contact with the great "lords" of prehistory, who were as spectacular as the archaeological relics of human civilizations. In *La Lanterne* the journalist and writer Jean-Camille Fulbert-Dumonteil was even more struck, describing the gallery in terms of phantasmagoria and orientalist imagery, noting, "I know nothing as strange as this tremendous resurrection of extinct animals, these prodigious skeletons, these horrible jaws, fantastic beaks, extravagant necks, these implausible profiles, all these apocalyptic beasts which seem torn from a Japanese screen or fallen from the brush of a Chinese Callot intoxicated with opium. This is not a museum, it is a nightmare."[25] The galleries, and the fossil past, could be represented in varied manners. While Gaudry understood natural antiquity through empathy and wonder, for other visitors the skeletons recalled other museum spectacles, like the productions of unfamiliar human cultures and decadent art, invoking strangeness, even horror. Yet the reports generally agreed on one thing: the ancient past, viewable in the modern city through skeletons and mounts, was awe inspiring, and the creatures were understood through idioms of art and antiquity.

Gaudry nevertheless cited problems with the new gallery, as the location for his hard-won specimens remained "insufficient." A major problem was that the specimens were arranged around the gallery in an ad hoc manner, without any sense of history or change across time. This went against Gaudry's view that links across the eras as the crucial lesson of paleontology. He ended his account of the galleries with the plea: "As I believe that life continues across all the ages, forming chains from the first manifestations up until the empowered forms of modern times, I would like the museum of paleontology to take the form of a long gallery that follows the series of fossil beings without interruption."[26] This "dream of the arrangement of a Museum of Paleontology" was a hope for the future, but if it could be realized "our spirits would be filled with sublime joy; and not only naturalists would be satisfied, but also, I think, artists and philosophers."[27] Paleontology needed new foundations, mixing art, science, narrative, and spectacle. A later chapter will show how Gaudry attempted to enact his dream, although in 1885 this was only a plan for the future. Still, paleontology had gained a new position as an independent subject, binding varied means of knowing the natural world to aesthetic appreciation of the wonders of creation.

The London Museum and the Elephants of Essex

Across the English Channel, the great metropolis of London was also a center of paleontological work. The natural history division of the British Museum, while not as nationally dominant as Paris's Muséum d'Histoire Naturelle, was nevertheless a huge institution—and marked by similar tensions. Rather than a single multiheaded organization devoted to the natural world, these collections were part of the wider British Museum, which also contained antiquarian, historical, and classical collections.[28] Since 1856 the natural history collections were under the authority of Richard Owen, who reformed them into subdivisions of zoology, geology, mineralogy, and botany, each with its own "keeper," staff, and budget.

Unlike in Paris, where there were conflicts over where fossils should be kept, in the British Museum geology was recast as primarily about life's history and fossils. Owen noted, "Geology, indeed, seems to have left her old handmaiden mineralogy, to rest almost wholly upon her young and vigorous offspring, the science of organic remains." The subject told the vast history of the earth and its inhabitants, showing how "the globe allotted to man has revolved in orbit through a period of time so vast, that the mind, in the endeavor to realize it, is strained by an effort like that by which it strives to conceive the space dividing the solar system from the most distant nebulae."[29] Notably, this was a different vision from Gaudry's, where "stratigraphy" and the more philosophical questions of life's history were linked but distinct projects. In the British context, these two ways of knowing the deep past were inseparable.

In the British Museum the minerals and fossils were exhibited in six rooms on the upper floor in the North Gallery.[30] The fossil displays were organized along a rough chain-of-being schema, with successive rooms devoted to plants and fish, and then two rooms each for fossil reptiles and mammals. The sixth and final room contained the most dramatic specimens—the fossil edentates and elephants. After walking through the scale of life, visitors saw the composite *Megatherium*, the "corrected" skeleton of Koch's mastodon, Siwalik elephants, a cast of the *Dinotherium* skull, fragmentary remains of fossil marsupials, and the "skeleton of Guadeloupe," a headless human fossil of unknown antiquity.

Public descriptions of the geological galleries were mixed. To the journalist William Jerrold they showed that "of all animal types, man is the highest and the strongest—removed from the most powerful mammoth and megatherium—the bones of which he has refixed, that they may, as stones, tell the story of their wonderful characters when alive."[31] However, others regarded the displays as lacking. A popular guide published by Chambers in 1850 wrote that the "fossil mammalia . . . are at present distributed in a state of confusion . . . and some time must elapse before they can be put in order." The author recalled Cuvier's characterization of the Montmartre fossils as "'mutilated fragments of many hundred skeletons piled confusedly around me,'" a quote which came to mind when "glancing at the heaps of bones lying like lumber in room VI."[32] While paleontology ordered past life, the display fell short of what was fitting for a museum.

London was not just the site for centralizing collections. Like Paris with its gypsum quarries, the city itself was a site of paleontological discovery, and the urban metropolis simultaneously saw centers of accumulation in its museums, while also being a field site for labor and extraction. While the Paris mines had largely closed by the mid-nineteenth century, fossil discoveries in London stepped up a gear. London was subjected to building works like the sewers, metropolitan railway, and various schemes of urban renovation, which Linda Nead has discussed as being great public spectacles in themselves and shifting understandings of the urban environment.[33] As workmen and corporations dug, they reached layers of gravel and clay deposited during the "drift" period, which contained fossils. In 1883 the *Illustrated London News* featured a story on the "Remains of Extinct Mammalia at Charing Cross," noting how "it is not an uncommon occurrence for the workmen, when digging deeply into the gravels and brick-earths which underlie London . . . to exhume the fossil teeth and bones of animals now extinct, or of the early ancestors of others which still survive, either in a wild or semi-wild state, in this country or on the European and African continents."[34] These remains included mammoth, aurochs, deer, hippopotamus, and rhino, and were displayed at the shop of the taxidermist Roland Ward, before being returned to Drummond's banking establishment, on whose property they had been found. Many fossils, following British property law, were kept by their excavators or the institutions owning the land where they were extracted. However, many were sold or donated to museums and associations devoted to the deep past, including the British Museum. Museum collections and urban renovation were brought together by the Pleistocene fossils of the Thames basin.

Some of the most productive fossil sites were on the eastern fringes of the city, in the brick pits of Ilford and Loxford. This region had historically been farmland, forest and marsh with a scattering of small towns, but in the early nineteenth century it rapidly became an industrial agglomeration.[35] Manufacturing and dock work proliferated along the River Lea, making this an economically important but also impoverished landscape, with overcrowded housing and poor sanitary conditions. The area was also a source of raw materials for the growing city, with clay and gravel deposits exploited for construction materials. Large teams worked the pits and, while digging the "brick earth," frequently found bones and tusks of Pleistocene beasts.

The writer Manley Hopkins (father of the poet Gerard) described these finds in a column for *Once a Week* in 1860, "Essex Elephants." After a tour through east London, described as "so dull, so flat, so poverty-stricken, and so redolent of odours," he moved through the threatened Epping and Hainault Forests, where "the deer-trodden thickets are fast disappearing before modern improvements," before settling on the developing suburban conurbation and the fossils found there. In "the long, grey, old church of West Ham . . . leaning against an altar-tomb, two immense bones rested—one being a shoulder-blade, three feet in length, and the other a rib—concerning which relics the inquirer was shortly answered that they were *mammoth bones*."[36] The modern city contained relics of the deep past. Older traditions of placing large bones from the earth in churches continued into the nineteenth century. However, these practices now conceived them as the bones of ancient elephants, which showed local antiquity.

The fossils in the Ilford pits attracted numerous collectors. the most prominent was Antonio Brady, an upwardly mobile clerk in the navy's victualing division and resident of nearby Stratford.[37] Brady was involved in London scholarly society (being a fellow of the Geological Society), local politics (involved in establishing the West Ham & Stratford Dispensary, and Plaistow and Victoria Dock Mission), and social networks with London naturalists. This local elite status gave opportunities to collect fossil bones. Brady remembered how these confluences came together in the 1840s, when, during a dinner with William Buckland's son Francis, he received a letter from his "dear friend Mrs. Curtis," the owner of the Uphall brick pit, stating "that the workmen in digging for Brick-earth had again come upon some more large bones, and knowing my geological proclivities, placed them at my disposal." The bone turned out to be "the femur of a large elephant, luckily in a very fair state of preservation," and was the first of many such acquisitions.[38]

Over the next thirty years Brady amassed a vast amount of material—over a thousand fossils, all stored in a private museum in his home. A procedure developed where, if workmen discovered fossil bones, they would alert Brady by letter. Brady would then visit the site and pay the men's wages while the fossils were excavated (something enabled by Brady's £1,000 annual salary). While for the workmen this seems to have primarily been an economic arrangement, for Brady this was a mission of national pride and religious interest, as "England, I am proud to feel, is the original home—nay birth-place—of this noble science, noble alike to the physicist and the divine," a science which could "trace the history of the world which we inhabit, written by the finger of God in the rocks."[39] As in France, paleontological knowledge was a patriotic project (with England also being presented as the originator of the science). Yet Brady also persisted in the project of natural theology, with knowledge of nature and knowledge of the divine being inseparable.

The Ilford fossils gave Brady significant inroads into British natural history. He corresponded with Henry Woodward, one of the geological curators at the British Museum, and entered into an agreement with the head preparator, William Davies, who would prepare and preserve fossil bones. This was particularly important because (as with the elephant of Durfort) excavating and preserving fossil material required specialized techniques. While Brady's first complete elephant femur was unusually well preserved, most bones from the site "when wet, are in most instances so soft as not to bear the slightest touch, and when dry are so friable as to be equally unmanageable. To exhume bones so situated manifestly requires great skill and care . . . [M]any otherwise beautiful specimens crumbled to decay and were destroyed from this cause."[40] Brady and his unnamed field assistants developed ways of securing fragile specimens with plaster and bandages, but this was still a complex task.

Visiting the Uphall Pits became a popular activity for members of learned societies, mixing leisure with knowledge of the deep past. In 1871 Henry Woodward described an excursion to Uphall by the Geological Society, which included a lecture, collecting specimens in the pits, and then a dinner at Brady's home to view his "magnificent Collection of Mammalian Remains."[41] The Manchester geologist William Boyd Dawkins was taken by Brady to the site and reported on the "most remarkable accumulation of bones carefully left *in situ* by the workmen," including "horse, *Rhinoceros hemitoechus*, Mammoth, urus, either brown or grizzly bear, and wolf." He wrote how "the bottom of an ancient river with all its contents lay before our eyes—a river in which all these animals had been drowned, and by which they had been swept to the exact position which they then occupied."[42] Ancient floods and disasters persisted in this imaginative reconstruction, but it was now a more localized event in the lands that were to become Ilford, rather than a universal Deluge.

Another account of visits to the Ilford pits was given by Henry Walker to the Essex Field Club in 1880, describing an excursion of "elephant hunting" by over sixty people. Fossil "hunting" was rational recreation, allowing educated publics to partake in scientific work. It was also comprehended through colonial tropes of exploration, hunting, and engaging with "native" inhabitants—illustrating the strong hold of imperial tropes over the imaginations of the British middle classes and how paleontology bound imperial dynamics and science. Walking from the Ilford train station was an adventure, as "the flats along which we have walked have reminded us of the rice and paddy fields of Ceylon, but another vegetation here confronts us. In Indian file we thread our way through ranks of well-hoed potatoes." Descending into the brick pits on wheelbarrow planks illustrated that "elephant hunting in Essex, in these modern days, is an underground sport."[43] This "hunt" also gave opportunities for acquisition. The participants immediately dug in the pits, finding the shells of mollusks who "lived in the waters which the mammoth frequented," having "since shared his grave for thousands of years."[44] Collection and acquisition were therefore important parts of engaging with the paleontological site.

While digging, the party interacted with the local working-class population, also described through strongly colonial language:

> Suddenly an alarm is given. We are not to invade these sacred haunts of ancient life with impunity. The aborigines of the country have been gradually closing in upon us unseen. They now appear, some of gigantic form, looking down upon us exultingly from the brink of the pit. We are fairly caught—outflanked and surrounded by a wily foe. Not an instant is to be lost. With great presence of mind Sir Antonio, our leader, advances with dignified mien to parley with the chief. It is an anxious moment. Happily, he speedily returns with good news. The natives are not hostile, but amicable. They are inclined to trade and barter. Better than all, their wares consist of the very spoils we are in search of. They carry with them, wrapped in textures of evidently European fabric, some of the enormous stone-like teeth of fossil elephants, and various gigantic bones. A brisk exchange is soon set up.[45]

The account displays unfamiliarity with and fear of the local inhabitants by these bourgeois visitors. However, the source also indicates that fossils were collected by the working-class people of the area. It is difficult from this or other sources to gain a sense of how the people of Ilford understood the fossils, although it raises an important parallel with the way local people were always crucial for paleontological work—in both European and extra-European settings—while often being belittled in scholarly accounts. And while Walker presented the collecting as a primarily economic activity, with the locals selling cloth-wrapped specimens to tourists, the acquisition of fossil bones and teeth required careful preservation, suggesting they had considerable expertise in extracting fossils.

The prize specimen from the Uphall Pits was a complete mammoth skull unearthed in 1864, which became known as "the Ilford Mammoth." The size of the specimen meant Brady did not attempt the excavation himself, but instead called for support from the British Museum. The skull's extraction was a delicate operation, supervised by William Davies and described by Henry Woodward. The skull was damaged when it was found, "resting half exposed in compact brickearth, requiring a spade or trowel to remove it,

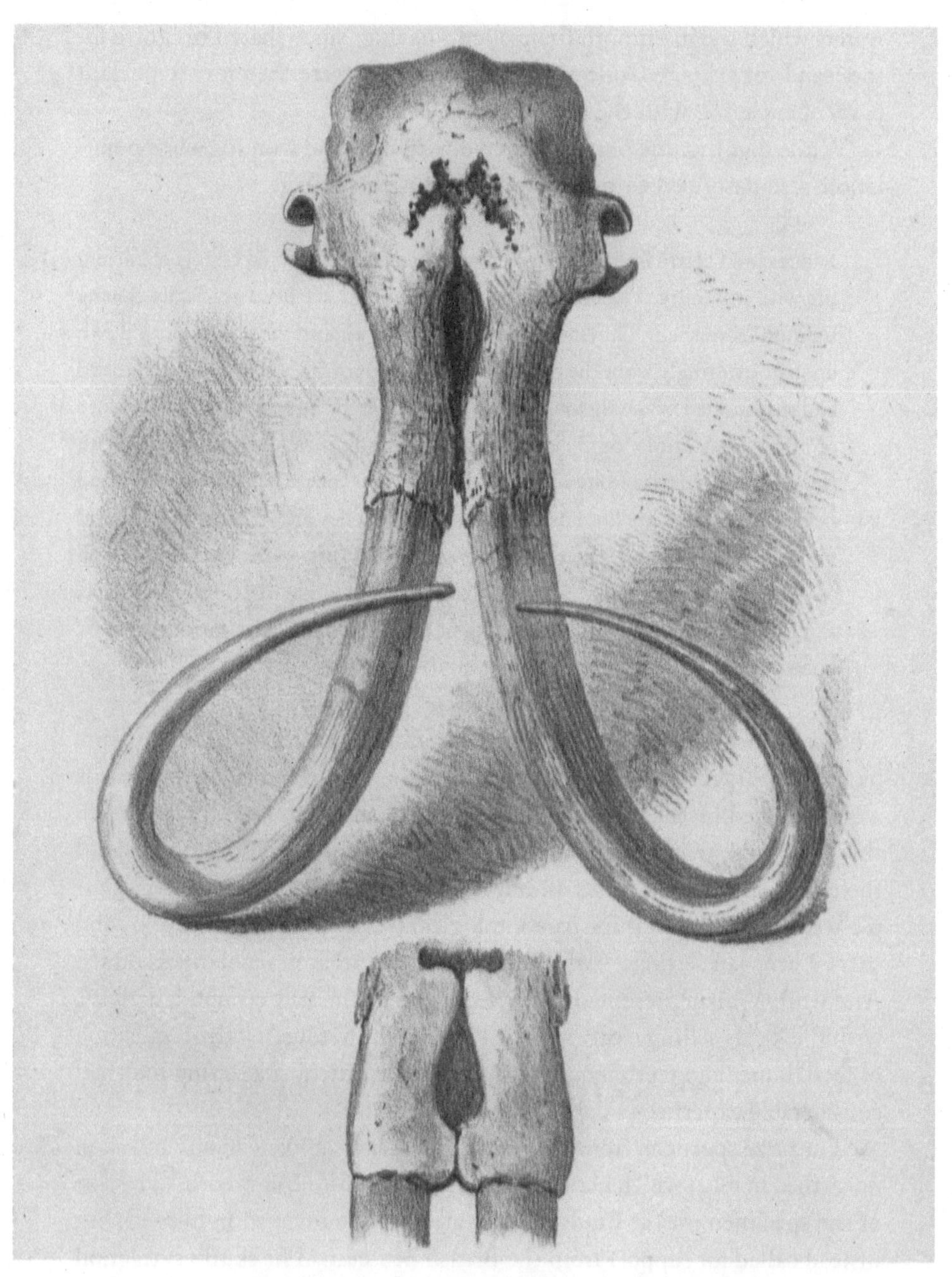

FIGURE 5.2. The skull of the Ilford Mammoth. H. Woodward, "On the Curvature of the Tusks in the Mammoth," fig. 1. Courtesy of the Staatsbibliothek zu Berlin-Preussischer Kulturbesitz, Shelfmark: Mf 2267–5.1868.

but the fossil itself as friable as decayed wood or tinder, the ivory of the tusk being equally soft and shattered." The tusks were supported by iron bars and plaster, and "the whole was swathed with bandages of canvas, hay, and cord, like a mummy. When thus secured, six men turned it gently over from its matrix and placed it upon a long plank prepared for it."[46] This was a delicate and skilled operation, requiring time and labor. Extracting the skull and tusks from the pit took almost a week. It was then prepared by Davies in the British Museum's workshops and proudly described by Henry Woodward as "a nearly perfect cranium" and the first complete mammoth skull found in Britain. Woodward gave a speech on the Ilford Mammoth at the 1868 International Congress of Prehistoric Archaeology and Anthropology in Norwich, drawing attention to the impressive specimen, which gave credence to the idea that the tusks attached to the mount of "the Adams Mammoth" in Saint Petersburg were not the actual tusks belonging to the specimen, and were incorrectly mounted, curving outward rather than inward.[47]

Rather than being kept in Brady's collections, the Ilford Mammoth was donated to the British Museum and placed in the public galleries—possibly in obligation for the museum's role in extracting the fragile specimen, but also reflecting how collectors entered into alliances with established institutions by offering material with an eye for posterity. By the 1870s Brady sought a permanent home for his collection, tapping into the educated life of the city through his paternalistic philanthropy. However, the British Museum was not his first choice. Brady had originally hoped to sell his fossils to the Bethnal Green Museum, set up in the East End of London in 1872 with surplus material from the South Kensington Museum and Richard Wallace's collections of Renaissance art. Brady justified this bequest in terms of the need "to preserve local collections in local museums," and to assist in "elevating the tastes and habits of the working men, my neighbours, at the East End."[48] However, the Bethnal Green Museum passed on the opportunity to purchase the fossils, citing lack of space and funds.

Brady therefore used his connections to sell his collection to the geology division of the British Museum as his second choice. The sale was a coordinated affair. In 1874 Brady drafted a *Catalogue of the Pleistocene Vertebrata* in collaboration with Henry Woodward, documenting a thousand specimens, costing around 1,000 guineas to acquire. The collection was offered for £550—half the implied cost—on the condition that it should not be broken up and should continue to bear Brady's name. This was important

FIGURE 5.3. The overcrowded Room 6 in the British Museum's geology galleries. The Ilford Mammoth can be seen in front of the *Megatherium*. Photograph by Frederick York (1875). Copyright British Museum Images.

not just for personal prestige, but for the values around paleontology. Brady wrote how the completeness of his collection bolstered its importance, as it contained "many varieties and series of comparatively recently extinct animals, not only showing slight difference of form in the several species (some doubtless owing to difference of sex), but connecting as it were, the past with the present orders of things, and what is more interesting still, showing the differences, etc., in the forms of life which existed, when these extinct animals roamed in forests and marshes, now the site of London, from those of similar, or nearly similar species, which now inhabit various portions of the globe."[49] Brady's collection allowed the life of the past to be understood in its diversity and variation. Moreover, the animals showed the history of the land now occupied by the great Victorian metropolis and connected it with parts of the world where similar animals still lived.

Acquiring the collections bolstered the natural history division of the British Museum, but also caused problems. Adding to the difficulties highlighted by the Chambers guide, the 1860s and 1870s saw the museum acquire even more fossils. Demands for additional space were constant features of Owen's departmental reports, noting how "offers of collections for sale and opportunities of acquisitions by purchase are declined or postponed, which, with due exhibition space, might have had claims for submission to the consideration of the Trustees."[50] While this statement partly reflected Owen's desire for more status, photographs of the galleries bear these comments out. The sixth room in particular was overstocked, with bones, tusks, and cabinets crowding one another out.

This need for space, and Owen's continued attempts to expand the natural history collections, led to a major controversy.[51] Owen and his supporters lobbied for the creation of a new and separate museum of natural history. One statement of Owen's ambitions was an 1862 pamphlet, *On the Extent and Aims of a National Museum of Natural History*, intended as a riposte to the House of Commons's rejection of the plan to move the natural history collections to a new site in South Kensington. Owen argued not just in scientific terms, but (like d'Orbigny) through national and imperial prestige, noting how "every European nation now possesses its National Museum of Natural History,"[52] and that "the greatest commercial and colonizing empire of the world can take her own befitting course for ennobling herself with that material symbol of advance in the march of civilization which a Public Museum of Natural History embodies."[53] The museum would have "the humanizing and ameliorating effect of such mere opportunity of contemplating the extent, variety, beauty, and perfection of Creative Power upon the people of a busy and populous nation."[54] Paleontology was crucial in these plans, as "no specimens of Natural History so much excite the interest and wonder of the public, so sensibly gratify their curiosity, are the subjects of such prolonged and profound contemplation, as these reconstructed skeletons of large extinct animals." Great beasts and their skeletons inspired interest in the natural world. Paleontology could also "be gratified at comparatively small cost" through the use of casts, as "a fossil bone and a coloured plaster-cast of it are not distinguishable at first sight."[55] Fossils were crucial for building appreciation of the world, mixing learned contemplation of specimens with wonder and curiosity.

However, as in Paris, calls for separation were often opposed. Opponents

FIGURE 5.4. The paleontological galleries in the British Museum of Natural History at South Kensington in 1882. The Ilford Mammoth can be seen toward the back of the gallery. Courtesy of the Trustees of the Natural History Museum, London.

queried the amount of space Owen requested, the cost of moving the specimens and constructing a new building, and also that the proposed site at South Kensington was too far from London's center to be easily visited. Meanwhile, many naturalists were opposed on intellectual grounds. Separating "nature" and "culture" was anathema to many scholars, as it implied a division between objects created by God and those created by humans. Other scholars opposed Owen's ambitions to maintain the collections under his control. Thomas Henry Huxley, along with many other scientific naturalists, advocated splitting the natural history collections into separate museums devoted to zoology, geology, and botany, with clearer focuses, more space for research laboratories, and more employment opportunities for naturalists.

Owen's supporters eventually won out, although the institution remained formally part of the British Museum, with the official title of British Museum (Natural History), although often referred to as the British Museum of Natural History (the name it would eventually adopt). A huge new building, designed by Alfred Waterhouse, was constructed in South Kensington on the model of a cathedral of nature, dramatizing Owen's views of the natural world. The architecture made a clear division, with the east wing devoted to extinct organisms and the west to extant creatures.[56] The building was also constructed with a sense of spectacle, decorated with carvings of extinct animals, including pterosaurs, saber-toothed cats, and paleotheres. The fossil collections were the first to be moved to South Kensington. In the exhibition space, mammals and reptiles dominated, with the mammal displays running the length of the geological galleries and extending into a further pavilion. While wall cases in a chain-of-being arrangement lined the galleries, the central corridor contained iconic specimens on individual stands, with visitors able to circulate in the space around them. The *Megatherium*, glyptodon carapaces, and two skeletons of *Dinornis* dominated the end. The main gallery was filled with a parade of proboscideans: first the skull of *Mastodon andium* from Chile; then the whole skeleton of the Ohio mastodon; the *Dinotherium* cast; the skull of *Elephas ganesa* from the Siwaliks; and as the final elephant, "in the centre of the floor is placed the fine skull, tusks, and lower jaw of the Ilford Mammoth."[57] The Ilford Mammoth gained its pedestal, but this depended on the dense networks of local interest and scholarly contestation that defined paleontological work in large cities and museums.

CHAPTER 6

Beasts from the West

Science at a Distance in North America

"KING OF LIVING THINGS AT THAT DAWN OF TERTIARY TIME, I NAMED HIM *EOBASILEUS,* and the species from its peculiar horns, *Eobasileus pressicornis.*"[1] This portentous statement was made by Edward Drinker Cope in an 1874 article in the *Pennsylvania Monthly*, "The Monster of Mammoth Buttes," describing fieldwork in the Bridger Basin in Wyoming Territory. While Cope stated that "the extraordinary scenery of this ruined world cannot be depicted in a short article, and indeed, language almost fails in the attempt,"[2] he nevertheless wrote a florid account of mystery, death, and horror. "Black Butte," where the fossils were found, was a "dark mass [which] stands like a ruined robber castle, guarding the great highway," while nearby Bitter Creek was "like a gathering of the exudations of Hades . . . its sources trickle down horrid gorges of the bad lands, and from beds of bones of the dead of former ages."[3] The land itself showed ancient transformations, as rock layers containing the bones of the dinosaur *Agathaumas* gave way to strata lacking great reptiles and containing extensive mammal fossils—a transition whose causes Cope could only guess. These "desolate forums" looked like "the dining table of the gods," filled with "bones of the titans they had slain; huge limbs, piles of vertebrae, fragments of great hips that once swung capacious bellies."[4] The most impressive specimen was the two-hundred-pound skull of

FIGURE 6.1. Imagining *Loxolophodon* in its ancient environment. Cope, "Monster of Mammoth Buttes," 521. Courtesy of the American Philosophical Society.

Eobasileus—"the invaluable relics of this ancient king"—carried out as paleontological plunder to waiting wagons through "the Mammoth Buttes, more perennial than the tomb of Cheops, more vast than the labyrinth of Minos."[5]

Cope described the West as a land of death and antiquity, interpreted through imagination and technical knowledge. The paleontologist was simultaneously a scientific worker, supervisor of labor extracting the specimen, traveler in a colonized region, and imaginer of the beasts themselves. *Eobasileus* symbolized the territory and deep time. It was from early in the Age of Mammals, possibly of similar antiquity to the creatures of the Paris basin. Yet unlike the diminutive *Palaeotherium* and *Anoplotherium*, the "Dawn King" approached an elephant in size, and possessed fearsome tusks and a massive horned skull. This beast showed the dramas of past life and American science's claims over natural wonders. Cope included a drawing of how a related animal, named *Loxolophodon cornutus*, may have lived. Gregariously assembling in herds, with trunks and large flapping ears, they showed changes in the landscape, the strangeness of the ancient fauna, and the scale of primeval American life.

However, Cope's Dawn King was problematic. His expedition was part of a rush by East Coast scholars to the North American West in the early 1870s. Fossil hunting followed the US government's invasion of Native American lands, which were subjected to surveying, settlement, and commercial exploitation. Career-building scholars like Cope, Joseph Leidy, and Othniel Charles Marsh either personally worked in this region or (more usually) supervised teams of workers at a distance. They used western fossils to build scientific reputations, and judge the paleontological past and modern environments. They also focused on similar specimens: the skulls (and less frequently vertebrae and limb bones) of large, horned herbivorous mammals. Cope called these *Eobasileus* and *Loxolophodon*, Leidy named them *Uintatherium*, and Marsh had *Dinoceras* and *Tinoceras*. Who had authority to describe these animals, and what they showed about deep time in North America, became a central issue for paleontology in the United States and beyond.

The "Bone Wars" between the Pennsylvania naturalist Edward Drinker Cope and his New England rival Othniel Charles Marsh have been understood as crucial to the development of US science.[6] While popular literature has seen the Bone Wars as a mainly dinosaurian affair, the first major explosions were over large fossil mammals. While this was a conflict over leadership within American science, Cope and Marsh were not the only workers in the West. Networks of collectors and collaborators were crucial, and other scholarly institutions moved into the territory. While public presentation revolved around control and prestige, excavation often depended on more prosaic, but no less significant, issues of access, communication, and transportation. Although Leidy, Cope, and Marsh did visit the Bridger Formation at various times, the overwhelming majority of fossils were gathered by military, geographic, and geological surveys, student parties, and collectors drawn from frontier society. Work in the West depended on local expertise and long-range communication. Science worked at a distance through the new technologies of railway and telegraph.[7]

Creating "the Badlands"

The term *badlands* was promoted by French traders, who called the region the "mauvaises terres à traverser" (bad lands to travel across), and most likely derived from the Lakota name for the region of "mako sica," meaning bad

FIGURE 6.2. A conventionalized depiction of the badlands. D. Owen, *Report of a Geological Survey*, 1:196. Reproduced by kind permission of the Syndics of Cambridge University Library.

lands.[8] These landscapes were important for Indigenous people, and while not traditionally used as extensively as the more open and watered grasslands, they had important roles in healing rites and vision quests. There were also numerous traditions about the region. A Dakota story saw the badlands as made from a water monster named Unktehi, who was killed after declaring war on humans and whose bones were still visible in the landscape.[9]

US geologists and surveyors meanwhile represented the badlands as a dry, inhospitable, and uninhabited region. David Dale Owen a member of a US surveying party in the late 1840s, wrote an account that built the formula for describing the badlands. After crossing "the uniform, monotonous, open prairie, the traveler suddenly descends, one or two hundred feet, into a valley that looks as if it had sunk away from the surrounding world." Within this were "deep, confined, labyrinthine passages, not unlike the narrow, irregular streets and lanes of some quaint old town of the European Continent." The region was like "some magnificent city of the dead, where the labour and the genius of forgotten nations had left behind them a multitude of monuments of art and skill."[10] Human and natural antiquity were linked, and descent into the canyons represented entry into a ruined past.

The inhabitants of the "city of the dead" were also visible. Travelers described "fossil treasures" weathering out of the rocks, adding to the morbid antiquity. Fossils also circulated through trading networks. In the late 1840s a fossil jawbone was acquired by the trader Hiram Prout from "a friend residing at one of the trading posts of the St. Louis Fur Company, on the Missouri River," who had in turn had found it at the White River "one hundred and fifty miles south of St Pierre."[11] Prout thought it the remains of a "Gigantic Paleotherium," whose thirty-inch jaw massively exceeded that of Cuvier's *Palaeotherium magnum*.[12] Owen wrote of the antiquity and strangeness of the bones, stating how "every specimen as yet brought from the Bad Lands, proves to be of species that became exterminated before the mammoth and mastodon lived, and differ in their specific character, not alone from all living animals, but also from all fossils obtained even from cotemporaneous geological formations elsewhere."[13] He described finding complete remains of what seemed to be Prout's gigantic *Palaeotherium* "eighteen feet in length, and nine in height," which "must have attained a much larger size than any [animals] which the Paris basin afforded." However, "the crumbling condition of the bones rendered it impossible to disinter them whole."[14] The fossils could only be observed in situ, or taken as fragments.

Western fossils were often sent to East Coast institutions, which maintained comparative collections and were linked with centers of scholarly and political power. In the 1850s and 1860s, many fossils excavated in the West were analyzed by Joseph Leidy, professor of anatomy at the University of Pennsylvania. Leidy has often been described as a "gentleman scientist," initially training as a doctor and studying various fields of natural history, especially fossils from the western territories sent to him by travelers, traders, and settlers. A particularly important collaboration was with the surveyor Ferdinand Vandeveer Hayden (an original member of the Smithsonian's Megatherium Club). Hayden's expeditions prospected western territories and, as discussed by James Cassidy, his reports conceptually transformed the region from "The Great American Desert" into a new imagined land suitable for settlement and agriculture.[15]

Leidy wrote several pieces on Hayden's fossils, most notably a large monograph titled *The Extinct Mammalian Fauna of Dakota and Nebraska*

(1869). There are some important points about Leidy's works. First, they were highly technical, arranging fossils along the chain of being, with extensive descriptions, measurements, and illustrations. Leidy opened *The Extinct Mammalian Fauna* with the explanation that it was "a record of facts, in paleontology" and "a contribution to the great inventory of nature. No attempt has been made at generalizations or theories which might attract the momentary attention and admiration of the scientific community."[16] This partly reflects the contemporary status of paleontology as a science balancing rigor and claims to "objectivity" against accusations of speculation. Leidy's cataloging style also meant he focused on the most numerous specimens, rather than the most dramatic. The fauna as a whole was "not remarkable for large size,"[17] and in contrast to Prout's monster *Palaeotherium*, Leidy emphasized the more numerous smaller creatures.

Leidy's works were descriptive, but nevertheless contained imagination and conjecture. This mixture is seen in Leidy's interpretations of the most numerous fossil animals, a "peculiar and extinct" family named the oreodons. The oreodons were small to medium-size herbivores, with hornless skulls having "somewhat the form of that of the peccaries; the cranial portion especially resembles that of the camel"; "well-developed" canines and incisors; and molars like those of the deer. In sum, "the character of the genus cannot probably be better expressed than when it is called a genus of ruminating hogs" that linked ancient forms. Leidy reconstructed their lifestyles and habits, living in "large herds, which once roamed over the extensive prairies and through the dense forests of ancient Nebraska, as the Peccaries do in our own times in South America."[18] Behind the typology was an imagining of the West once covered by forests and filled with numerous animals.

Leidy's comparisons oscillated between animals in the modern West and the fossil fauna of Europe, especially France. He aligned the oreodons with American animals like peccaries and (assumed) Old World animals like camels. He placed some carnivore bones within the genus *Amphicyon*, coined by Lartet,[19] but described it as between a prairie wolf and red fox in size, using modern North American animals as comparators. More carnivore fossils were linked with *Hyaenodon* (a creature named by French scholars from fossil jaw fragments and interpreted as a marsupial predator "about the size of a Thylacine"[20]). The largest American species, *Hyaenodon horridus*, surpassed its European counterpart, and was "probably the most sanguinary and dreaded enemy of its numerous ruminant associates, the

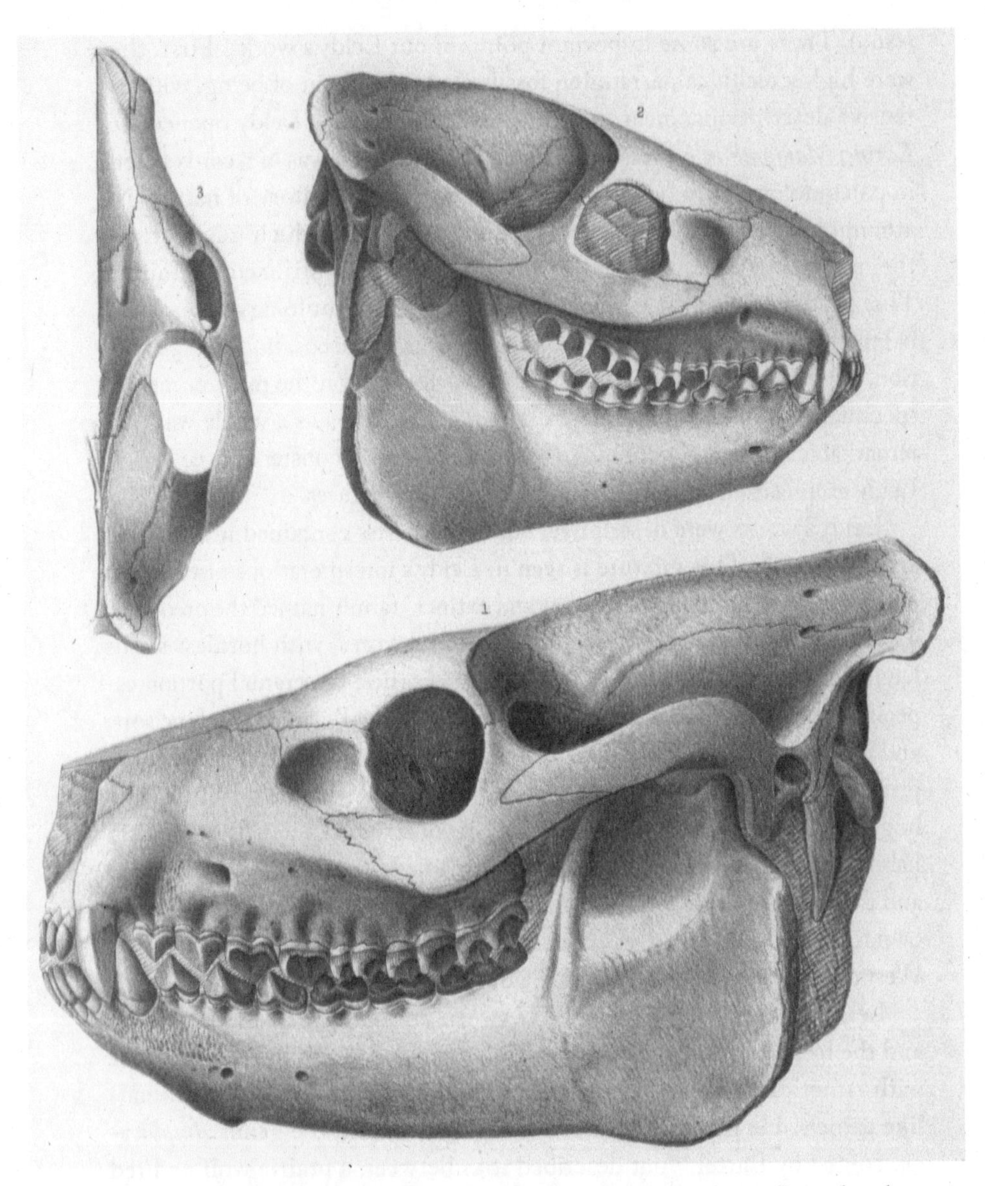

FIGURE 6.3. Oreodon skulls. Leidy, *Extinct Mammalian Fauna*, plate 6. Reproduced by kind permission of the Syndics of Cambridge University Library.

oreodons, etc., [and] greatly exceed in size any of the described European species; its skull fully equalling that of the largest individuals of the Black Bear, *Ursus americanus*."[21]

Leidy's choice of comparisons undoubtedly partly owed to access. In addition to French publications being available in Philadelphia, Leidy had studied in Europe during the revolutionary year of 1848, and sent some casts of the Paris basin mammals to the Philadelphia Academy of Natural Sciences.[22] The prestige of Cuvier's and Lartet's studies meant French works remained crucial. However, this was not simply deference, as Leidy was confident enough to differentiate the American fauna from the European. He reanalyzed the teeth of Prout's gigantic *Palaeotherium* and wrote that it was probably a "different genus . . . and should the suspicion prove correct, *Titanotherium* would be a good name for the animal, as expressive of its very great size."[23] The American fauna was different from the European, and worthy of study in itself. Its distinctiveness stretched back into geological time, and understanding American antiquity required comparison of fossil animals from all over the world, as well as modern western creatures.

The Rush to Fort Bridger

While Leidy relied on relatively small-scale networks, there was massive economic, scholarly, and territorial expansion into the northern plains in the 1860s. Science was still undertaken at a distance between western field sites and East Coast centers of interpretation, but was reframed, as the fossils of the West were looked on as prestigious objects, exploitable through transportation, communication, and political and military control. These expansions drove the feud between Othniel Charles Marsh and Edward Drinker Cope, memorialized in the paleontological literature.

Despite their rivalry, Cope and Marsh followed parallel careers, using their independent wealth and scholarly connections to control fossils and field sites and to expand influence within US government agencies, scientific associations, and the media. Marsh, hailing from New England, gained support from his uncle George Peabody, the prominent businessman and financier. In 1866, with Peabody's sponsorship, Marsh was appointed professor of paleontology at Yale, and a new natural history museum was established under Peabody's name.[24] Cope was from a rich family of Philadelphia Quakers and was apparently inspired by fossils at the age of six after seeing Koch's *Hydrarchos* exhibited in Philadelphia.[25] As a student he studied anatomy and zoology, taking Leidy's course around 1861. After a brief stint as a professor at Haverford College in Philadelphia, he worked through his own private collections and the social world of US scientific associations. Both Cope and Marsh had significant private means, and secure East Coast bases from which they could seek control over North American paleontology.

The turning of paleontological attention to western territories beginning in the late 1860s was not simply scientific conquest by Cope and Marsh, but drew from changing political and economic conditions in the West itself, especially in the fossil-rich "Bridger formation" in Wyoming Territory. While the oreodons of the White River seemed to have lived sometime before the mammoths and the mastodons, the Bridger fossils were placed in a more definite period. As already demonstrated from Cope's reveries at Black Butte, this place showed the very beginnings of the Age of Mammals, with the rocks grading seamlessly from strata containing dinosaurian reptiles to those containing mammals. Work in this region could therefore demonstrate an epochal transition from the Age of Reptiles to the Age of Mammals.

Beyond the privileging of the landscape, political changes drove exploitation. In prior decades Indigenous power over the northern plains was well established, and incursions by settlers and US government agencies were small scale, through trading stations and geographic surveys. However, the completion of the transcontinental railroad meant white settlement increased dramatically. As the military force of the US government was directed against the Plains tribes, transportation networks, geological prospecting, and telegraph networks expanded. Paleontologists followed in the wake of economic and political annexation, taking advantage of the extension of military control.

These dynamics can be seen in Fort Bridger itself. This settlement grew out of a trading and telegraph post established by the fur trader James Bridger and his partner Louis Vasquez in 1843. After being burned down during the Mormon Wars, it was rebuilt as a military station. Fort Bridger also benefited from the transcontinental railroad, with the Carter Station depot being ten miles north. Carter Station itself was named after Judge William A. Carter (often called Mister Fort Bridger), simultaneously local judge, director of the trading post, and post office agent. Carter was an important patron, assembling a library, sponsoring schoolhouses and district courts, owning a sawmill, and having interests in mining and cattle ranching. In particular, Carter worked with the army surgeon J. K. Corson, and James van Allen Carter, who both married two of Carter's daughters. The store at Fort Bridger was at the center of trade networks, with civilians, military personnel, Indigenous people, settlers, farmers, and surveyors all relying on this node.[26]

In the early 1870s the Bridger basin saw no less than three paleontological collecting efforts, as Leidy, Cope, and Marsh all followed the well-worn tracks of travel and exploration. The railway was particularly crucial. After the 1868 American Association for the Advancement of Sciences meeting in Chicago, Marsh took the railroad into Wyoming to observe the fossil landscape and organized collecting teams from Yale in 1870. Having prominent East Coast scientists in the territory was a boon to local elites. Judge Carter wrote to Marsh, "You will find no more interesting and satisfactory fields for investigation than this portion of Wyoming Territory, one that is so entirely free from the dread of hostile Indians and whose silence has so seldom been broken by the foot of Science,"[27] and recommended a photographer and suppliers. Meanwhile, Leidy traveled to Fort Bridger in 1872, on his first trip to the West, and was shown sites by van Allen Carter and Corson. Cope used his personal connections to be named the official paleontologist on Hayden's survey (this did not come with a salary, but facilitated travel to the region). The three scholars showed distinct ways of opportunistically taking advantage of US invasion and economic exploitation, working through governmental and informal networks.

The paleontologists described the landscape in various ways, some conventional, others quite new. Cope's florid description has already been discussed. Leidy also portrayed Wyoming as a place of dead antiquity. He wrote to the Academy of Natural Sciences how "the country is the most remarkable that I have ever seen," with the buttes resembling "great earthworks or huge railway embankments," and "their eroded sides give them the appearance of a vast assemblage of Egyptian pyramids flanking the plains above." The fossils weathered out of an earth which seemed "parched" and of "overwhelming silence." The large numbers of fossils and ruinous landscape were depicted as "the wreck of another world which was once luxuriant with vegetation and teemed with animals."[28]

As well as communing with the landscape, Leidy described its former inhabitants. There were few full skeletons, but many fragmentary bones and teeth. The fossil turtles ("the most abundant vertebrate fossils") were skipped over quickly, most attention being focused on the "largest and most extraordinary mammals." The first were the jaws, crushed cranium, teeth, and bones of "a tapiroid animal exceeding in bulk of body and limb the living Rhinoceroses," which Leidy named *Uintatherium robustum*. Beyond this large herbivore, "the most exciting incident" was "Corson's discovery of the upper canine teeth, apparently of the most formidable of Carnivores, the enemy of the *Uintatherium*." These were almost nine inches long, resembling those "of the sabre-toothed tiger," known from Europe and Brazil. Descending into another reverie, Leidy described how "these canine teeth terminating in lance-like points must have proved most terrific instruments of slaughter. Their possessor was no doubt the scourge of Uinta, and may therefore be appropriately named *Uintamastix atrox*."[29] The animal remains gave knowledge of the lost world and projected violence and combat.

While Leidy and Cope became dreamlike communers with the ancient landscape, another image of science in the West developed at this time: the paleontologist as fossil hunter and scientific frontiersman. This was the view presented by Marsh's teams, with one of the most famous historic images of paleontology being a photograph of the Yale College Expeditions of 1872 with the team decked out in prospector outfits. The image was highly contrived, and though rarely noted, official photographs of the other Yale expeditions showed the participants in formal urban clothing without the "western" accoutrements. However, it nevertheless became a defining image, connecting paleontology with a masculine frontier identity. An article by one participant, Charles Betts, in *Harper's New Monthly Magazine* in 1871, described the Yale expeditions through dramatic western incidents: traveling under military escort (and accompanied for part of the journey by Buffalo Bill), encountering Lakota and Cheyenne war parties, and enduring the climate and wildlife.[30] Apart from an initial vignette imagining the interior of North America in distant geological times as being a gigantic seaway, paleontological research was barely mentioned.[31] Far more important was the transition of the Yale students into rough frontiersmen. As they traveled home, Betts recalled: "Our ruffianly appearance created consternation among sober railroad tourists. Months of hardship, labor, and adventure had made many a rent in our well-worn clothes; and the buckskin breeches and army blouses of several members gave to the party a wild and warlike character, in keeping with the open display of revolver and bowie-knife, and bronzed faces covered with the un-trimmed stubble of a season."[32] Adventuring in the West and becoming a tanned masculine explorer of dangerous regions were an important part of their scientific persona.

FIGURE 6.4. The Yale College Scientific Expedition of 1872. YPMAR.001445, YPMA. Courtesy of the Yale Peabody Museum of Natural History.

Science at a Distance through Telegraphs and Collectors

The drama of paleontological fieldwork was paralleled by the specimens. Research in the Bridger Formation focused on the remains of a large herbivore whose limb bones indicated a bulky, almost elephantine body, while the huge skull had multiple horn-like protuberances. Even more "extraordinary" (in the language of the reports) were saber-teeth, which Leidy had assigned to a terrible carnivore, but which later seemed to belong to these herbivores. Leidy, Cope, and Marsh collected and promoted these animals through new communications and transportation technologies. Fossils were packed in sacks stuffed with paper and cotton, carried by horse and wagon to the railhead, and then by train to New Haven or Philadelphia. Initial descriptions of the fossils, such as Leidy's letter on *Uintatherium robustum*, were sent by post and telegraph to scientific institutions. Cope moved at even greater speed. At the American Philosophical Society in September 1872, "the Secretary announced that he had received a telegram from Prof. Cope . . . announcing the discovery of *Lefalophodon dicornutus*, *bifurcates*, and *excressicornis* Cope,"[33] although it transpired that the telegraph operator had garbled Cope's desired name *Loxolophodon* in transmission. Meanwhile Marsh named his specimens the Dinocerata (terrible horns).

Rivalry over these names fed into one of the most bitter instances of conflict during the Bone Wars, with the disputes playing out in the scientific press in 1872 and 1873.[34] Leidy, Cope, and Marsh acknowledged that their *Uintatherium*, *Eobasileus*, *Loxolophodon*, and Dinocerata were similar animals, difficult to interpret owing to the fragmentary fossils and strange features, and could be the same genus. Developing conventions around scientific naming, and particularly the "principle of priority," meant that the first name given to a species needed to be retained over any later names, if any of them were found to be synonymous.[35] The principle had been proposed by French naturalists in the early nineteenth century and was endorsed by the British Association for the Advancement of Science in 1842. However, conventions were still shifting and uncertain, and debates over borderline cases were common in botany, zoology, and paleontology.

Paleontological terms were open to revision, but rapid research on fragmentary fossils meant it was easy to create new genera from initial findings, and then make strenuous efforts to maintain those names. This was a core part of the Cope–Marsh dispute, with the two rushing to name as many specimens as possible. Priority was even more difficult to judge, as the letters and telegrams from Leidy, Marsh, and Cope were all delivered within a few weeks of each other. Also, what constituted the "publication" of a name was uncertain, especially whether a brief telegram to a scientific journal counted. The specimens were important for debating rules of priority, particularly the rapid-fire modes of publication occurring during these field projects.

The dispute also concerned how scientists could understand these animals. While paleontology required imagination and conjecture, there was a countervailing trend: that speculation needed to be reined in to maintain the field's credibility.[36] Cope and Marsh were (at least rhetorically) at different ends of the spectrum, with Cope engaging in imaginative reveries on the ancient world, and Marsh stating his commitment to "objective" and sober research—although as shall be seen, matters were more complex. Marsh accused Cope of overly fanciful reconstructions and lack of anatomical knowledge. He identified sixteen errors with Cope's analysis of the animals. Cope's misinterpretation of the teeth as incisors rather than canines and misplacement of the horn-cores showed his inability to understand fossils.

Meanwhile, Cope's suggestion the animal had an elephant-like trunk was disproved by basic principles of anatomy: its short limbs and long skull "rendered a proboscis unnecessary, as the muzzle could readily reach the ground" and it had a "small nasal opening—smaller even than that of the rhinoceros or tapir." Marsh argued Cope's reconstruction "belongs in the Arabian Nights and not in the records of modern science."[37]

Cope and Marsh sparred with one another in the *American Journal of Science* in increasingly vociferous, but also increasingly marginalized, manners. Their letters were eventually relegated to an appendix, which they needed to fund themselves. While Marsh claimed that the animals themselves were "comparatively of little consequence,"[38] this was not accurate for their status in the paleontological imagination. The beasts of the Bridger—whether called uintatheres, Dinocerata, or *Loxolophodon*—were fossil wonders of the American West, and Marsh and Cope continued to work on them. However, fieldwork operated at a distance, as East Coast scholars supervised local collectors to gather material.

While records of Cope's collectors are difficult to find, Marsh's correspondence contains a wealth of material relating to his western field-workers, showing attempts to manage teams remotely and the views of the fossil collectors themselves. Marsh's initial forays into the Bridger Formation were conducted by Yale students (with considerable guidance from military personnel and local notables), but from 1873 he preferred to work with people already in the West. Van Allen Carter requested of Marsh: "Can't you come alone that is not with an 'expedition.' I think a great deal of you Professor, but with due respect I don't 'go a cent' on the Yale boys as helpers to science."[39] While these comments are potentially dismissals of eastern interlopers, Marsh himself increasingly used the expertise of local collaborators rather than students. His workers included Sam Smith, a sometime store owner (and previously one of Cope's collectors), and John Chew and Ervin Devendorf, both former Civil War soldiers.

Additionally, interaction between paleontologists and Native Americans is gaining attention in the secondary literature, often taking inspiration from Adrienne Mayor's work on Native American geomythology,[40] and emphasizing fossil collecting as an act of treaty-breaking, with fossils becoming yet another resource extracted from Indigenous territories by the United States.[41] These issues are significant and deserve further dedicated studies, but what is notable is how absent Native Americans are from

Leidy's, Cope's and Marsh's accounts, except as either threats to be avoided or people to be outwitted, befriended or dealt with in the pursuit of fossils. Marsh was known for his support of Indigenous rights and his friendship with Red Cloud, and he worked Indigenous stories into the terminology of natural history, naming one beast similar to Prout's "Giant *Palaeotherium*" as *Brontotherium* (thunder beast) after a Lakota mythic creature.[42] However, while relying on Pawnee scouts for his expeditions, he preferred to collect through white settlers rather than use his links with Native Americans. Meanwhile Cope argued for the inferiority of all nonwhite peoples and occluded Indigenous people in his writings. Fossil collecting often drew from Indigenous knowledge of the landscape, especially for locating fossil-bearing formations. However, this dependence was increasingly erased from the 1870s on, with fossil hunting becoming another instance where paleontology was tied with extractive programs.

Rather than composing reveries on the deep past or tales of western adventure, the collectors saw themselves as organizing manpower and surveying. Marsh organized fossil collecting in the 1870s through letter and telegram. Sam Smith frequently argued for increased wages, in 1874 demanding one hundred dollars a month, arguing "I can not make as much as a common herder at seventy five per month."[43] Smith seems to have been effective at negotiating, although matters came to a head in 1883, when Marsh was appointed paleontologist for the US Geological Survey, with an annual budget of $15,000. Smith requested a considerable increase in wages, and when this was initially refused (with an implied accusation from Marsh that Smith was going to pocket the extra money and not hire additional collectors), Smith wrote, "I have always tried to do the best I could for you but since the work has become dissatisfactory & I shall quit and do something else."[44] This was followed by a hasty panicked telegram from Marsh: "Letter misunderstood. Another mailed. Proposition accepted. Prepare for work immediately."[45] While clearly showing Marsh's dependence on his collectors, this was the end of the collaboration. One account suggests Smith was murdered while on a trek shortly after.[46]

The field-workers negotiated with Marsh to secure as much money as possible for their skilled labor; this also reflected the difficulties of working in Wyoming Territory. Devendorf wrote it was hard to hire workers, as "it is rather a difficult matter to keep men in the country working, for they are mostly of a timid nature & are afraid to stay there and I cannot blame them much for it."[47] Partly this was due to continued conflicts on the northern plains, as the Lakota and Cheyenne were hostile to prospectors. Equally pressing was the environment itself. In one letter Smith promised, "I am a going to make it rain Fossils in New Haven . . . that is if I don't get alkalied."[48] Indeed, in Smith's accounts it was the environment itself and the danger of himself or his horse being poisoned by bad water or injured that were of concern.

Despite these difficulties, the collectors became skilled in identifying fossils and appealing to Marsh's wishes. They were aware of Marsh's interest in large skulls, although knowledge of the animals themselves was variable. In November 1873 Devendorf referred to his specimens as "mammoth heads," or simply "heads," and asked whether Marsh could "inform me of what species is this head & also how you like the specimens."[49] Ten months later he wrote with pride that he had developed his knowledge "so now that by treading over a country now I can tell nearly exactly where I can find Fossils also the different formations from the Primary, Transition, Secondary, Tertiary, & the different Pliocenes," and asked Marsh for scientific papers.[50] Smith also discussed technical details, writing to Marsh with pride and curiosity on how "Dinoceras No 37 is the skull of a young Dino., if not it is a new thing altogether the tusk where it comes out of jaw is no larger than Paleosyops and the head is about 2 feet long with posterior and anterior horn cores very small."[51] After working for Marsh for almost ten years, Smith wrote with curiosity, "If you have a woodcut of the Dinoceras as a whole skeleton please send me one so that I can see what kind of a Fellow he is."[52]

While Marsh has been presented as a domineering organizer—developing his "fifteen commandments" for the conduct of fieldwork in 1875[53]—he often needed to defer to his workers, and his instructions were often refused. On receiving Marsh's command to pack fossils with hay and grass, Devendorf wrote, "I could get no grass nor hay this summer with one exception it would have taken the entire party more than a week with pincers to get grass for to pack fossils."[54] Working through local intermediaries was difficult, and in later years Marsh preferred to employ specially trained fossil hunters like John Bell Hatcher, rather than informal networks around people like Smith and Devendorf. Despite their claims to have mastered the territory, eastern collectors were often at the mercy of their employees.

A number of other people became involved in western paleontology, inspired by fossils and frontier mythology. Despite van Allen Carter's misgivings, western expeditions became something of a rite of passage for geology students in many US universities. Princeton maintained a particularly strong presence through the geological program built by Arnold Guyot, a Swiss radical who left Europe in 1848. Guyot, who lectured in theology as well as geology and physical geography, connected the deep past with providential development derived from Protestant natural theology, and successfully navigated the complex debates at Princeton over materialism and evolutionary thinking.[55] This enabled him to develop a strong paleontological and geological program. The artist Benjamin Waterhouse Hawkins was employed from 1876 to 1878 to paint a series on the prehistoric life of New Jersey, and Guyot's students read the papers of Cope and Marsh as well as popular reports on fossil adventures in the West.

A group of Guyot's students were sufficiently inspired by Betts's *Harper's Monthly Magazine* account to organize their own geological trip to the West with Guyot's support. The prime movers were Henry Fairfield Osborn (son of the railroad baron William H. Osborn), William Berryman Scott, and Francis Speir. The group went to the Bridger Formation in 1877 as the paleontological division of the wider Princeton Scientific Expedition to the West. The arrival of the Princeton expedition was regarded with excitement by local collectors, with van Allen Carter writing to Marsh about their arrival,[56] and both Chew and Smith hoping to work with them.[57] However, rather than hire these established collectors, the Princeton expedition worked independently. The iconography around these expeditions was telling, linked with frontier myth, but in a tongue-in-cheek manner. A photograph of "The Triumvirate" of Scott, Osborn, and Speir showed them dressed in "western" garb, similar to the more elaborate images of the Yale expeditions. Scott's memoirs indicate this was at least partly playful. The photo was taken in New York following the expeditions, when Scott and Osborn were about to go study in Europe. Scott wrote, "It was a very Wild Western picture, but the effect was rather spoiled by our small and neatly blacked shoes, while our clean-shaven faces and my smooth hair" gave the impression of "as mild a mannered man as ever cut a throat."[58]

As was now conventional, the Princeton expedition aimed to collect great fossil herbivores, with Guyot being extremely pleased when they returned with a "fine *Dinoceras* head which I so much coveted."[59] In 1881 Osborn wrote up the main results of the expeditions as "Memoir on *Loxolophodon* and *Uintatherium*: Two Genera of the Sub-Order *Dinocerata*," a title that

FIGURE 6.5. The "Triumvirate" of Osborn, Speir, and Scott. Image #116113, American Museum of Natural History Library.

carefully balanced the names bestowed by Cope, Leidy, and Marsh. While the piece was technical, it reached poetic flourishes when describing the landscape and creatures. The opening followed the idiom of writing about the badlands, describing "the noble and grotesque forms which centuries of wind and rain have carved out of the sandstone bluffs," where "the records of an ancient land are written in characters which cannot be mistaken."[60] Further effort went into understanding the lifestyle of *Loxolophodon*. Its "spreading feet" indicated it lived in an ancient marsh, and "taken altogether, the long body, wide thorax, and comparatively short limbs with spreading toes indicate an animal rather of the habits of the Rhinoceros than of the Elephant," and "the long canine tusks" would have given the creature "a very formidable appearance."[61]

Following the Princeton expedition, Speir went into business, but Scott and Osborn continued studying geology and biology including in London under Thomas Henry Huxley and Germany under Karl Gegenbauer. While American paleontology was coming into its own, finishing their studies in Europe was still crucial for the formation of American scientists. After returning to the United States, Osborn and Scott would become important figures in American paleontology. Scott succeeded Guyot as the leading geologist at Princeton, and Osborn built the paleontological division at New York's American Museum of Natural History into the most well-resourced paleontological collection in the world.

The Great Creatures Live Again

Osborn's and Scott's studies in Europe indicate that, despite its national importance, American paleontology developed within an international context. Paleontology maintained definite centers of authority, and many of the oldest collections and type specimens were in Europe. But European scholars looked on developments in the United States with interest, especially as fossils from the western territories were impressive in their extent and difference from European fossils. The Munich paleontologist Karl Alfred von Zittel reported on the "progress" of American science, which countered stereotypes his German readers might have had of "America as a country where money-making suppresses all other interests." In literature and branches of knowledge "connected with practical life," America was moving ahead. Zittel cited Marsh's collection in New Haven, which was "not inferior in value to the collection of the British Museum in London. It is infinitely more than all the material ever seen and studied by Cuvier during his whole life," containing "twenty-five gigantic skulls of Dinocerata in the professor's laboratories."[62] Cope's collections, displayed in his Philadelphia home (and also the house next door, which he purchased to accommodate this expanding collections) was an "improvised museum, where almost all the rooms are filled nearly up to the ceiling with cases, shelves, drawers, trunks, and boxes, where one finds piled on the floor, or along the walls, enormous skulls of mastodons and Dinocerata." While not open to the public, it was as impressive as the older Philadelphia Academy of Sciences. European reports also played on the associations between paleontology and the frontier. Zittel wrote how "for months Professor Marsh and his assistants were camping in the reservations of the Indians, protected by an escort of cavalry. With the great chiefs of the Sioux, 'Red Cloud,' 'Red Dog,' he used to smoke the pipe of peace: against others he had to defend himself, revolver in hand."[63]

Over the next few years Marsh dominated discussion of the Bridger fossils, through a combination of frantic preparation, collection, publications, and international connections. By the 1880s Marsh claimed to possess the remains of more than two hundred Dinocerata, and his *Dinocerata: A Monograph of an Extinct Order of Gigantic Mammals*, partially funded by the US Geological Survey, was a long and lavishly illustrated account of the animal, deducing its habits, diet, and lifestyle. The animals were valued as unique to the region, and "gigantic beasts, which nearly equalled the elephant in size, [and which] roamed in great numbers about the borders of the ancient tropical lake in which many of them were entombed."[64] However, despite the quantity of material, understanding the creatures was difficult. The fossils were fragmentary, and Marsh built his Dinocerata as composites, reconstructing skeletons of particular types using the remains of many individuals. Marsh identified three distinct genera: *Dinoceras*, *Tinoceras*, and *Uintatherium*, keeping one of Leidy's names, but consigning all of Cope's to oblivion.

Most of the early part of the monograph was descriptive and technical, going through individual bones of the Dinocerata and assigning them to the three genera. Detailed typological and anatomical work was essential for building authority and offseting accusations of speculation. Yet while Marsh was averse to reconstructing extinct creatures "in life," and frequently accused rivals of overspeculation and poor knowledge of anatomy, his work was full of illustrations of bones and skeletons. Marsh wrote how "the text of such a Memoir may soon lose interest, and belong to the past, but good figures are of permanent value in all departments of Natural Science."[65] The centerpieces of the monograph were full skeletal illustrations of *Dinoceras* and *Tinoceras*, which were also worked up as sculptures in papier-mâché and eventually plaster. To avoid accusations of speculation, the composites were detailed to show exactly which bones came from collected fossils and which

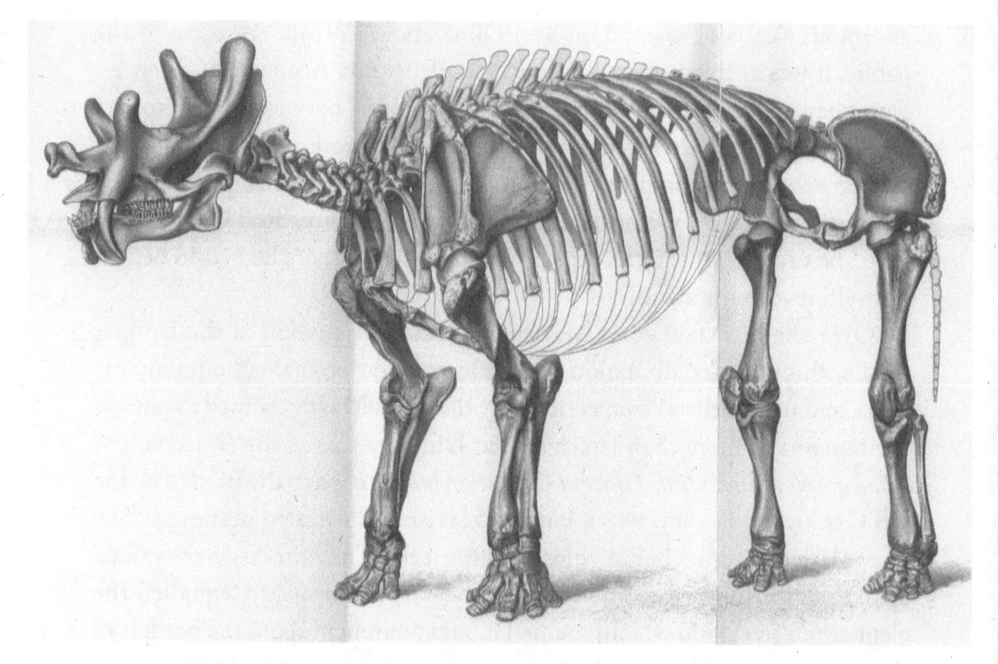

FIGURE 6.6. The full skeleton of *Tinoceras ingens*, the larger of Marsh's two *Dinocerata*. Marsh, Dinocerata, plate 56. Reproduced by kind permission of the Syndics of Cambridge University Library.

were built through comparative conjectures. Yet these were still showy specimens, being almost as much works of imaginative reconstruction as Cope's reveries, but veiled in the rhetoric of sober typology.

Marsh gave a long account of the possible lifestyles of the animals. Analysis of anatomy, discussions of the ancient landscape and vegetation, and comparisons with modern creatures allowed Marsh to go beyond his self-consciously "objective" style and into more imaginative forms. It was a social herbivore, and "that the animals lived in herds is also suggested by the position in which the remains are found. Their favorite resorts would seem to have been around the borders of the great Eocene tropical lake described in the Introduction of the present volume. Here, they found an abundance of food, which was evidently the soft succulent vegetation which flourished, then as now, in such localities." Their physical features meanwhile mixed different animals. "*Dinoceras mirabile* when standing at rest would have a general resemblance to a very large rhinoceros. When walking, the movement of the hind limbs would at once suggest the elephant, as we know it to-day." However, "the head of *Dinoceras* must have had some resemblance to that of the hippopotamus, but was very different from that of any known animal, living or extinct."[66] The Dinocerata were comparable to living herbivores—although Marsh was at great pains to argue that it was not a direct ancestor of any of these, being much stranger. But in all features defining mammals, the creature was "primitive," with simple teeth, a very small brain (almost "reptilian" according to Marsh), and flat five-toed feet. The anatomy explained the animal's extinction, as "the small brain, highly specialized characters, and huge bulk, rendered them incapable of adapting themselves to new conditions, and a change of surroundings brought extinction. Smaller mammals, with larger brains, and more plastic structure, readily adapt themselves to their environment, and survive, or even send off new and vigorous lines."[67] The implications of this will be discussed in chapter 8, but the Dinocerata were not just fossil wonders—their anatomy indicated larger trends in life's history.

Despite Marsh's confident reconstructions, there was trouble behind the scenes in his fractious workshop. He gained a reputation as a domineering figure, and his realtions often exploded into conflict (as occurred with his western field collectors). When Erwin Hinckley Barbour, one of Marsh's assistants, broke with him, he wrote some "Notes on the Paleontological Laboratory of the United States Geological Survey under Professor Marsh" for the *American Naturalist* (notably, Cope was one of the editors of this journal). Barbour argued that Marsh's reconstructions were full of "deceits and falsehoods."[68] The Dinocerata were singled out for being "plump with plaster,"[69] with their size and completeness exaggerated by manufactured bones. Moreover, the lavish publications and lithographs were overindulgent wastes of public funds. Barbour accused Marsh of "wholesale misuse of Government men and money for his own personal benefit,"[70] having used Geological Survey resources for his private work and publications.

Accusations of misusing government money stuck, and Marsh's work as a paleontologist for the Geological Survey was subjected to Congressional scrutiny, tied to wider criticisms of federal funding for noneconomic science. Marsh's position deteriorated rapidly. Marsh donated his collections to the Peabody Museum in 1898, although with the condition, that they remain under his "supervision and control" and that only he, or scholars approved by him, would be able to study them.[71] However, after Marsh's death, the collections made when he was paleontologist on the Geological Survey

FIGURE 6.7. "Professor Marsh's Primeval Troupe: He Shows His Perfect Mastery over the Ceratopsidæ," *Punch*, September 13, 1890. Author's collection.

were classed as government property and transported to the US National Museum in Washington, DC. A huge amount—estimated at 592 boxes requiring five railway cars to transport—formed the core of the museum's paleontological collection.[72]

Barbour's criticisms of Marsh's practices do not seem to have been taken up internationally—something notable given that Barbour had addressed his criticisms partly "to those scientists in foreign lands, especially Germany, who have marveled at the exceptional beauty and perfect preservation of Prof. Marsh's specimens."[73] Plaster modeling and "artful" reconstructions were widespread in European collections (as in Richard Owen's comment on plaster casts being "indistinguishable" from actual bones), so the criticisms may not have been accepted. Marsh's Dinocerata monograph was well reviewed in the European press. William Henry Flower wrote on receiving the work, "How can I thank you enough for such a splendid copy of your great work on the Dinocerata? I have read it all through—it is wonderfully complete—the great creatures live again in your descriptions and illustrations."[74]

In the 1880s there were many attempts to secure models of Dinocerata from Marsh's laboratory. The British Museum of Natural History was the first to acquire one (which arrived along with Flower's copy of Marsh's monograph), and it was placed in the gallery of fossil mammals. Not to be outdone, Gaudry met Marsh at the International Congress of Geology in London in 1888 and persuaded him to send a copy to Paris. While this went against the convention in the Paris Muséum of exhibiting only "original pieces," it was justified, "for the *Dinoceras* is a beast so different from the other mammals that—doubtlessly—all naturalists will want to be able to examine it." Moreover, *Dinoceras* bolstered Gaudry's evolutionary theories: "The mighty King of the Eocene had an ephemeral reign, and seems to have died without posterity." It was "a parvenue type reaching the summit of its development and flourished in strength and magnificence," but then disappeared as it lost evolutionary force and energy.[75] In return Marsh was unanimously elected a correspondent of the Paris Muséum in 1889, "in consideration of your beautiful works of paleontology." More privately, Gaudry, whose wife had recently died, wrote to Marsh that "in the deep sorrow I have been thrown . . . your gift was the first occasion where I had for an instant forgotten my misfortune."[76]

Paleontological work and networks were not just about scientific prestige, but also reflected personal relationships. The dramatic rivalry between Marsh and Cope is of course an example, but paleontologists worked across local, national, and international contexts. While Marsh ran into difficulties in the United States, he was feted abroad—and the Dinocerata were key to this. When he visited Britain to participate in the British Association for the Advancement of Science meeting in Leeds in 1890, he was depicted in *Punch* taming his prize specimens, *Dinoceras* and *Brontotherium* in a circus ring, while lecturing on the newly discovered dinosaur *Triceratops*. Henry Woodward recalled that "he was delighted when he saw the image."[77] Marsh's mastery of the beasts of the past was aimed as much at overseas observers as those in the United States. The rise of American paleontology depended on a variety of locations—the western field sites, the East Coast collections, and the broader transatlantic networks of paleontology, all essential for developing the field and privileging the "terrible horns" and "dawn rulers."

CHAPTER 7

Narratives of the Tertiary

Time, Geology, and the Difficulties of Progress

BY THE MID-NINETEENTH CENTURY THE DEEP-TIME SCIENCES HAD ELABORATED A VARIETY of ancient mammals and their associated environments. These included the waterlands of the Paris gypsum with its anoplotheres and paleotheres, the cold territories of the European drift with mammoths and hyenas (and the apparently equivalent periods in Australia and South America, with great edentates and marsupials), the geologically dramatic environments of Eppelsheim and Sansan, the abundance of ancient Pikermi and Siwalik, and the "ruins" of the badlands of the United States. Fossils from these localities showed the history of creation and a series of lost worlds, distinct from one another. Ordering them into a coherent narrative was difficult and required different modes of reasoning.

The construction of a narrative history for the earth was a dramatic instance of paleontology mixing imagination and rigor. It mixed different ways of knowing the natural world, including studies of rock strata; fossil animals and plants; evidence of forces like water erosion, volcanic eruptions, and the scouring of ice; and observations of modern environments and animals. It also brought in narrative models of time and history. The nineteenth century has often been defined as a period when historical consciousness, and uses of the past, shifted considerably.[1] Nations, peoples, classes, and civilizations were all inscribed with dramatic histories, marked by change and development. New geological narratives gave a similar history to life and the earth, which was as dramatic and dynamic as human history. It was also interpreted in the same manner, as a story of ages and events marked by progress, but also by decline, crisis, and cyclical rise and fall.

Stratigraphic geology was central to defining the earth's history. A defining point within contemporary geology was that different layers of the earth could be thought of as representing distinct periods of time, with the more recent layers of rock being laid over the older.[2] Cuvier and Brongniart's column of the Paris basin was an early example of this, which set the model for similar columns in other parts of the world. Metaphors abounded that the rocks could be "compared to the pages of a book, each containing a history of a certain period of the past."[3] The fossils in each layer also showed changes in form and were argued to move through the scale of life, with the bones of mammals in upper layers, reptiles below, and only invertebrates below that. Life's history therefore moved through the scale of nature across geological time (although the nature of this progress, and its regularity, was uncertain).

By the mid-nineteenth century, after several contested debates,[4] the structure of the book of the earth was broadly agreed on. After a long period lacking any obvious life came the Primary rocks showing the Paleozoic era of "ancient life," a germinal world with invertebrates, corals, and early fish, with amphibians and reptiles appearing at the end. Next were the Secondary rocks and the Mesozoic era of "middle life," the time of the great reptiles increasingly known from Britain, Germany, and North America. Then came the Tertiary rocks, the "recent life" of the Cenozoic, which encompassed the Age of Mammals. Finally, whether the drift, now christened the Pleistocene, should be a new Quaternary period marked by the appearance of humans was a controversial issue.

This stratigraphic chronology was important in several ways. It produced a narrative of earth's history which defined time through a sequence of eras. The disorienting and dramatic nature of the discovery of deep time has often been highlighted, as the earth sciences created an uncountably vast chronology, destabilizing humanity's position in the world and the older scriptural ideas of a relatively young earth.[5] However, disorientation was mollified, as deep time was not incomprehensible, but could be understood as a series of worlds, each passing (by admittedly mysterious forces) into the next. Just as layers of the earth graded into one another, so did the different phases in earth's history. This again operated in similar manners to human history, which could be interpreted through the dominance of particular civilizational or social forces. The nineteenth century was a key time for "stage" models of human development, whether these be the stadial theories of Enlightenment thinkers of progress through savagery, barbarism, agriculture, and commercial civilization (which maintained a strong hold over concepts of human civilization across the century); the Three-Age model of archaeological development through stone, bronze, and iron ages; or the more granular developmental theories of Auguste Comte or Lewis Henry Morgan.[6] These progressive models often jostled with historicist ideas, seeing particular periods as having distinct qualities and needing to be understood on their own terms, rather than as mere stepping stones on a developmental road.[7] The earth had a similar logic of passing through distinct ages, which could reinforce ideas of progress, but could alternatively emphasize the distinctiveness of particular eras.

Sometimes, the metaphor between human history and earth's history was quite explicit: Bommeli's *Geschichte der Erde* included a table depicting "Antiquity," "Middle Ages," and "Recent" periods of earth's history, linking the frameworks used to understand the human past with the history of life. Meanwhile William Henry Flower wrote, "In geology we know nothing about centuries. We have no kings' reigns, as in political history, to mark the course of time, so we speak of 'Miocene' much in the same vague kind of sense in which we speak of the 'Middle Ages' in our chronology of the historical events in Europe."[8] Yet as well as being general approximations of time, the ages of prehistory were definable through dominant forms of life, and these potentially linked with debates over human history (where values given to particular eras were often contested). The germinal and creative periods of human antiquity could be connected with the first stirrings of life in the primary ages. The Middle Ages—possibly presented as a time of obscurantist superstition and violence, or alternatively as an age of origins and dynamic heroism—could be likened to the power and force of the Age of Reptiles. More recent periods, valued as a time of progress and moral change, could be connected with the rise of the elevated mammals over the Tertiary.

The concept of earth history as a successive series of worlds ensured that defining it could not be solely, or even primarily, the work of elite gentleman scholars discussing rock strata and fragmentary fossils. As Ralph O'Connor and Martin Rudwick have shown, literary and artistic techniques were

Geologische Zeitentafel.

Zeitalter	Zeit	Periode
Die Neuzeit der Erde	Diluvialzeit	Periode des Mammuth und des Urmenschen
	Tertiärzeit	Periode der Mastodonten und Dinotherien
Das Mittelalter der Erde	Kreidezeit	Periode der Dinosaurier und der ersten Laubhölzer
	Jurazeit	Periode der Fisch- und Flugsaurier und der Urvögel
	Triaszeit	Periode der Panzerlurche und Krokodile und der Riesen-Schachtelhalme
Das Alterthum der Erde	Carbonzeit	Periode der Kryptogamen Sigillarien Lepidodendren Calamiten
	Devonzeit	Periode der Panzerfische und der ersten Landpflanzen
	Silurzeit	Periode der Trilobiten und der Tange

essential to making deep time comprehensible.[9] Numerous systematizing, and often self-consciously popularizing, works outlined the narrative of earth history. Especially since the 1860s what had been termed the Epic of Evolution became a set genre, particularly in Britain, France, the German lands, and the United States—the places where the earth's deep history was most widely researched.[10] Some works were written by professional popularizers, such as Louis Figuier (whose *World before the Deluge* was a defining example of the genre), Camille Flammarion, Wilhelm Bölsche, and Henry Neville Hutchinson, while others were by leading paleontologists like Oscar Fraas, William Boyd Dawkins, and Karl von Zittel. These works mixed forms, drawing from literature, popular science, and poetry. Many were copiously illustrated, depicting specimens, landscapes, and often entire tableaux of past eras. Evolutionary epics narrated the depths of the past and ordered fragmentary and dispersed relics into graspable scenes.

Myth and religion also helped structure narratives of life and the earth. As in earlier studies of caves and the drift, plunging into "the Primeval world" was self-consciously dramatic, retrieving strange creatures likened to folkloric entities. Popular works of geology often expanded these analogies.[11] In the preface to the German work *Die Wunder der Urwelt* (Wonders of the primeval world), the author wrote, "We know that thousands of creatures once lived that have now disappeared from the earth—creatures which were frequently more wonderful and outrageous than ever conceived by the writers of fairy-tales."[12] As much of the literature around relations between religion and science should lead us to expect, this new deep past did not necessarily unsettle religious ideas or lead to secular views of earth's history. Rudwick has even argued that scriptural narratives were "one major source—even arguably *the* major source—for this new vision of nature as historical."[13]

The deep past could be used to debate theological ideas about nature and revive older concepts of the earth. Indeed, Figuier's *La terre avant le déluge* and Oscar Fraas's *Vor der Sündfluth!* explicitly referred to the Deluge in their titles, although with differing meanings. For Fraas, who initially trained as a pastor before becoming involved in paleontological work and being

◄ FIGURE 7.1. The eras of earth's history. Bommeli, *Geschichte der Erde*, frontispiece. Courtesy of the Staatsbibliothek zu Berlin-Preussischer Kulturbesitz, Shelfmark: Mp 1304.

appointed a curator at the royal Naturalienkabinett in Stuttgart, the Deluge was primarily metaphorical, used to explain complex geological concepts. He argued, "If natural sciences are to be popularized, then geology has the decisive advantage above all others in that it can tie into the concept of the flood of the Deluge, which has become, so to speak, a fundamental concept in the consciousness of the people." The period of the drift remained a clear "boundary between the primeval world and the present world."[14] Figuier was more literal, noting that, while geology had sometimes been regarded as "suspect," it in actuality "far from undermines Christian religion[;] . . . the antagonism which might have existed in the past has now given way to the happiest agreement."[15] He went on to argue that "nothing makes the thought of the unity of God and his omnipotence penetrate into the spirit of youth than the study of the successive evolution of our world, and the generations of living beings that have preceded and prepared the way for man."[16] Life's history showed progress, evidence of "deluges," and possibly the sudden appearance of humanity, which could all reinforce Christian concepts.

The evolutionary epic represented earth's history through progress and development. The history of life showed improvement up the scale of creation and unfolding complexity. These progressive qualities are readily apparent from the language of the texts themselves. The author of the *Vestiges of the Natural History of Creation* argued, "We see, from what remains have been found in the whole series, a clear progress throughout, from humble to superior types of being."[17] According to Flammarion, "the history of the earth carries in itself the most magnificent and eloquent witness in favor of the law of progress which is accessible to our observations. It is progress incarnate in life, from the mineral up until man."[18] Meanwhile, Richard Somerset has described how even ostensibly "facts only" authors like Figuier masked a central progressive narrative.[19]

The importance of "progress" in the nineteenth century is widely recognized as a value cutting across politics, economics, and the natural sciences, serving as a legitimizing force for reforming agendas, colonial policies, scholarly work, and numerous other programs.[20] However, it is important to qualify exactly what progress in life's history meant. In Britain, David Page wrote that while the story of life showed advancement, this should not detract from the beauty and comprehensiveness of creation as a whole, or mean "that one form or family is less perfect than another, either in its nature or in the functions it was designed to perform."[21] Progress often meant variation, as all the world's inhabitants became suited for their positions. Additionally, analogies between the deep history of the earth and human history, scripture and mythology emphasized struggle, difficulty, and rise and fall. Progress in deep time was often crooked and interspersed with great disasters. Some eras—such as the harsh diluvial period—seemed far more chaotic and unstable than the harmonious earlier worlds of the Paris basin or Pikermi. Life's history was as conflict-ridden as human history or biblical narratives, and saw fall from grace and instability, as well as advancement and salvation. Similarly, evolutionary epics presented each period as having its own distinct personality and value. Time was made up of integrated epochs, with characteristic faunas, floras, landscapes, and climatic conditions. Like the eras of human history, these needed to be understood holistically, as much as being links on a chain of development. This was not simply linear time, but movement across a series of worlds.

Mysteries in the Early Ages

The Tertiary and the Age of Mammals were conceptualized as a late period in a longer history of development. It is therefore important to initially discuss assumptions of the earlier ages in order to understand the framework underlying the Tertiary. Accounts of earth's history often began with the formation of the earth, linking cosmology and geology. Across the deep histories there was general consistency in following the "nebular hypothesis," that the earth originated as a heated gaseous mass, which slowly cooled and solidified. While Simon Schaffer has noted that the nebular hypothesis was often dismissed in astronomical circles as "low" knowledge (often owing to its progressivist character),[22] it was incorporated quite unproblematically into evolutionary epics. Figuier's *La terre avant le déluge* opened with a description of "the current earth as an extinct sun, as a cooled star, as a nebula passing from the state of gas to a solid state."[23] The earth was not eternal, but connected different ways of understanding nature.

The nebular hypothesis implied that the earth had cooled across its history. While the cooling earth was rejected by some scholars, most notably Charles Lyell and other uniformitarians, the model was useful for narrative writers. A cooling model enabled the age of the earth to be estimated, from calculating how long it would take a ball of molten rock or metal of the earth's size to cool into a solid form. However, there was tremendous variation in these calculations. Buffon estimated the earth at around 75,000 years old, but nineteenth-century estimates were rather longer. W. F. A. Zimmermann noted that Professor Karl Gustav Bischoff in Bonn calculated 353 million years was required.[24] In 1863 the British physicist Lord Kelvin estimated that the earth would be between 20 to 400 million years old, with a best estimate of 98 million years, which he revised down to 24 million years in the 1890s.[25] As a result, the length of earth's history could be supposed, but never conclusively stated. Zittel wrote that questions like "how old is the earth? . . . are always freshly addressed to the geologist, and are never satisfactorily answered."[26] This was one of the great mysteries in the history of life.

The cooling earth conditioned development. As the earth cooled, it was assumed to become better suited for particular animals and plants. In contemporary European thinking, there was often a stereotype that organisms from colder regions were more active, dynamic, and progressive, needing to exert themselves within difficult environments. Meanwhile, hot tropical climates were regarded as home to eccentric and degenerated forms, while also being exuberant and abundant.[27] These climatic stereotypes had implications for prior eras of earth's history. Hot periods, like the Mesozoic, could have been the equivalent of modern tropical faunas. Meanwhile more recent cooler periods could temper more "progressive" animals, like mammals. These ideas also valorized divisions in the human world. The idea that "races" and civilizations were conditioned by climates was also widespread, and often mapped onto views of the natural world—with hot climates seen as leading to decline and degeneration, while cooler ones led to civilization and progress (unless these were too cold, as in the Arctic regions).

Cooling did not just affect life and its moral values, but the structure of the earth itself. Although this was a period before theories of continental drift, the earth was not assumed to be static. The continents and oceans had altered across geological time (indeed, evolutionary epics would frequently include maps estimating ancient landmasses in different periods). The presence of marine or freshwater fossils in regions currently on land, and similarities between animals and fossils in different parts of the world, showed great changes. Sometimes these were assumed as caused by forces derived from uniformitarian or catastrophist geology, with either small changes or dramatic upheavals altering the land and seas. These ideas also drew on the model of the earth cooling from a heated state, which would lead to cracking and contracting as heat dissipated, causing volcanic eruptions and earthquakes, and mountains and oceans to rise and fall. Changes in the landscape would also affect the environment, as already seen by Gaudry's assertion of the dying fauna of Pikermi, or Marsh's ideas of drying Eocene lake beds.

Senses of mystery pervaded deep time. Far from expressing themselves in a language of certainty, accounts of earth's history often emphasized the unknown. This was an important rhetorical device, building senses of the "scientific sublime," as discussed by Allan Gross in relation to twentieth-century evolutionary and cosmological narratives (demonstrating their long persistence).[28] Mysteries in life's history could be a call to action, with further research required to establish "truth," "bring light," and progress knowledge. However, mysteries also emphasized the difficulty in knowing the past. The vast stretches of time and immense changes were often presented as unfathomable, dealing with forces that could not easily be understood rationally but had to be imagined or felt. Paleontology, linking rational and imaginative ways of knowing the past, almost required mystery for its conceptual power.

Mystery being both a call to action and an acknowledgment of the limits of human comprehension is reflected in one of the greatest unknowns of the paleontological past—the origin of life itself. This was almost universally regarded as unknowable by current scientific means, and whether natural science could ever answer this question was an open issue. Marsh stated that "the primal origin of life . . . may perhaps never be known," but that "it is certainly within the domain of science to determine when the earth was first fitted to receive life, and in what form the earliest life began."[29] However, David Page stated that the geologist could study the "dawn of life" but "he knows of nothing beyond this primordial zone, and the spirit of true philosophy forbids him to substitute conjecture for fact, or hypothesis for reality."[30] The origin of life nevertheless saw tremendous research, most notably in the 1860s, with controversies over whether Canadian fossils named *Eozoön canadense* were the first life-forms.[31]

While the origins of life were mysterious, accounts of earth's later development were generally consistent. Early eras saw life move up the chain of being. The Paleozoic began with "ages" of invertebrates in the Cambrian, Silurian, and Ordovician, where the trilobites became the dominant life-forms in primordial seas. The next stage, the Devonian, was an age of

FIGURE 7.2. The "ideal landscape of the Eocene" period. Figuier, *Terre avant le déluge*, fig. 271. Author's collection.

fish. Then followed the Carboniferous, a period of great interest which saw the first appearance of amphibians and also great forests, which eventually formed the coal driving the Industrial Revolution.[32] Finally, the Permian, named after Russian sites, saw the development of terrestrial animals.

Next came the earth's "medieval" period, the Mesozoic, dominated by the giant reptiles that stunned public audiences and scientists across the nineteenth century. In Europe, much of the attention focused on marine reptiles like the ichthyosaurs and plesiosaurs, whose fossils were found in England and Germany, as well as terrestrial dinosaurs like *Megalosaurus* and *Iguanodon*. The Mesozoic earth was often portrayed as a watery world, prior to the formation of modern continents. In the United States, where the remains of huge terrestrial dinosaurs were increasingly excavated during the late nineteenth century, the imagery differed, with land, or at least swamp, dwelling dinosaurs dominating the landscape. Other creatures, particularly small early mammals, were highlighted in the Mesozoic, but this was nevertheless a reptilian world, characterized by size and strength.

What happened at the end of the Mesozoic to cause the demise of the great reptiles was another mystery. Sometimes environmental change was cited, with the cooling earth becoming less suited to giant reptiles. At other times, vague providential forces were highlighted, with the Age of Reptiles passing to allow more dynamic mammals to take over. Scott presented this as a general transition, as in the "Mesozoic, medieval features already began to show signs of yielding to more modern ones which are continually coming. In the new era Reptiles dwindled, Mammals increased & became the dominant race."[33] Whatever the cause, there was a shift from a physical and brutish era to one based on improvement and higher qualities.

The Dawn of the Age of Mammals

With the mysterious disappearance of the great reptiles, the history of life moved into a new era. This was the Tertiary, the Age of Mammals. Oscar Fraas described how "after the nightly darkness in which the creation of the first ages of the world is shrouded, human spirit joyfully welcomes the dawn of modern times."[34] The Tertiary's "brightness" had a double meaning. First, it was more knowable than earlier eras. The Tertiary rock strata were well surveyed (in Europe and North America at least), and Tertiary fossils were more abundant than those of earlier periods. Moreover, the long-standing privileging of the mammals meant that much effort had already gone into investigating the Tertiary. It was conceptualized as "light" for a further reason: as Fraas described, these were the "modern times" of the earth, showing progress and development after the dark and terrifying world of the great reptiles. Following concepts of rise and fall, the world now passed to the mammals, with their intelligence, sociability, and diversity. This was an unambiguously progressive era, moving to light and improvement.

This sense of light was enshrined in the name of the Tertiary's first era: the Eocene, or Dawn Period, conceptualized as a new day of creation, bright, germinal, and full of potential. The Eocene was the period of some of paleontology's most iconic animals, including the fossil mammals of the Paris basin, and the Dinocerata, uintatheres, and *Loxolophodon* of the American West. The Eocene environment was presented as lush and verdant, with fossils of palm trees, ferns, and reptiles in addition to early mammals. Flammarion spoke of a world of "carnivores, rodents, many types of birds, crocodiles and tortoises,"[35] and Oscar Fraas referred to the period's "mammals, birds, amphibians and a rich tropical flora."[36] The cooling earth indicated this was still a hot world. The tropical Eocene had a dual implication, entangled with the Janus-faced image of the tropics in Western culture. These ancient forests and lakes were originary and germinal, and a promising cradle for animal life. However, modern regions which retained aspects of the imagined Eocene—crocodiles, palm trees, and tapir-like mammals—were primitive, trapped in early phases of development. The Eocene and the tropics were simultaneously productive yet antique.

This tension, of the Eocene being both promising and primitive, conditioned descriptions of its fauna. Systemizing works devoted much attention to Cuvier's Montmartre beasts and the "remarkable abundance and great variety of diverse genera of pachyderms, which are entirely lacking among the quadrupeds of our days."[37] William Boyd Dawkins used a range of analogies to re-create the fauna and flora of Eocene southern England: In the palm-tree forests lived "the varied mammalian fauna" including "animals (*Palaeotherium*) like the tapirs of tropic Asia and America." "Anchitheres, have been proved by the researches of Professors Marsh and Huxley to have been the ancestors of the horse. They were about the size of Shetland ponies, and possessed three distinct hoofs on each foot, reaching to the ground." "Overhead on the trees there were opossums," and there was also a predator, the *Hyaenodon*, "a carnivore, which to the ordinary characters of a placental mammal united the marsupial attribute of three sectorial molars in each jaw, arranged as in the marsupial *Thylacinus* or Tasmanian wolf, which it rivalled in size."[38] The descriptions referred to animals regarded by Victorian naturalists as "primitive" and undeveloped—most notably tapirs and marsupials—and those potentially ancestral to modern animals. This was a time of origins, with ancestors of familiar creatures and some groups relegated as primitive.

Mid-nineteenth-century reconstructions of the Eocene drew on a European fossil norm. However, the increased prominence of American fossils meant the earliest eras also showed regional variability. In his *Succession of Vertebrate Life in America*, Marsh imagined the Eocene of the North American interior, consisting of swamps and lake beds created through geological changes after the Cretaceous: "As these mountain chains were elevated, the enclosed Cretaceous sea, cut off from the ocean, gradually freshened, and formed these extensive lakes, while the surrounding land was covered with a luxuriant tropical vegetation, and with many strange forms of animal life."[39] Large mammals—such as the Dinocerata and *Coryphodon*—revealed the expansion of life and coexisted with smaller creatures similar to those of Europe. Opossum-like fossils showed the existence of marsupials, small horselike animals demonstrated the possible origins of this important lineage, and tarsier-like fossils represented early primates. The main branches of mammalian life could be found in their earliest forms in Eocene sites.

The Eocene in both Europe and North America represented harmony and abundance, with only small numbers of predators (namely crocodiles and a few mammals like *Hyaenodon*). The Eocene was a time of origins, but also one which needed to be moved beyond. This was an almost Eden-like state, an initial stage of peaceful harmony within nature, but which later pressures would expel. The history of the mammals, like biblical narratives and human history, could not be locked in early abundance. The Eocene was crucial for understanding origins and major mammalian groups; but the earth needed to change considerably from its lush ancient environments for further progress to occur.

The Apogee of Mammalian Life in the Late Tertiary

After the verdant wetlands and generalized fauna of the Dawn Period, the world shifted. The late Tertiary was divided into a series of periods, named by Charles Lyell as Miocene and Pliocene (and possibly both preceded by an Oligocene period), each denoting the increasing "newness" of the era. Despite the chronological subdivisions, the general narrative around the late Tertiary was fairly consistent as times when landscapes and creatures became more familiar. The world moved toward distinct biological provinces, presaging modern faunas. Flammarion wrote that in the late Tertiary "nature approaches modern appearances more and more."[40] However, the mid-Tertiary was also a time of drama and grandeur, as expressed in the reveries of Gaudry and Falconer on the lost environments of Pikermi and the Siwaliks (two regions which were now placed in these eras). While the animals may have been less strange, they existed in tremendous variety, with numerous species of ungulates, proboscideans, and carnivores, and with many of the animals, like *Dinotherium*, exceeding modern mammals in size. The Miocene and Pliocene were therefore the apogee of the Age of Mammals.

Once again, understandings of the fauna rested on landscape and climate. Unlike the tropical islands and waterways of the Eocene, the Miocene

Figure 7.3. The "ideal view" of the Miocene. Figuier, *Terre avant le déluge*, fig. 290. Author's collection.

and Pliocene were thought of as more continental. The late Tertiary saw steady drying and cooling, and the world became less lush. The Miocene was often (in Europe at least) depicted as filled with warm forests and marshlands, while the Pliocene was a time of open plains and foreboding mountains. Page wrote of great transformations, as "the Eocene sea is being gradually elevated into shoals and islands" and "volcanic energy gives birth to new mountain-chains, which interrupt the former currents of the air and ocean, and new external influences begin to prevail."[41] Figuier similarly spoke about the decline of palm trees and reptiles in the northern hemisphere in these eras.[42] The environment moved slowly away from tropical "exuberance" toward variation.

Climatic change was partly due to the cooling earth. However, it also drew from changing geography. Unlike the Eocene islands, the Miocene and Pliocene saw land rise and connect to form continents. These created migration routes, and animals could more easily move between different parts of the world. Sometimes these geographic changes were expressed through the geological sublime, as caused by tremendous volcanic eruptions—given

FIGURE 7.4. Animals fleeing erupting volcanoes in Miocene France. Flammarion, *Monde avant la création de l'homme*, 693. Author's collection.

credence by Miocene formations in southern France which seemed to be huge extinct volcanoes. Flammarion imagined that these volcanic mountains rose almost two thousand meters higher than modern equivalents, and "the dinotheriums and hipparions witnessed these eruptions and left their remains in graves that were covered by basalt eruptions."[43] The abundance of the era was punctuated by terror in the face of such vast geological processes.

Alongside these dramas the Miocene and Pliocene were the apogee of mammalian life, in diversity and size. Flammarion wrote, "During the Miocene period, all orders of mammals are represented, pachyderms, carnivores, bats, rodents, proboscideans, marine, ruminants, insectivores, primates."[44] These were less "aberrant" creatures than in the Eocene or Age of Reptiles, and could be presented as "progressive" forms persisting into the present. The most well-studied late Tertiary locations in Europe and the Americas were understood as mixing animals now associated with different continents. Animals defined as African, Asian, European, and North American mingled with one another, and with extinct forms like dinotheres, chalicotheres, and saber-toothed predators. Fraas highlighted that in the Pliocene "the rich development of herbivores and ruminants . . . resulted in a corresponding abundance of predators . . . which exceed in strength and size those that still exist in Asia and Africa."[45] The Miocene and Pliocene were eras when "the economy of nature" was at its most varied, bountiful, and well balanced.

Miocene and Pliocene landscapes were depicted as filled with elephants, rhinos, and hippos. Page wrote that "this preponderance of bulky frameworks, and the number of intermediate forms that serve as connecting-links between species now widely separate, are perhaps the most notable features."[46] Particularly striking was the rise of the Proboscidea. As already noted, elephants were iconic animals within natural history and museum displays. Their fossils also seemed to suddenly appear throughout the world during the Miocene. As well as ancestors of modern elephants, there were dinotheres and mastodons in varied combinations in Europe and Asia. In North America interest in the American Incognitum transferred to a range of Proboscidea, showing the continent as home to a varied elephantine fauna. The sudden flowering of elephants in the Miocene showed it as a time of majesty, when the most noble quadruped had reached its zenith. However, the suddenness of the appearance of the elephants was also problematic, raising mysteries as to where exactly they had come from.

Primates also developed. The Eocene saw forms classed as tarsier-like, or "Lemuroid" (marking out modern tarsiers and lemurs as "primitive"), but Miocene Europe and India saw the first recognizable fossil apes and monkeys, reported from Sansan, the Siwaliks, and Pikermi. Apes were potentially (although controversially) creatures linking the animal and human. Flammarion described early primates as representing "progress on progress, in the organic kingdoms which people the earth, working toward the preparation of the human kingdom."[47] The fossils also indicated apes were more widespread in the geological past, having appeared early in the Tertiary. However, what this implied about human origins was unclear. Despite developing debates over possible human descent from an apelike creature, none of these specimens was regarded as a potential human ancestor (even by those who were willing to countenance such an idea).

While European and North American fossils were often presented as the norm, Miocene and Pliocene sites were also identified in other regions. As mentioned, the Siwalik fauna of India was dated to these periods, and research on non-European fossils reconstructed a diverse ancient world. Many fossil faunas were understood as presaging modern animals, such as edentates in South America and fossil marsupials in Australia. There was no one universal fauna, but different regions of animal life, each with its own history. A common idea developed that the late Tertiary was when modern biogeographic zones acquired their particular stamps. David Page discussed how "as the miocene and pliocene epochs advance, the more and more do their fossil forms assimilate to those now peopling the same geographical regions."[48] Much as mammal groups diversified, so different regions saw new complexes of animal life arise—and, as will be seen from the next chapter, understandings of geographic differences in animal communities were key preoccupations within late nineteenth-century natural history. These were rooted in their deep histories.

The Rupture of the Pleistocene

If the late Tertiary was an abundant period in earth's history, the next period—christened the Pleistocene—was an age of conflict and drama, going against any claims that geological history was simply a steady story of progress. The Pleistocene not only represented a new period in earth's history, but potentially a whole new era. The current geological age was beginning, a new phase named the Quaternary, which Flammarion called "the modern era, the current state of creation of our planet."[49] The Pleistocene layers were the formations traditionally understood as "drift" and "diluvium." Many works, especially popularizing ones, continued to propose a great deluge. However, as the discussions from Fraas and Figuier should illustrate, the nature of this deluge, and whether it should be regarded literally or metaphorically, was open to debate. Over the mid-nineteenth century an alternative view of the drift developed, defining it as a glacial period marked by the spread of ice and a freezing climate. Tobias Krüger has shown how the Ice Age was debated internationally. Theories of gigantic ice sheets across the northern hemisphere developed through the glaciological studies of Karl Friedrich Schimper and Louis Agassiz in North America and the Alps.[50] That the drift had been deposited by immense movements of ice was simultaneously novel but also maintained elements of older ideas of the Deluge. While not a scripturally deduced global flood, it was still a great cataclysm (with an inundation of frozen rather than liquid water), causing massive changes in the earth and sundering past ages from the present.

The glacial period was another area of controversy. Some rejected the idea entirely, preferring to argue for a large-scale deluge as the single mechanism for recent changes in landscapes and the deposition of fossils. A notable example was *The Mammoth and the Flood* (1887) by Henry Howorth, a British MP and active participant in London learned societies. Howorth described "the extreme views" of "Agassiz, Croll, James Geikie, etc. etc." as "a glacial nightmare,"[51] based on overeager desires to separate geology from religion, and cling too tightly to uniformitarian geology. Howorth argued that legends, geological studies, and the distribution of mammoth fossils could only be explained by a sudden universal change in the earth, for which a single great flood was the best explanation.

Those arguing for an ice age also faced problems. Figuier (who imagined both a deluge and a glacial period in the recent past) could not satisfactorily answer the question of why the "so unforeseen and intense phenomen[on]" of an ice age might have occurred. The available hypotheses of solar energy, oceanic currents, changing distribution of the earth's landmasses, alterations of the earth's axis, movement through a cold region of space, or a generalized cooling of the planet's core all seemed unprovable.[52] In the 1870s James Geikie wrote how there was no easy way of explaining an ice age, but argued that the climatic history of the earth was more complex than a cooling earth, going so far as to say "this cooling of the earth may be safely disregarded."[53]

Whatever the environmental changes of the Pleistocene, it was regarded as a harsh and threatening world, full of dangerous creatures. In Europe the fauna included large herbivores like mammoth, woolly rhinoceros, and hippopotamus. Smaller herd animals, including aurochs, reindeer, horse, bison, boar, goats, and sheep, were also abundant. Carnivores were widespread too, including the hyena, cave lion, cave bear, wolf, and wolverine. The animals themselves also indicated the climate. The hairiness of the elephants and rhinos and the robustness of the bears and big cats indicated that this was a cold period, where animals needed protection from the elements.

Other parts of the world continued the animal diversification begun in the Miocene and Pliocene. Zittel split his narrative geographically in the Pleistocene, as "with the ongoing differentiation . . . each country has its own history and requires special attention."[54] The life of the Americas was extensively examined. Some Pleistocene fossil animals, including bears, wolves, pronghorns, and buffalo, were still present in North America. However, others were not, like the mammoths and mastodons, which were often understood as reaching their largest size in North America. South American forms like ground sloths and glyptodons also appeared and were understood as migrants from the now connected southern continent. Predators included the short-faced bear (even larger than European cave bears), dire wolf, and saber-toothed cat. Life in the Pleistocene Americas paralleled Europe, representing a larger, more abundant, and more ferocious version of the present, with numerous unexpected migrants. In structure, if not in form, the life of the "old" and "new" worlds followed similar courses.

The mixture of animals also fed into other debates on whether there had been one single period of glacial cooling followed by steady warming, or more cyclical patterns, where frozen climates were punctuated by periods of warmth, before plunging into ice and cold again. Pleistocene Europe seemed to mix animals now associated with distinct regions of the earth, not all of which seemed suited for cold conditions. How, for example, the hippopotamus could have survived in a glaciated environment was a major problem. Charles Lyell suggested these animals may have undertaken seasonal migrations, and "the geologist, therefore, may freely speculate on the time when herds of hippopotami issued from North African rivers, such as the Nile, and swam northwards in summer," basing this partly on reports

FIGURE 7.5. Depiction of the Severn Valley in Pleistocene Age by M. Rowe. PDW: 5/4. Courtesy of the Manchester Museum, University of Manchester.

from travelers in Africa of the long travels of these animals.[55] Others used animal evidence to reconstruct climatic trends. In Britain a dispute erupted between William Boyd Dawkins and his Scottish rival James Geikie. Dawkins thought the animals all lived together in a land of cold winters punctuated by relatively mild summers, which then slowly warmed to the modern climate, implying a single glacial period. Geikie meanwhile argued these animals were not contemporaries at all, and their mingling in Pleistocene layers was an artifact of a pulsing climate that was warm for many thousands of years before cooling again, leading to cycles of glacial and interglacial periods which the animals then reacted to. These deductions from the geological formations and animal fossils had important implications for the modern world—had it arisen in a steady progressive manner after a single disjunction, or was it a product of alternating processes of freezing and thawing? Linear and cyclical visions of development jostled and persisted.

An image commissioned from an artist named M. Rowe by William Boyd Dawkins aimed to imagine the Pleistocene fauna of western Britain. This depicted a wild craggy landscape inhabited by a variety of animals. Some were posed in a fairly awkward manner, with straight-tusked elephant, mammoth, bison, hippo, and deer all lined up for viewing, while lions scattered herds of wild horses. The image also hinted at something else. In the background were small figures, tending fires in front of caves. Humans were also part of this landscape, separated from the other animals, although with their precise condition difficult to identify. In an article titled "The British Lion," Dawkins reflected on how ancient humans had been immersed in the struggles of the natural world. "The river-drift hunter, armed with his roughly chipped stone implements, doubtless had great difficulty in making good his place in the struggle for existence among the beasts of prey . . . and sometimes, when he had the chance, he would be likely to eat the lion, and at other times the lion would certainly eat him."[56] Early humans were a part of this wild environment, but were not necessarily its masters.

The Human Ages and the Quaternary

Rowe's illustration indicates that by the 1860s and 1870s the most important change in the Pleistocene was thought to be a new form of life: humans. The drift was even further detached from biblical schemas by what has been described as "the establishment of human antiquity" or, more grandly, "the time revolution." These developments have seen a large literature, arguing that in the years around 1859 a series of cross-disciplinary endeavors, particularly between scholars in France and Britain, systematically studied caves and recent geological formations to demonstrate that ancient humans equipped with stone tools lived alongside the already well-known drift animals.[57] The timescale of human existence was stretched far beyond the six thousand years deduced from biblical chronology, but firmly positioned within geological antiquity. While earlier epics would often have humans appear at the end as special creations, they were now presented as developing within the natural world, in their own series of ages. These ideas were widely presented through specialist texts and popularizing works, such as John Lubbock's *Pre-historic Times*, Louis Figuier's *L'homme primitif*, and Dawkins's *Early Man in Britain and His Place in the Tertiary Period*,[58] works synthesizing ancient humanity within the paleontological eras.

The historiography has noted strong links between the elaboration of Stone Age humans and ideologies around racial difference and social evolution, as human prehistory shifted ideas of the "primitive."[59] Lubbock's *Pre-historic Times* and the theories of French scholars like Gabriel de Mortillet were predicated on what has been termed "the comparative method," that peoples classed by European ethnologists as primitive were similar enough that analogies could be made between them—whether they lived in the deep past or in the present. Modern peoples judged low on the scale of civilization or racial hierarchies, were marked as living representatives of the deep past, just as how the platypus and tapir were seen as representing ancient mammals. These analogies were often contested in terms of how far they could be pushed, and whether they should be understood in an analogical sense or in literal terms of modern and prehistoric peoples being biologically identical. In either case, they were prevalent and connected the fossil past of humans and animals into a single frame.

Dawkins proposed an extreme version of the comparative method, with a dynamic biogeographic landscape that shifted with climatic change and migration. Pleistocene animals were divided into three distinct faunas moving around Europe. First was a "southern group of animals"[60] including lions, hippos, elephants, and hyenas, associated with Africa and moving north during warm periods. Second was a northern group adapted to colder Arctic climates, including "animals which are now only to be met with in the colder regions of the northern hemisphere,"[61] such as marmots, lemmings, wolverines, and reindeer, which moved south as temperatures dropped. Finally, there was a temperate group, which consisted of "those still living in the temperate zones of Europe, Asia, and America,"[62] namely beavers, rabbits, bison, bears, and otters. Across the Pleistocene the animals migrated and intermingled depending on temperature. This was expressly discussed in terms of territory being "invaded" as animals "pushed forward," with their migrations mirroring human conflict. Competition was critical for the development of animal life.

These climatically associated groups included humans as well as animals. For Dawkins, linkage between ancient and modern "primitives" was conceptualized literally and racially. The first group, the "river-drift" people with chipped-stone tools, were part of the southern mammals, and linked with the San and the Aboriginal Australians. They were "driven out" by "the cave men," adapted to cold environments and taken to be literally the same as the modern "Eskimo." Finally, the Neolithic "Iberians" migrated from the east at a later date with the temperate fauna, bringing agriculture and domesticated animals. Early human development operated in lockstep with animal life, with particular human "races" being included within animal faunas. These patterns of climatically induced migration, conflict, extermination, and extinction were therefore naturalized—and used to marginalize particular peoples in the modern age, by casting them as prehistoric relics in a world which had since moved on.

The human and animal pasts were linked in a different way by Edouard Lartet, who from the 1860s moved away from researches on the early Tertiary and became a prime proponent of human antiquity, mobilizing material from sites in southern France. While many scholars preferred evolutionist models illustrating the progress of material artifacts or "racial" change, Lartet linked human prehistory with animals. "Primitive humanity" moved through "the Age of the Great Cave Bear, the Age of the Elephant and the Rhinoceros, the Age of the Reindeer, the Age of the Aurochs." This was a localized chronology primarily focusing on southern France (and Lartet himself argued for uneven change, as "the age of the Aurochs persists today in Lithuania, and the Reindeer still lived in the Hercynian forest in the time of Caesar").[63] The animal ages depended on paleontological concepts. First was the idea that recent geological history saw a decline of large beasts, with first the great carnivores like cave bears, then elephants and other large herbivores, and then medium-size herbivores disappearing in sequence. The animal ages were structured on the notion that human development depended on the decline of the animals. Whether this was due to triumph in a struggle for life, a slow retreat in the face of a changing climate, or a "fading away" in the light of providential change, the Pleistocene was when the beasts retreated and humans inherited the world.

Ancient human interaction with the Pleistocene fauna was not just interpreted in terms of hunting and extinction. The human and animal worlds dramatically intersected in artistic objects retrieved from many European Old Stone Age sites, usually pieces of horn, ivory, or bone carved with images of animals, abstract designs, and occasionally human figures. The realistic renderings of Pleistocene animals showed the capacities of Stone Age humans. These were not just brutish "savages," but potentially (following one of the captions in Louis Figuier's *L'homme primitif*) "the precursors of Raphael and Michelangelo."[64] Artworks showed ancient humans engaging with the natural world through spiritual, religious, and artistic sensibilities.

FIGURE 7.6. The Mammoth of La Madeleine. Reproduced in Dawkins, *Early Man*, 105. Author's collection.

One of the most discussed of these artifacts was the Mammoth of La Madeleine, five fragments of mammoth tusk which—when pieced together—showed an engraved figure of a living mammoth, found by Lartet's workmen in surveys of Perigord in 1863, along with animal bones and other artifacts of carved "mobile art." The engraved mammoth was initially controversial, due to assumptions of the "low" capacities of early humans and concerns about whether the object might have been a forgery (especially as it was found at the same time a controversy was raging over the authenticity of a human jaw found at Moulin Quignon). However, the undisturbed nature of the site indicated it must have been deposited during the Pleistocene. Hugh Falconer authenticated this depiction, and "the practised eye of the celebrated paleontologist, who has so well studied the Proboscideans, at once recognized the head of an Elephant" and "a bundle of descending lines, which recall the long shaggy hair characteristic of the Mammoth, or Elephant of the Glacial Period."[65]

The Mammoth of La Madeleine became one of the most widely reproduced Stone Age artifacts. Stahl, the preparator at the Paris Muséum, made a number of plaster casts, and illustrations were ubiquitous in works on human prehistory, ancient art, and the life of the past. The mammoth demonstrating human existence alongside the beasts of the Pleistocene was just one aspect of its importance. The artifact was also compared with mammoth fossils and with the frozen remains that were still being found in northern Siberia. Karl Ernst von Baer, aligning his cast of the artifact with reports of Siberian mammoths, wrote how the piece is "undeniably a mammoth," indicated by its curved tusks, long coat, and hump, marking it as quite different from modern elephants.[66] Following the promotion of the Mammoth of La Madeleine, the image of the mammoth began to shift. Prior to the La Madeleine carving, mammoths were often depicted as hairy elephants, modeled on living proboscideans observable in zoos, circuses, and stuffed in museums. After this find, mammoths were increasingly depicted as humped, with manes, and with a gait matching the carving. Understandings of prehistoric animals now drew on the cultural productions of humans living in far-off periods, who were increasingly difficult to dismiss as "savages," low on the scale of "progress." Meditations on ancient art linked humans and animals across vast stretches of time. The emergence of human creativity was almost the culmination of the deep processes of development traced across the history of the earth. But how it had emerged, and what it signified, was another of life's mysteries.

CHAPTER 8

Development, Origins, and Distribution

Theory and Mystery in the History of the Mammals

THE NARRATIVE HISTORIES OF LIFE TRACED IN THE LAST CHAPTER CONSTRUCTED GRAND visions of the development of the earth. Life's history was a purposeful drama, leading toward improvement, but also affected by rise and fall, decline and crisis. Ideological notions of progress and human dominance, but also contestations over what progress and human dominance actually meant, were naturalized. Behind these narratives were conceptions of the forces driving changes in the earth and powering the development of life. These led to various debates, as points of mystery were used to enshrine the importance of paleontology to a suite of biological and social questions and acted as "prophecies" and calls to action.

This chapter examines how paleontology became an important contributor to the charged debates on evolution and variation in the late nineteenth century, drawing on the fossil world, and the fossil world of mammals in particular, to contribute to three important research problems: how life changed and developed; how important lineages originated; and the distribution of animal life, as the field of "biogeography" linked the past and the present while valuing different parts of the world according to their assumed evolutionary pasts. These all tied paleontological understandings of origins, progress, and diversity to modern nature. In these debates, reflections on mammals were again critical. The valuing of mammals as diverse and superior made them crucial for understanding development. How mammals changed over geological time, and how modern mammals were stamped by their histories, was crucial for paleontological discourse. It also had wider social and political implications. Tracing the origins and evolution of mammals conditioned understandings of humans, as they became tied with this lost animal world.

Late nineteenth-century paleontology was a confident science, contributing extensively to the debates on development that were raging across European and American society in those years. There used to be a stereotype that nineteenth-century paleontology was largely descriptive, at most supplying evidence for debates on evolution. This idea was typified by Stephen Jay Gould's comment that "paleontology as an evolutionary discipline either languished in intellectual poverty or rushed up blind alleys with exuberance during most of its history since 1859. Darwin himself viewed paleontology more as an embarrassment than as an aid to his theory."[1] However, this idea has been deconstructed over the past few decades. Most notably, Peter Bowler's *Life's Splendid Drama* provides a detailed account of the intellectual projects around investigations of the fossil record, which involved extensive theoretical debates.[2] Studies of fossils frequently contained warnings against overspeculation, citing the incompleteness of the fossil record. However, the synthetic models of paleontological research, mixing forms of knowledge and linking technical and popular ideas, meant paleontologists could take a leading role in debates on the development of life. Areas of uncertainty were reworked as calls to action, mysteries that future generations could uncover, or hypotheses and "prophesies" based on knowledge of deep time.

As discussed in the last chapter, evocations of mystery were important rhetorical techniques within paleontological discourse, presenting the limits of current knowledge (which could nevertheless be tested), and raising senses of awe and inquisitiveness at unknown issues. Paleontological engagement with origins, evolution, and distribution drew off these mysteries, exacerbated by the assumed "poverty of the fossil record," namely that there were still many gaps in evolutionary lineages, many periods of geological history still known only from a few sites, and large regions of the world, especially outside Europe and North America, not yet subjected to paleontological research. This was not necessarily a flaw in the field. The construction of lost organisms like *Megatherium* and *Dinotherium*, the building of a narrative of earth's history, and the working out of a few particularly well-known lineages were used to argue that if paleontology were conducted with sufficient ardor and vigor, these gaps would be filled. In this way the poverty of the fossil record was turned on its head: it showed not that paleontology was a flawed science, but that it had tremendous potential to add to the "store of knowledge" and inscribe the modern world with insights from the deep past.

Progress and Diversity in Life's Development

The link between paleontological theorizing and mystery can be seen through the often uncertain relationship that leading paleontological scholars had with one of the most widely studied theoretical developments in the nineteenth century: Darwinian evolution. Indeed, one major reason why paleontology has often been regarded as a nontheoretical field is due to its contested relationship with Darwinism. Darwin's own career shows an ambiguous relationship with fossils. The fossils Darwin brought back from the *Beagle* voyage illustrated the diversity of South American mammals. They also linked the continent's ancient and modern fauna, indicating that particular regions had been inhabited by related "types" throughout geological history (a significant point for developing ideas of biogeography).[3] However, fossils were often problematic in Darwin's later publications. The rarity of transitional fossils between classes of animal was a problem for evolutionary theory, as was the relatively young earth deduced by geological models of cooling, which afforded nowhere near enough time for the gradual processes Darwin argued for. Likewise, the centerpieces of paleontological science, museums, and fossil collections were more an embarrassment for Darwin than a demonstrator of evolution. He lamented that "now turn to our richest museums, and what a paltry display we behold!! . . . [T]hat our palaeontological collections are imperfect is admitted by every one."[4]

Darwin's own works of course were only part of the promotion of evolution. Many proponents of Darwinian models highlighted the current paucity of fossil collections not as a problem but as a call to action. Thomas Henry Huxley transformed many of the flaws outlined by Darwin into agendas for future work. In "The Rise and Progress of Paleontology" (1881), he argued against the "impoverishment of the fossil record": "Although paleontology is a comparatively youthful scientific speciality, the mass of materials with which it has to deal is already prodigious. . . . [I]n many groups of the animal kingdom the number of fossil forms already known is as great as that of the existing species."[5] The promise of paleontology was especially shown by research on the mammals, including "the labours of Gaudry, Marsh, and Cope" and "the marvelous fossil wealth of Pikermi and the vast uninterrupted series of tertiary rocks in the territories of North America." These fossils, and especially those of ungulates and carnivores, showed "gradually increasing specialization of structure," particularly in the key markers of mammal anatomy, "limbs or dentition and complicated brain."[6] Paleontology was not a flawed subject, but an essential science that needed to be conducted with more effort.

Paleontology reinforced particular narratives of evolution. As illustrated by Huxley's quote, much focused on increased specialization affecting particular parts of the anatomy. These were often the same features that earlier generations of natural historians had used to illustrate mammalian superiority and variation: teeth, limbs, and brains. The last were especially significant. Cope and Marsh both highlighted that the brains of Eocene mammals were small and "of a low, almost reptilian type."[7] This meant the true force driving evolution over the Tertiary was not growth in the body (the pinnacle of which had already been reached by the immense dinosaurs), but the growth of the brain and mental power—to such an extent that Cope and Marsh (despite their rivalry and different views of evolution) argued for similar "laws" of development, with brain size increasing over the Tertiary.

Evolutionary progress did not necessarily go in a single direction, as the mammal lineages showed specialization and variation. Even if the general track of progress led to greater intelligence, other parts of animals' bodies developed more variably. Divergent tracks among mammal groups could signify improvement within their own class, but not necessarily wider hierarchies. Each mammal group existed on its own branch, with its own pattern of development. The evolutionary history of mammals was both branching and progressive, with divergence reflecting improvement. In this respect, the common distinction made in histories of evolutionary thinking between linear "ladder thinking" and divergionist "tree thinking" somewhat broke down[8]—the branching of the mammals indicated they were at the summit of progress.

Tracing divergences and links between fossil and living organisms became an important research program. One of the earliest proponents was the Russian anatomist Vladimir Kovalevsky, who visited collections in France, Britain, Italy, and Germany, investigating as many herbivore fossils as he could find. In papers in English, German, and French in the 1870s, Kovalevsky discussed his work "diligently studying the organization of the fossil forms." He closely analyzed the bones and teeth of many herbivores, showing a movement from *Palaeotherium* through Tertiary ungulates like *Anchitherium* and *Hipparion* to the modern horse. Making an "apology for the somewhat minute osteological details," he argued that these animals could be arranged into series, moving across geological time with increasing anatomical specialization. This went beyond "the splendid osteological investigations of Cuvier, [who] had revealed to science a glimpse of a new mammalian world of wonderful richness," but which primarily cataloged new forms, "multiplying the diversity of this extinct creation." Instead, Kovalevsky drew on "the wide acceptance by thinking naturalists of Darwin's theory" by following "a deep scientific investigation of forms allied naturally and in direct connexion with those now peopling the globe."[9]

Notions of increased specialization and variation across geological time were significant, but raised problems. The first was what drove it. For some scholars, Darwinian processes of natural selection were enough. That natural selection, especially in the steadily harshening environment of the Tertiary and Quaternary, would encourage specialization in locomotion and dentition and require greater intelligence seemed logical for many—including Huxley and Marsh. However, natural selection was not always thought to fully explain all development. One of the most significant issues around Darwinian theory was the lack of a clear explanation of how new characteristics arose or were inherited. In studies of fossils, this was particularly important, as changes in lineages sometimes seemed dramatic and sudden. Paleontologists often saw Darwinian models as interesting, but not enough in themselves to create the variety within the fossil record.

Albert Gaudry presents an important example of the flexibility and extent of paleontological notions, mixing together several mechanisms behind poetic and metaphysical concepts. As professor of paleontology at the Muséum d'Histoire Naturelle, Gaudry maintained a strong hold over French studies of fossils. He drew attention to the impact of Darwin's work, and especially Clémence Royer's progressivist translation of *The Origin*.[10] He wrote that he had "always been distant in certain respects from the philosophical ideas of Charles Darwin," but nevertheless "read his book on the *Origin of Species* with a passionate admiration. . . . I took it in slowly, in small sips as one would a delicious liqueur."[11] Drawing on the French transformist tradition, Gaudry argued evolutionary principles were deeper than Darwinian ones, stating that "discussions of the doctrine of evolution must have paleontology as the foundation. This doctrine does not consist of theoretical views, but the patient comparison of beings who have succeeded one another over geological time."[12] Any accusations of paleontology as just descriptive were turned around; it was based not on abstract theorizing, but empirical proof. The fossil record was not—as for Darwin—a weak and impoverished thing, but the true basis for reflections on the natural world.

Gaudry proposed numerous mechanisms to understand the transformation of species in the fossil record. He saw natural selection as potentially a force, but more significant was a development toward harmony, as "the geological world is not a theater of carnage, but a theater of majesty and tranquility."[13] Gaudry had imagined the Pikermi fauna as a majestic, peaceful community. In later works he applied this to the entire natural world. Rather than a war of all against all, nature showed order, with every element having its place. Natural harmony and progress also had spiritual implications, demonstrating a divine figure mysteriously although perceptibly guiding improvement: "The history of past beings reveals a succession of indefinite nuances: the Divine Wisdom knows how to coordinate these nuances."[14]

Edward Drinker Cope also conceptualized development as driven by mysterious teleological processes. Cope's paleontological views were fairly conventional, seeing life progressing as a whole, but with regressions and destabilizations in particular lineages. Most of his texts were typological, focusing either on anatomical reconstructions or descriptions of geological formations. However, he also theorized on the mechanisms of evolution, most notably in *The Origin of the Fittest* and *The Primary Factors of Organic Evolution*. Cope has been presented as a "Neo-Lamarckian," and has been considered the foremost figure in an entire school of "American Neo-Lamarckianism" influential across the life sciences.[15] It is possibly more accurate, however, to discuss Cope building a complex synthesis mixing varied notions, including Lamarckian development, other evolutionary models, and fossils, neurology, and comparative anatomy. Darwinian natural selection was a supplementary factor, but could not cause creative evolution. Cope noted, "Nothing ever originated by natural selection." It explained "the gutters and channels which conduct the water," but not "the pump and the man who pumps it."[16] As with Gaudry, natural selection played only a destructive role in life's history, not a creative one.

To understand "the pump and the man who pumps it," Cope developed a series of laws driving development: "The study of phylogeny shows that the evolution of life-forms has been from the simple to the complex, and from the generalized to the specialized. These two forms of expression are not identical."[17] Growing complexity and specialization could have divergent results; they could potentially lead to more efficient organisms better suited to their lifestyles and environments, but could also create monstrosities lacking the ability to change or adapt beyond specialized conditions.

Development for Cope was driven by a mysterious evolutionary energy termed "bathmism," which was responsible for an organism's adaptations. It was a finite, fluid-like growth energy, potentially originating in the nervous system, directed to develop particular organs, while others receded. Cope wrote that "a given animal organism can only convert a given amount of force, and that capacity must remain uniform so long as the machine or structure remains the same. . . . When, then, a useful organ is added, subtraction from some less important locality must result, and, as a consequence, the latter must become still less prominent in the general economy."[18] This was later refined into the theory of "kinetogenesis," that movement and action in an organism's life led to an accumulation of bathmism and the development of particular characteristics, which were transmitted to later generations. Once again the teeth as organs for processing food, the limbs as organs of locomotion, and the skull as the seat of the brain and intellectual capacity were all prime locations for the accumulation of bathmism and the working of kinetogenesis. The results also differed when applied to these organs. The accumulation or reduction of "growth-force" in the limbs and teeth led to either increased specialization or decline, as they became "perfected," developed in overspecialized directions, or atrophied from lack of use.

While kinetogenesis acting on the limbs and the teeth led to diversification, its impact on the brain was more one-sided. The pressures here could lead in only two directions: increased brain size and therefore intelligence; or conversely, that lineages that used their growth force to develop limbs, teeth, and other organs at the expense of the brain may have gained short-term advantage, but were doomed in the long run against more intelligent animals. A key example was the primates. Early in the Tertiary they were dominated by "more powerful rivals; the ancestors of the ungulates held the fields and the swamps, and the Carnivora, driven by hunger, learned the arts and cruelty of the chase." The primates meanwhile devoted themselves to "ceaseless vigilance," and the "quality of inquisitiveness and wakefulness was stimulated and developed which is the condition of progressive intelligence," allowing them to triumph and eventually form humans. Intelligence was one area where Cope allowed natural selection to play a role, as "the 'survival of the fittest' has been the survival of the most intelligent, and natural selection proves to be, in its highest animal phase, intelligent selection."[19] Brain growth was the true progressive path within evolution and would prevent creatures from being swept aside by destructive forces.

Cope's model sought to square the diversity of animals in the fossil record and the present, with a broad if uneven narrative of progress. This integrated metaphysical and material forces around growth energy, effort, and exertion. It was also a moralized model, showing an ordered, regulated, and disciplined natural world generating progressive forms across geological time. Despite the great diversity of life, the line of progress moved in one direction: toward brain growth. Intelligence formed a single hierarchy on which all life—human and animal—could be positioned. This followed Cope's wider social and political commitments, predicated at placing white males at the summit of development, which has been well described by Kyla Schuller. Writing extensively on the inferiority of women and people racialized as nonwhite, Cope saw hierarchies as ordained by geological time.[20]

Cope and Gaudry argued for the primacy of paleontology within evolutionary discussions. Discussions of mammalian development consolidated around branching progress and arguments that some form of creative energy had constructed the natural world (although whether this came from an external creator or an inner force was unclear, which was a distinction between Gaudry and Cope). Their separation from Darwinian mechanisms was potentially extreme, although this should be considered from the other side: these two scholars had the confidence to reject the creative potentials of natural selection, arguing for the importance of paleontological work. Potentially more significant is that their evolutionary models inscribed the natural world with harmony and progress, deduced from the fossil record. Directional evolutionary models posited the unity of life and its movement toward diversity and hierarchy. This not only explained how life changed, but bolstered hierarchies in the modern world by judging animals and people as higher or lower on the chain of being. Morality and intelligence were key factors in life's history, whether through the work of a creative and driven evolution or the vitalistic and moralized visions of kinetogenesis. Those who understood these forces were therefore placed at the highest scale and could claim authority over the entirety of life, both human and animal.

The Unclear Origins of the Mammals

Questions of development opened another important issue: Where exactly had the mammals themselves originated, and what were their first forms? Much developmental and historical thinking in the nineteenth century was concerned with searches for origins, whether of animals, social groups, civilizations, or nations. This could work on various levels, with the original ancestors sometimes seen as presenting a group's inner core and essence, or as low and primitive forms that needed to be superseded through later development. These tensions were marked in discussions of early human societies, where debates veered over whether prehistoric people were genius-like precursors of modern humans, savages and barbarians who needed progress and "tempering" by civilization to reach their potential, or uncomfortable mixtures of these two possibilities.[21] Similar debates emerged over the origin of the mammals, and the mystery of their origins was one of the most significant, but also inconclusive, calls to action in nineteenth-century paleontology. While the mammal ancestors were often presented as strange and primitive, their early history also potentially gave keys to understanding the entire group.

The narrative of mammalian life from the Eocene on was increasingly mapped out, but its earlier history was more mysterious. By the mid-nineteenth century hierarchical arrangements of the mammals, presented by Blainville and accentuated by figures like Richard Owen, presented the monotreme, marsupial, and placental mammal divisions as an ordering from inferior to superior. However, what this hierarchy represented was uncertain, especially when the scale of life became refracted into more developmental modes. Some scholars—most notably Ernst Haeckel in Germany—argued that monotremes "are clearly the last surviving remains of a formerly diverse group of animals, who were the sole representatives of all the classes of mammals in the old Secondary period, and developed into the second subclass, the Didelphians, apparently in the Jurassic."[22] The three mammal groups therefore showed movement through distinct evolutionary phases, with monotremes transforming into marsupials, which then evolved into "superior" placentals. The idea that marsupials and monotremes represented

FIGURE 8.1. A Mesozoic scene, with early mammalian "duck bills" hunted by prehistoric reptiles. Illustration by Carl Whymper. Knipe, *Nebula to Man*, 69. Reproduced by kind permission of the Syndics of Cambridge University Library.

earlier stages of mammalian evolution was carried on in other media, such as the 1905 evolutionary epic poem by Henry Knipe, *Nebula to Man*.[23] The work included an image by Charles Whymper of the mammals of the Mesozoic, as scurrying, platypus-like creatures terrorized by the era's more dynamic flying and ambushing reptiles.

However, an evolutionary scale from monotreme to placental was by no means agreed. Some scholars saw the egg laying of the monotremes and "premature" births of the marsupials as marks of isolation or degeneration or special adaptations to the Australian environment. Thomas Henry Huxley wrote that while "the Ornithorhynchus and the Echidna are thus the representatives of the lowest stage of the evolution of the Mammalia," they were still highly specialized and could not represent generalized first ancestors.[24] Another current of thought did not include the monotremes within mammals at all, but saw them as either sui generis creatures linking different classes, or strange offshoots from the tree of life. Henry Neville Hutchinson called them "a great puzzle to naturalists, and one hardly knows whether we ought to call them reptiles or mammals."[25]

Whatever the character of marsupials and monotremes, debates over mammalian origins revolved around two major hypotheses. The first drew directly from chain-of-being reasoning, positing that mammals evolved from an unknown reptile. However, this idea was often criticized. First, there were persistent misgivings against too literal readings of the chain of being, and so this clear echo could be presented as tradition rather than rigorous scientific research. An alternate hypothesis was that mammals had a completely separate line of development. Thomas Henry Huxley for example argued that amphibious Labyrinthodonts showed enough mammalian features, and also potential for transformation and plasticity, to possibly illustrate that mammals' ancestors derived directly from amphibians, rather than progressed through reptiles.

To solve the mystery of mammals' origins, much attention focused on fossils from the Permian period at the end of the Paleozoic. These were particularly well known from South Africa, where Alexander Geddes Bain, a road engineer in the Cape Colony, excavated fossils in the Karoo region from the late 1830s.[26] Bain's fossils included creatures that seemed broadly reptilian, but with some mammal-like features, particularly two large and protruding tusks (which led Bain to christen them as "Bidentals"). Most of these remains were sent to Britain, and Richard Owen renamed them

"Dicynodons." He wrote that while they were essentially reptilian, "the Dicynodon seems to have borrowed its peculiarities from a higher class, and to have engrafted some mammalian characteristics upon the upper jaw,"[27] mixing forms, as dolphins and marine reptiles like ichthyosaurs had fishlike characteristics. While Owen himself was famously opposed to evolutionary transformation, thinkers with more evolutionary ideas could easily take these characteristics and present Dicynodons as not just "mammal-like," but mammal-ancestral.

Another important fossil was one Bain initially termed "the Blinkwater Monster," but was eventually transformed in scientific publications into another beast: *Pareiasaurus*. The animal was given its new name by Harry Govier Seeley, a British paleontologist who often focused on groups regarded as "interstitial." His work *Dragons of the Air*, for example, concluded that pterosaurs, the flying beasts of the Mesozoic, were not "flying reptiles," but mixed forms like monotremes, merging characteristics from different lineages.[28] Seeley also conducted several projects in South Africa, excavating vertebrate fossils while being escorted around Karoo sites by Andrew Bain's son, Thomas.

Most of Seeley's attention focused on the *Pareiasaurus*. In 1888 he gave a report to the Royal Society on the skull and vertebrae *Pareiasaurus bombidens*, secured from the Karoo in the late 1870s. It represented a series of transitions. Most features of the skeleton were judged reptilian (placing the animal as a whole in this class), but its skull showed similarities with the amphibian labyrinthodonts. Even more significantly, as well as incorporating amphibian and reptile features, it also showed some characteristics Seeley regarded as distinctly mammalian, particularly in the pelvis and sacrum, which it shared with many of the more mammalian South African vertebrates. These were "evidence of affinity," indicating "a common origin for these mammalian and reptilian structures by inheritance from Amphibian ancestors."[29] *Pareiasaurus* showed that both reptiles and mammals could have evolved independently from great amphibians.

Seeley brought back a more complete pareiasaur, named *Pareiasaurus baini*, after a further trip to South Africa in 1889. This expedition depended on European settlement and mining networks, with further assistance from Thomas Bain and another naturalist, William Guybon Atherstone. Seeley described his main roles as having been "to direct and control the quarrying operations of ten men" and reconstruct the heavy and fragile specimen in the

FIGURE 8.2. *Pareiasaurus baini*. "The Royal Society's Conversazione," *The Graphic*, May 16, 1891. Author's collection.

British Museum of Natural History with help from the casting workshop. This turned "an unprepossessing heap of rock, among which were some indications of bones," into a reasonably complete skeleton.[30] The animal was exhibited to the Royal Society in 1891, highlighted by Seeley as a mixed creature with reptilian, amphibian, and mammalian affinities, potentially showing the origins of reptiles and mammals in the depths of the Permian. A vivid illustration of the specimen being pored over by scholars of the Royal Society was published in *The Graphic*, along with an illustration of "our artist's dream" of the living animal. As well as showing the need for imagination in understanding prehistoric creatures, the illustration dramatized Seeley's theories, with the "giant reptile" having webbed feet like an amphibian, a mixture of reptilian sprawling and mammalian upright legs, and a slavering bulldog-like head.

Seeley's discussions of the South African fossils were contentious, with both Cope and Gaudry raising problems with his analyses. The period from the 1890s saw even greater research on Permian vertebrate fossils, as more were uncovered in South Africa and connected with similar remains from North America and Russia.[31] While debates on the origins and first characteristics of the mammals were therefore uncertain, they raised important points around paleontological work. They reinforced ideas that the search for new fossils—and particularly new fossils from regions that had not been extensively studied geologically, but were now subjected to European colonial authority—would allow hitherto unknown dynamics in life's history to be understood. Paleontology was therefore further tied with the global expansion of colonial and commercial systems. However, debates over creatures often defined as "mammal-like reptiles" from the Permian or strange and unsettling like *Pareiasaurus* also had implications for the origins of lineages. At the base of the tree, types could be deduced, but these were often tangled and interstitial and mixed characteristics defined as primitive and specialized. The history of life therefore was not an easy ladder of progress (even if some aspects could be made to fit this narrative), but combined forms in wondrous and monstrous ways. Successful creatures emerged unpredictably from this disruption.

Teeth of "Dwarfs" and the Mesozoic Mammals

Whether mammals originated from an early reptile or a separate amphibian stem, they originated before the Age of Reptiles. Therefore, how the mammals developed between their origins and sudden flowering to dominance in the Eocene was a live question. Fossils ascribed to the mammals of the Mesozoic added to these mysteries, generating interest and disquiet. The first of these were two jawbones collected by a stonemason gathering fossils in the Stonesfield slate for William Broderip in 1812. One of these was in turn sold to William Buckland, who eventually classed them as the remains of an animal like a small opossum living at the same time as *Megalosaurus* and other great fossil reptiles.[32] This was in itself controversial, especially among French scholars like Blainville, who doubted the presence of mammals in these reptilian ages.[33] Other Mesozoic mammal fossils were subsequently found and identified, but these were almost invariably small and fragmentary, primarily consisting of teeth and jawbones. While there were not many specimens, and they tended to be concentrated in large collections in western Europe or North America, there were enough to lead to some general studies. Richard Owen wrote a monograph in 1871 describing the twenty specimens he had access to, and Henry Fairfield Osborn produced his own study of forty specimens in European collections in the 1880s.[34] The tiny and fragmentary fossils required special techniques and claims to authority to study them, with magnification and illustration being essential for discerning their small features. Debates over Mesozoic mammals were also ways of claiming status and power over the past.

That fossils of Mesozoic mammals were small, and came from an era defined as the Age of Reptiles, dominated by giant dinosaurs on land, marine reptiles in the sea, and pterosaurs in the air, was a difficult issue. It reinforced one of the key tropes of paleontological narratives: that earth's history was a movement through a series of eras, each dominated by progressively higher forms of life. Mammals may have existed during the Age of Reptiles, but they were subordinate to the true "rulers" of the age. The Mesozoic mammals were therefore often described not as originators of progressive qualities, but as stunted and lacking in high features. Richard Owen ended his monograph with a sense of disgust: "Mesozoic Mammalian life is, without exception . . . low, insignificant in size and power, adapted for eating insect-food, for preying upon small lizards, or on the smaller and weaker members of their own low mammalian grade." This confounded the assertion of mammalian superiority. Mesozoic mammals were "rat-like, shrew-like, forms of the most stupid and unintelligent order of sucklers. The results of Neozoic paleontology sometimes move one to exclaim, in regard to Mammals, 'there were giants in those days!' but, descending to earlier periods, we find only dwarfs."[35] For mammals to truly reach their potential, the earth itself had to move beyond the Age of Reptiles, into a new age of creation. These ages were therefore driven as much by vague providential forces as they were by the characters of the organisms themselves, which reinforced the mystery and epic qualities around life's development.

Mesozoic mammals were still significant for what they demonstrated about origins. The fossil jawbones and teeth opened the way for paleontological analysis, as teeth were regarded as critical for understanding diet and increasingly for developmental history. The teeth of Mesozoic mammals became one of Cope's most strident illustrations of kinetogenesis and the reptilian origins of mammals. He argued that the teeth of the earliest mammals were all tritubercular, having three cusps, and developed logically from the single cone teeth of reptiles. Through minute studies of mammal teeth from the Mesozoic to the Eocene and across the Tertiary, Cope argued that all varieties and complexities of mammalian dentition derived from the tritubercular form, which differentiated as bathmism and exertion led early mammals to specialize. Step-by-step evolutionary models showed how different mammal molars, premolars, canines, and incisors could have developed from the original three-coned teeth of Mesozoic mammals, all sharing a common origin.

Osborn supported Cope's theories, and his work on the teeth of Mesozoic mammals also provided an opportunity to contest domineering authorities. One of the most heated controversies in late nineteenth-century American paleontology was between Osborn and Marsh over issues which seemed technical, but were highly significant. In 1888 Marsh announced that his field-worker, John Bell Hatcher, had found almost a hundred remains of Cretaceous mammals (again, almost entirely jaws and teeth) in the Laramie Formation in the American West. Marsh argued that they represented twenty-seven species—a large trove demonstrating "the rich mammalian fauna that lived during Cretaceous time"—and promised a memoir under the auspices of the US Geological Survey.[36] The animals were all small, ranging from the size of a mouse to a rabbit, and rather than represent a gradual movement from marsupial to placental forms, Marsh's fossils seemed similar to Jurassic mammals, as well as monotremes and opossums. This meant there was "a great faunal break between the time in the Cretaceous when they lived and the earliest known Tertiary."[37] Life changed dramatically between the Age of Reptiles and the Age of Mammals, with the "higher" placentals suddenly appearing in the Eocene.

However, Osborn wrote a scathing review of Marsh's work which, he claimed, "almost entirely eliminated the work of the author."[38] Osborn argued that most of Marsh's species were defined on poorly preserved individual teeth (some of which may have been fish or reptile, rather than mammal), and the number of new genera was seriously inflated. Marsh expectedly attacked Osborn in a similar multifaceted manner, as he had Cope in the conflicts over the Dinocerata, writing that Osborn had made numerous technical errors (including egregiously getting the title of Marsh's paper wrong), had not seen the original specimens, and had "never collected any Mesozoic mammals" himself.[39] In response Osborn brushed Marsh's misgivings aside with a more significant issue: he was not permitted to see Marsh's jealously guarded specimens, which despite "belonging to the government" (owing to Marsh's funding by the United States Geological

Survey) "are not accessible to American paleontologists," unlike those in "the various foreign museums" which Osborn was allowed to study.[40] Marsh was censured for not living up to conventions of scientific openness, and as he was at the time undergoing scrutiny for his use of geological survey funds, this was a major threat.

The dispute eventually fizzled out, but Osborn continued reinforcing the reptilian origin of the mammals. Of course, personal issues were highly significant. Osborn later wrote to Robert Broom that he "could not resist the opportunity of getting even with" Marsh.[41] Osborn's increased confidence indicates that American paleontology was beginning to shift into a new period marked by codes and negotiations, rather than the heated disputes of the Cope and Marsh era. The earliest mammals were still uncertain, but conventions around their research were solidifying.

The Imperial Science of Biogeographies

Paleontology became a science of origins and development, which defined modern nature. Possibly the clearest example of how paleontology was used to explain the present was in biogeography, the science of animal distribution. Biogeography was one of the most prominent topics within the late nineteenth-century life sciences, but has been relatively neglected by historians, with the main contributions being by Janet Browne for earlier periods, plus discussions in relation to Alfred Russel Wallace.[42] The neglect of biogeography is surprising, given the prominence of the subject. Biogeography was not just a major research problem for scholars like Wallace and Haeckel, but contemporary encyclopedias and textbooks were replete with maps of the world divided into zoological provinces, defined by types of animal. There were also numerous popularizing works like Wilhelm Bölsche's *Tierwanderungen in der Urwelt* and Richard Lydekker's *Geographical History of Mammals*.[43] As in much scientific work, the boundary between the public and the scholarly was either porous or nonexistent.

There are two things about late nineteenth-century biogeography which should be emphasized. The first is that it was conditioned by imperial structures—and indeed Janet Browne described it as "one of the most obviously imperial sciences in an age of increasing imperialism."[44] The empires of the late nineteenth century provided a framework in which biogeographic studies were undertaken. The classic images of biogeography, dividing the world into provinces marked by "natural" boundaries which animals and plants did not cross, drew closely from images of empire, which associated particular parts of the world with particular climates and products. Imperial systems also formed the material basis for biogeography, as large numbers of animals were killed, preserved, and shipped to centers of calculation in Europe to define the fauna of particular regions. Biogeography also engaged with questions around commercial and economic development, connecting with attempts to acclimatize animals and plants to expand agricultural and pastoral production in colonized territories, while considering the fundamental qualities of landscapes, regions, and environments.

The second issue is that biogeography was increasingly based around the premise that the current distribution of animals depended on their history, which could be understood only through paleontology. Biogeography did not just focus on the "what" and "where" of the distribution of animals, but also the "why." Answering this depended on understanding life's history. Richard Lydekker wrote, "To understand rightly the present distribution of animals, it is, however, essential to study their past history," and "without such history it would be quite impossible to grasp the reason of many apparent anomalies in their present distribution." He specifically brought up the persistence of two groups of "primitive" mammals—tapirs in South America and southeast Asia and marsupials in Australia and the Americas—as points which could not be understood "without the aid of paleontology."[45] These animals had once been widespread (with fossils of marsupials and tapirs found across Eurasia and North America), but now existed only in regions where nature was still akin to earlier geological periods.

The mix between imperial natural history and paleontological explanation meant biogeographic studies linked older traditions of the animal world and new knowledge. Many concepts drew from early modern precedents and the assumed characteristics of particular environments among European or North American scholars. That animals, plants, and people were stamped by climates was a long-standing idea, whether thought to be due to direct influence, or more indirect processes of adaptation. Ideas that tropical climates led to strange, aberrant, and possibly degenerate forms; that open continental climates created more dynamic and migratory organisms; and that colder places forged hardy and ferocious creatures were generally maintained through biogeographic thinking. However, while biogeography drew from older ideas that particular climates conditioned life, it also needed to explain why life in geographically separate regions with similar climates was so different. In Wallace's words, climate could not be the sole factor in biogeographic distribution, as "countries exceedingly similar in climate and all physical features may yet have very distinct populations."[46]

Climate therefore interacted with other factors, most notably migration and its counterpart, isolation. A central assumption in nineteenth-century biogeography was that animals would—if there was sufficient food and no geographic or climatic boundaries—migrate to fill all available space, and higher forms would displace the lower. Conquest and migration were features of the animal world, just as they were conceptualized as crucial to human history (in another parallel with contemporary imperial thinking). This resonated with Darwinian ideas of competition, and spread to writers who doubted natural selection was sufficient to explain life. These assumed trends in the animal world meant superior types would conquer large territories. Creatures regarded as primitive, aberrant, or isolated were marked as prehistoric relics and worthy of study, but were also most likely doomed in the present.

Some of the earliest nineteenth-century biogeographic models dealt with distributions of birds (which Kirsten Greer has connected with British military collecting networks).[47] However, as birds were able to fly, they were generally not seen as the best candidates for understanding general biogeographic principles. As in much nineteenth-century natural history, mammals were again taken as the most indicative group. William and Philip Sclater wrote how "mammals, as the most highly organised and altogether the best-known group of the animal kingdom" were ideal, especially as (excepting bats and marine mammals) "land is the means by which they extend their ranges, and seas and rivers form their restraining boundaries."[48] For Wallace too the "preeminent" mammals "possess the additional advantage of being the most highly developed class of organized beings, and that to which we ourselves belong," having "so much power of adaptation as to be able to exist in one form or another over the whole globe."[49] These were of course the same reasons why mammals absorbed attention across the nineteenth century, as the "highest," and most diverse and widespread, animals.

That most mammals could travel only on foot had wider implications for the earth's history. The distribution of mammals potentially illustrated the ancient distribution of the world's landmasses. While there were debates over floating and mobile continents, particularly prompted by Wegener's hypotheses of continental drift in the 1910s, most geologists and geographers thought the position of the earth's landmasses was relatively fixed. However, while the position of the continents may have been fixed, their surfaces were not, and were subjected to huge and violent changes, as mountains and other landforms rose and fell, and as the sea encroached on and receded from the land. The distribution of land animals, and how they were either similar to or distinct from animals in other parts of the world, served as a proxy for imagining ancient continents, land bridges, and rising and sinking landmasses. Biogeography was not just about distribution, but about understanding a historicized earth conditioned by geological change.

What then did the division of the world according to mammals look like? There were several models, but by the late nineteenth century the most popular—promoted by Huxley, the Sclaters, Haeckel, and Wallace—were to first separate Australia and South America as the two most distinctive biogeographic provinces. These regions were held as unique, but also isolated and "inferior" in important respects. Australia, with its marsupials and monotremes, was an iconic continental isolate. The "Neotropical" region of South America, possessing large rodents, edentates, and even stranger extinct animals, was also conceptualized as separate, even though "northern" animals like big cats and bears now had a strong presence. The rest of the world was then divided into four further major regions: a "Paleoarctic" region, consisting of Europe, North Africa, and northern Eurasia; "Ethiopia," consisting of sub-Saharan Africa; the "Oriental" zone of south and southeast Asia; and the "Nearctic" region, representing North America. This created six biogeographic provinces, of roughly comparable size and with various degrees of mixture, migration, and interchange among them. The islands of New Zealand and Madagascar were often presented as their own zones, with Madagascar in particular sometimes seen as a relic of the lost continent of Lemuria.[50] These were all distinct "worlds" of nature, where development had occurred at different rates and along distinct trajectories.

These biogeographic regions were not conceptualized as equal. While the southern zones were distinctive and marked by particular faunas, the northern ones were thought more mixed and "progressive." The mixture of forms in Europe was presaged in the writings of Lyell and Boyd Dawkins, who saw combinations of arctic, temperate, and "African" animals in recent Pleistocene history. The history of the north was marked by the ebb and flow of animals that migrated, competed, and followed changing climates.

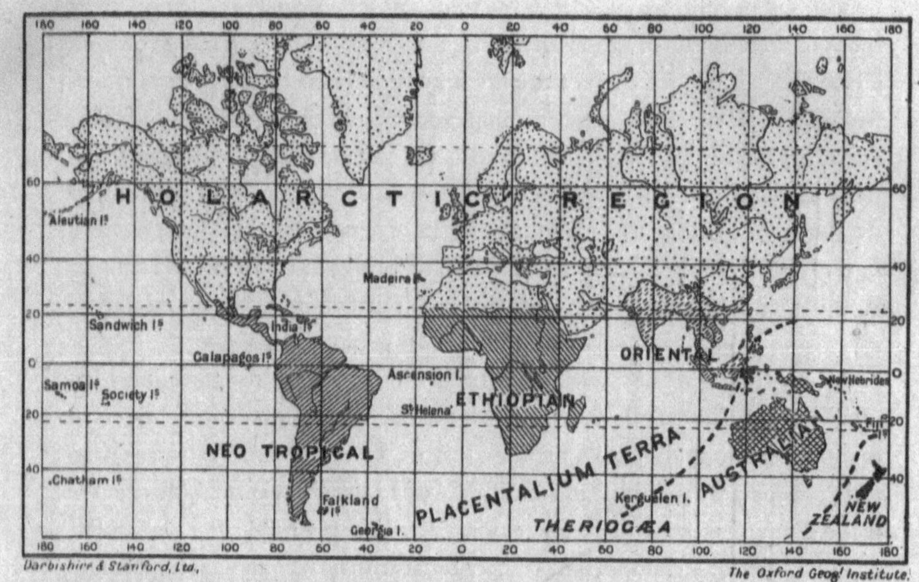

Fig. 42.—Map of the World, showing its division into great provinces and regions characterised by the presence of different kinds of animals. The first division is into (*a*) New Zealand and (*b*) Theriogæa. Then Theriogæa is divided into Australia and the Placentalium Terra. Lastly the Placentalium Terra is divided into the four regions, viz. (1) the Holarctic, (2) the Neo-tropical, (3) the Ethiopian, and (4) the Oriental.

FIGURE 8.3. A division of the world into conventional biogeographic "provinces." Lankester, *Extinct Animals*, 63. Author's collection.

Unlike the isolated southern continents, the northern hemisphere was a vast connected zone. American scholars in particular attempted to construct a northern region incorporating Europe and northern Asia and stretching into North America as being the original home of all the animals defined as the most dynamic, including equids, cervids, dogs, bears, rodents, cats, and (according to later theories) potentially humans. Henry Fairfield Osborn promoted the term *Holarctica* for this area and argued for alternating periods of exchange and isolation across the Tertiary, as North America and Eurasia became linked and unlinked by land bridges.[51] Holarctica was a dynamic region, where superior types could form and move across the northern latitudes, and potentially spread south when geographic connections permitted.

While biogeography emphasized the northern hemisphere as the norm, it also promoted the idea that research across the world was essential. Filling in gaps in the map or solving biogeographic mysteries motivated research expeditions to areas thought to be important in biogeographic history. Biogeographic ideas and searches for migration routes connected paleontology with colonial projects and rhetoric, as it became defined by wide-ranging searches. While North America, Europe, and parts of India and South America had been reasonably well prospected, and particular lineages—especially equids, cervids, and rhinos—were well studied, other parts of the world and other animals were less known. Paleontologists noted how few fossils were taken from Australasia, eastern and central Asia, and Africa, and so the universalizing collections of Europe and North America required specimens from these regions. The histories of several lineages, including elephants and marsupials, were also mysterious. Prophesies and calls to action gave an increased global vision to paleontological work.

These issues—hunts for origins, attempts to clarify biogeographic problems, and the impact these varying drives had on the balance of authority among scholars in different parts of the world—will be core themes in the next chapters. Biogeography further emphasized paleontology's global and colonial aspects, and reworked relations between central collections and field sites. While these agendas bolstered northern institutions, they also gave scholars based in understudied areas opportunities to gain authority, either by controlling access to sites and specimens, or by using their positions to promote alternative views of life's history. This simultaneously gave particular localities an important status, but also often led to more localized visions of paleontology and earth's history, moving in some respects away from the grand narratives discussed in the last chapter to more focused understandings of what particular places were like during distinct periods and what patterns had affected their development. These relations, between global exchange and local consolidation and between centers and field sites, were of course far from new in these years, having affected fossil work since its inception. However, they were now expressed with a new urgency, as the tightening global entanglements of the late nineteenth century expanded paleontological work and allowed a greater variety of places to be defined through their deep pasts and by new actors and individuals.

PART III

GLOBAL TRANSFORMATION AND NEW HISTORIES OF LIFE

FIGURE 9.1. The American Museum of Natural History's fossil preparation workshop in the 1900s. Image #46501, American Museum of Natural History Library.

CHAPTER 9

Building and Contesting Collections

New Museums and Other Institutions

THE FOSSIL PREPARATION LABORATORY IN NEW YORK'S AMERICAN MUSEUM OF NATURAL History (AMNH) was a hive of disciplined work. Fossils came encased in plaster from traditional field sites in the American West, and were accumulated in storerooms. Promising blocks, cross-referenced through field notes and catalogs, were selected for preparation. The workers used techniques from stonemasonry, dentistry, sculpture, and model making to extract fossils and turn them into comparable artifacts. The laboratory was organized to maximize productivity and output: its 1896 report noted: "Three complete skeletons have been mounted; 163 skulls, 111 jaws, 17 feet have been prepared for exhibition." This was driven by hard-working practices, as "Mr. Hermann," the laboratory's director, "is a rigid disciplinarian, and extracts a full day's work from every man in the room. No loitering is permitted, and every man takes a keen interest in the work of the department."[1]

Overseeing the institution was a pyramid of authority, with the department's curator, Henry Fairfield Osborn, at its summit. After finishing studies in Europe, Osborn became increasingly involved in university and museum management, but only after working in his father's railroad company. According to William Berryman Scott, "his father . . . was determined, that, whatever else his sons might do, they should acquire a knowledge of business and its ways."[2] Following this stint, Osborn was appointed to an unsalaried post at the AMNH, where he founded the museum's Department of Mammalian Paleontology in 1891, which was renamed Department of Vertebrate Paleontology in 1895, as equal attention became devoted to fossil reptiles. In 1907 he became the museum's president, overseeing a complex of collecting, research, and exchange. Ronald Rainger has shown how Osborn's links with New York high society, and particularly his uncle John Pierpont Morgan, allowed him to gather resources for museum work.[3] Osborn promoted the biologized politics of the Progressive Era, supporting the eugenics movement, urban educational reform, and immigration restriction. He aimed to develop scientific work, connect urban New Yorkers with the natural world, and show how evolution could lead to progress or degeneration.[4]

The AMNH's Department of Vertebrate Paleontology was far from simply Osborn's project, but maintained a complex division of labor. Indeed, the AMNH provides a central example of Lukas Rieppel's argument that American museums in the 1890s and 1900s often drew from US business practices, with their organizational structures and principles of vertical integration mirroring those of industrial trusts.[5] While Osborn was listed as the author on many papers, much of the work was conducted by others. Scientific staff like William Diller Matthew and William King Gregory worked on fossils, took measurements, prepared notes, and often did much of the writing. The men (and they were primarily men) in the preparation laboratory constructed specimens, while typists, printers, artists, illustrators, catalogers, and other workers all conducted essential activities.

In 1895 the new Hall of Fossil Vertebrates opened. While histories of the AMNH tend to emphasize its work on dinosaurs, Matthew was to write in 1902 that "the greater part of the specimens are *Mammals*."[6] The growing numbers of dinosaur fossils excavated during the "Second Dinosaur Rush" meant that a new Dinosaur Hall was opened in 1905, with the mounted skeleton of *Brontosaurus* as its centerpiece, and the larger Hall of Fossil Vertebrates was rechristened The Hall of Fossil Mammals.[7] The room depicted mammals' diversity and "progress" across geological time. In the southern galleries, visitors saw a stream of perissodactyls, including ancestors of the rhino and horse, and the extinct titanotheres, chalicotheres, and paleotheres. On the northern side, proboscideans held the center, with uintatheres, rodents, and carnivores to their left, and even-toed ungulates

FIGURE 9.2. The mammal galleries at the American Museum of Natural History, visited by a school party. Image #286913, American Museum of Natural History Library.

and South American mammals to their right. The specimens consisted overwhelmingly of mounted skeletons, accompanied by smaller fossils showing anatomical features, charts, and diagrams explaining evolution and geology. Illustrations, usually by the museum's artist Charles R. Knight, showed (following Matthew) "the *probable appearance* of the different extinct animals, according to our best judgement, as indicated by the characters of the skeleton, appearance of their nearest surviving relatives and the habits of life for which the animals seem to have been fitted."[8] The gallery brought the past to life, through numerous strategies and media.

The AMNH has often been the focus of histories of paleontology.[9] Indeed, this institution illustrates key changes overtaking the field in the late nineteenth century, as it became strongly connected with what has been termed the new museum movement.[10] Expanding collections like the AMNH, often funded by philanthropists or state governments, aimed to appeal to and "elevate" public audiences, while also conducting scientific and scholarly research. Within these collections, the life of the past was crucial. Paleontology was a field where prestige could be consolidated and built into public spectacle and ideological messages around evolution and development.

The programs of the new museum movement shifted paleontological work considerably. Its significance will be seen not just in this chapter, but in later ones, with institutions like the AMNH and Carnegie Museum becoming key players within national and international paleontological research. Their scale of funding, personnel, and collections led to a dramatic expansion of museum work, as well as particular visions of research and nature. They were also particularly prominent in dinosaur paleontology, as these institutions possessed the large systems of excavation and preparation, and huge display halls, sufficient to exhibit dinosaur specimens.[11] Well-funded, reforming collections like the AMNH of course did not exist in a vacuum, but interacted with and influenced other institutions also devoted to the deep past. This was especially the case in mammal paleontology, which was always conducted in diverse and heterogeneous places and gave scope for other voices.

Therefore, as well as examining the expansion of large collections in the late nineteenth century, this chapter also considers how other institutions attempted to maintain their positions within the expanding context of museum work, partly through emulating elements of the new museum program, but also following other strategies. In particular, two types of institution will be highlighted. The first are those based around universities, which were never as well funded and resourced as large municipal and philanthropically funded museum collections, but could nevertheless hold their own through adopting more flexible strategies and generating prestige and resources through education. The second type is traditional collections in established centers of authority, most notably the British Museum of Natural History and, especially, the Muséum d'Histoire Naturelle in Paris. These in some respects were part of the "new museum" movement, expanding their collections and reforming their displays and practices in similar ways. However, restrictions placed by their often more limited funding and opportunities presented by their inherited prestige and established collections led them to maintain their status through other means. Examining this range of new and reforming institutions shows how paleontological work spread into a range of places and was conducted through varied values and projects—all aiming towards asserting control over the deep past and its significance.

The New Museum between the Organic and the Bureaucratic

What then was the new museum movement, and how did paleontology become central to it? Across the world in the late nineteenth century, a wave of municipal, national, and imperial collections were founded (or heavily re-formed), aiming simultaneously for educational, research-based, and public roles. Natural history, including fossil displays, were crucial for these institutions, emphasizing narratives around progress, scale, and development. Moving away from the potentially "unrespectable" showiness of older forms of display, such as the penny museum in the United States or the public show in Britain and France, was a stated aim of these collections—but so was a move away from crowded older collections, such as the ad-hoc gallery of paleontology established in Paris in 1885, or the cluttered final geological room of fossil edentates and mammoths in the British Museum. A wide literature has noted how the architecture of these museums—often based on Gothic cathedrals or neoclassical temples—were intended to instill hushed reverence, as displayed objects became relics to be pondered and revered.[12]

There was a further metaphor for the new museum, that it was a constant work in progress, almost a living entity, expanding and growing, even as it reached gigantic proportions. The maxim by George Brown Goode, director of the US National Museum, that "a finished museum is a dead museum, and a dead museum is a useless museum" was constantly invoked.[13] William Henry Flower, director of the British Museum of Natural History reiterated that "a museum is like a living organism—it requires continual and tender care. It must grow, or it will perish."[14] Just as organisms in their evolutionary past either grew and innovated or risked falling into decadence, so the museum needed to expand and constantly change to survive.

The idea of the museum as a great organism emphasized its scale. Museum annual reports—directed to trustees, government officials, and often the general public—emphasized scale and expansion. These highlighted visitor numbers increasing year by year (or offering commiserations and excuses if they did not). They listed budgets, which needed to be maintained or increased. They emphasized the expansion of collections, through exchange, purchase, and the organization of expeditions. Personnel and employees grew in number. Publications poured out. Public galleries were constantly reworked. The museum could never be finished, but had to grow. "Finishedness" was not just the end of growth, but the end of vitality.

New museums were also connected with urban life and spectacle. Much contemporary and historical attention has focused on the expansion of new forms of metropolitan culture—such as the bar, cabaret, gallery, and daily newspaper—in this period.[15] The museum was another of these urban institutions, deploying new techniques from business and commerce. Many museums installed electric lighting to illuminate public galleries and enable longer working hours in storerooms and preparation labs. Museum research publications emulated those of universities and learned societies, and many collections organized public subscription clubs and popular magazines. Card indexes and increasingly complex storerooms made the cataloging of nature as much a bureaucratic and technical problem as an anatomical one.[16] Meanwhile, the public galleries were filled with vitrines, adopting visual strategies from department stores and expositions. The boundary between the commercial and scientific (an always tense dichotomy) could be blurred. The wonders of life were as dazzling, spectacular, and bewildering as the wonders of the modern metropolis.

Museum visitor numbers increased dramatically in almost every country. By 1900 the AMNH was receiving 523,522 visitors a year,[17] and the British Museum of Natural History 485,288,[18] and they were to increase even further in following years. Growing public interest required new ways of managing visitors. Turnstiles, ticketing booths, labeling systems, shops selling guidebooks and postcards, public lectures, security guards, and "guide naturalists" were all pioneered in these years. The increase in visitors was crucial to the new museums, which linked research with public education. The public role was also political, although in varied ways. Tony Bennett has argued for the ideological role of "New Liberalism" in the evolutionary and racial displays in British museums,[19] and in the United States, the new museum movement is often connected with Progressive Era social reform agendas.[20] Lynn Nyhart has drawn attention to the "biological perspective" prominent in German society, tied to debates over modernity and reform, and particularly presented through museums and other civic institutions.[21] The natural world could be invoked in political struggles between a range of positions, and the museum was an important arena, although often leaned more toward establishment views, owing to reliance on philanthropy and state sponsorship.

Within natural history museums, displays of fossil mammals were almost obligatory. What was striking about museum displays in this period is how formulaic the specimens often were. Across Europe and North America these generally recapitulated the "classic" creatures discussed in previous chapters of this book. Museums sought out casts of Cuvier's Paris mammals, a *Megatherium* (often paired with a glyptodon), a giant deer, *Dinotherium* skulls, a large fossil elephant of some sort, one of the Dinocerata (often papier-mâché from Marsh's workshops), and a collection of Pleistocene beasts from European caves. For example, the initially "comparatively small" paleontological collection in the Carnegie Museum in Pittsburgh consisted of specimens expected of a paleontological display, including a *Megaloceros* bought from London and "one of the most perfect skeletons of the mastodon which has as yet been discovered," purchased from the specimen dealer Henry Ward.[22] This canon of fossil mammals accentuated the authority of collections that held original fossils. The Paris Muséum, with the Montmartre and Pikermi fossils, and the British Museum of Natural History, with its *Megatherium*, *Dinotherium* and Siwalik fossils, could maintain positions even as their resources were eclipsed by better-funded American collections. Wealthy museums could also opportunistically expand by purchasing classic and important specimens, most notably the AMNH's purchase of Cope's fossil vertebrates and the Warren Mastodon (the latter bought with a special bequest from J. P. Morgan).

The new museum—a heterogeneous and expanding institution—required a large number of workers. At the summit were the "museum masters,"[23] people like Henry Fairfield Osborn at the AMNH, William Jacob Holland at the Carnegie Museum, Karl Möbius in Berlin, and William Henry Flower and E. Ray Lankester in London. Whether in charge of the whole museum as directors or at the head of well-resourced paleontology departments, these domineering figures set the tone for their institutions. Their activities were diverse: issuing publications; overseeing research staff; providing an overarching "vision"; acting as public faces for the institution; representing it in elite circles of politics; philanthropy and patronage (essential for securing resources); and producing popular books and magazine articles.

The varied role of the museum masters should of course not obscure that they were only one point in more complex and fragmented institutions. As the number of personnel increased, museums offered employment opportunities for research scientists. In addition to full-time salaried scientific staff, museums also contained numerous associated scholars who were experts in particular branches of natural history who sometimes held honorary or voluntary roles, or minor positions with a small salary or stipend. While this status might suggest junior positions, these scholars were often privately wealthy, and connected the museum to gentlemanly and voluntary societies and imperial networks. Richard Lydekker, one of the most prominent theorists of biogeography, had various informal roles in the British Museum of Natural History, and the AMNH frequently gave assistant positions to foreign collaborators. These roles were therefore crucial for spreading the reach of museum work.

Work in museums was also highly gendered, with relatively few opportunities for women, though educated women did sometimes become museum assistants and researchers. Important examples included Maria Pavlova, who worked in Paris under Albert Gaudry before returning to Moscow to build a large paleontological network, and Dorothea Bate, who in the 1890s gained a salaried position at the British Museum of Natural History by collecting fossils from the Mediterranean islands.[24] These examples were exceptional, however; most women in museums remained in technical, voluntary, and lower-ranked positions. While much of the historiography on paleontological artwork has focused on grand mural painters like Charles R. Knight,[25] much artistic work was performed by women, such as Alice Woodward (daughter of the geologist Henry Woodward), who produced artworks for the British Museum of Natural History. This followed some of the gendered conventions in natural history work, where "gentleman scientists" would often work in a family firm environment, publishing works under their own names, but with much of the research, illustration, and writing being performed by female members of the household.[26]

Preparation and casting remained central to museum business. Most fossils were incomplete and fragile and required intervention to become usable scientific objects (or just to prevent them from crumbling to dust). In the new museums fossil preparation became increasingly formalized. A. Hermann at the AMNH and Francis Arthur Bather at the British Museum of Natural History wrote articles describing technical changes in preparation up to the 1900s.[27] Merging the role of artist, sculptor, taxidermist, and conservator, preparators were critically important for fossil collections. Osborn was extremely proud of the AMNH's casting workshop, writing to Henry Woodward that they employed "a young lady artist" to color the casts, and "in Paris they admit that our casts are fully as good as theirs, an admission which I think tantamount to an acknowledgement when coming from a Frenchman, that the casts are actually a little better than theirs."[28] While drawing on stereotypes of French snobbishness, in Paris the *atelier des moulages* was also a pride of the museum, crucial for producing material for exchange.

A new category of paleontological worker was the "field man" or "fossil hunter" (usually specifically gendered as male). These were technical experts whose primary responsibility was taking part in expeditions extracting fossils. While the fossil hunters of the 1870s had been geological surveyors, university students, or members of frontier society, the trained "fossil hunter" merged these facets into a single identity. Fossil hunting required deep familiarity with landscapes and geological strata and expertise in supervising workers, excavation, and preserving fossils. The last involved developing techniques like encasing whole blocks of rock in plaster, which could then be transported and extracted by preparators in museum laboratories. The fossil hunters often worked seasonally, conducting excavations during the "season" (which varied by region, depending on climate) and then spending the remainder of the year either supervising preparation, studying fossils, or writing publications. As with the casting workshops, "field men" were often in demand, sometimes being museum employees, or working as independent specialists. Charles Sternberg for example, after initially being one of Cope's collectors, worked independently for hire for US and European museums. Meanwhile, John Bell Hatcher was employed by several institutions—first working for Marsh, then moving to Princeton, and finally to the Carnegie Museum—and trained a large number of people in this developing craft.[29]

The field man, simultaneously a bold frontiersman and scientific observer, was a particularly pronounced type in the United States, to the extent that Osborn called it a "distinctively American profession."[30] In European contexts, fieldwork tended to be conducted by assistant and curatorial scientists, who were involved in supervising excavations, writing up results, and often preparing and ordering fossils. These differences could potentially be due to the less structured nature of European museums, which did not create the same "vertically integrated" practices as US ones, so fossil researchers needed to adopt more heterogeneous roles. However, it also reflects cultural traditions linking US paleontology with the frontier. The American fossil hunter was as much a scientific prospector testing his strength against the environment as a museum-based scholar.

Fieldwork, casting, and exchange led to constant accumulation of material. One of the largest issues for museums was therefore space. Requests for more display and storage space were constant refrains across museum annual reports. Observing the displays in London around 1900, William Diller Matthew reported there was "an enormous exhibit of fossil vertebrata whose effectiveness is greatly decreased by placing too much material on exhibition."[31] Difficulties over space were especially marked in vertebrate paleontology, which relied on large specimens to show the drama of life. This could quickly turn into something of a curse, with mammoths, stegosaurs, and giant sloths cramming rooms.

Institutional histories and departmental rivalries also caused difficulties. A report by Arthur Smith Woodward in 1901 on possibly rearranging the London galleries in a more evolutionary manner and linking fossils with the remains of living animals demonstrates this. While Smith Woodward wrote that "an ideal arrangement would be, the classification of organisms in natural groups, mingling the fossils with each group to illustrate (i) its ancestry, and (ii) the variations in its geographical distribution at different periods," in practice this was impossible. "The Imperfection of our knowledge of extinct organisms" meant displays would need to be constantly rearranged. Fossils and recent remains also required different skills to preserve and display, which in some cases were diametrically opposed—"the gum of gelatine used for mounting and hardening fossils has never hitherto been poisoned and is often infested with mites, which might be detrimental to skins or other soft tissues placed with them." Collaboration between departments was also difficult, due to different cataloging systems, and so "chaos in administration would otherwise be the result."[32] The difficulties of the work, factional jealousies, and the need for "accurate" museum displays posed problems, and made work in the organism of the museum a tense affair.

Paleontology in the University

Large public museums were not the only significant institutions in paleontological work, especially mammal paleontology. A number of universities developed strong paleontological focuses, especially those with established departments of geology, which were often relatively well funded (owing to educational investment in the exploitation of mineral resources), and provided opportunities for those with paleontological interests to build careers. And university systems in these years were beset by similar drives to expansion and reform as museums.[33] While never as well resourced as national or philanthropically funded museums, university paleontology departments could nevertheless maintain strong positions. Indeed, emphasizing the importance of university work in paleontology presents something of a reversal to general trends in the history of science, where the importance of museums as research institutes has been overshadowed by an overarching focus on university- and laboratory-based research. In the history of paleontology meanwhile, the role of universities has often been neglected, with the subject being seen as a quintessentially museum-based discipline. But universities could simultaneously develop educational programs and their own independent research collections, and were persistently significant in paleontological work—although they needed to work in strategic and opportunistic manners to garner prestige and authority.

University paleontology was particularly established in countries where higher education systems expanded most dramatically, like Germany and the United States. The German example indeed provides a particularly clear case; if there was an international center of paleontological education, it was in the Bavarian capital of Munich, where the Bavarian State Collections were closely linked with the University of Munich. The Wittelsbach house patronized geology and mining, and scholars like Andreas Wagner had been important in paleontological networks. Bavarian paleontology grew even further under Karl Alfred von Zittel. In 1866 Zittel was appointed as a "young, very educated and eager scholar"[34] to be the professor of paleontology at the university and custodian of in the Bavarian State Collections of Paleontology (taking the position previously established by Wagner). Zittel's career integrated Bavarian paleontology with international links. He had studied in Paris and Vienna; conducted excavations in the French and Swiss Jura, Egypt, and Libya; and traveled in the United States. Through an organized program of collecting, education, and publishing, Munich became a central paleontological node.

The importance of Munich paleontology is noted in several works, most notably by Marco Tamborini.[35] However, writing its history is difficult, even within the often fragmentary records of nineteenth-century paleontology. The Bavarian State Collections generated a huge amount of material, and Zittel and his coworkers corresponded with scholars all over the world. However, the collections themselves were destroyed in 1944 during an intense bombing raid. A huge number of fossils and the wing of the institution's building that held "Zittel's worldwide correspondence, kept in numerous files" were lost.[36] From the surviving material, we can see that the Bavarian state collections had a substantial staff for its educational activities and growing collections, drawn not just from Bavaria, but all over the world. Combining research and teaching was central to Zittel's plans; when negotiating his position in Munich, he specifically asked whether the role of conservator of the collections would also make him a member of the university faculty, as "I would not like to completely part from the teaching, which I have come to love."[37] Zittel also built a position through writing systemizing works. These included the already discussed evolutionary epic *Aus der Urzeit* and his *Geschichte der Geologie und Paläontologie bis Ende des 19. Jahrhunderts*, outlining the expansion of geology and paleontology across the modern period.[38]

Another of Zittel's key works was the *Handbuch der Paleontologie*, whose five volumes were published between 1876 and 1893 (and was translated into French), which was followed in 1895 by *Grundzüge der Paläontologie*, a condensed textbook (although still 971 pages long with 2,048 illustrations), which saw several further editions and an English translation.[39] Zittel described paleontology as a meticulous science, systematically documenting the history of life through illustrations, numerical data, and textual descriptions. Zittel was ostensibly a Darwinian evolutionist, although it is difficult to get a sense of this from his works. Most typically, his studies showed a starkly typological and descriptive perspective. As Tamborini argues, this was critical to Zittel's approach. New technologies like microscopes and chemical preparation were used to analyze fossils in an empirical and "objective" manner, with the mechanisms of evolution being largely omitted.[40] Nevertheless, Zittel was clear that paleontology had important lessons for the modern world. This was particularly expressed in the final section of the *Grundzüge*, whose 212 pages covered mammal groups (fossil reptiles were dealt with in 97 pages). The account was conventional, emphasizing the mammals' superiority and diversity, and how they "take the highest rank among the vertebrates," and that "their functions are more specialized, nervous systems and sensory organs are more developed, their teeth and organs of movement more differentiated, than any other class of animals."[41] Rather than revise older ideas of mammalian superiority and differentiation, this systematic approach reinforced them.

Zittel's publications were only a small part of the activities of the Munich collections, which extensively promoted paleontological education aimed at both German and international students. A substantial proportion of people training as paleontologists from the 1880s to the 1900s spent time in Munich, and Zittel eventually supervised over fifty doctoral students, mainly from Germany but also other parts of Europe and the United States.[42] Paleontological methods and principles developed at Munich were widely dispersed and emulated through international networks. Even after Zittel's death in 1904, Munich maintained its importance, with Ferdinand Broili, Max Schlosser, and Ernst Stromer von Reichenbach continuing to build the collections, and produce new editions of Zittel's works.

When Zittel became director of the Bavarian state collections, the fossil mammal holdings primarily included Wagner's Pikermi fossils, some Pleistocene material from Franconia, and casts of *Megatherium* and the Siwalik animals (gifted and exchanged with British collections). However, the collections expanded considerably over his tenure. Schlosser later recalled that "the now so impressive collection of mammals has expanded mainly due to the rare prudence and activity of the late Professor Zittel, who raised it from small beginnings in less than fifty years." He drew much from international links, purchasing material from France "to acquire almost all the typical species of the European Tertiary" and making "exchanges with the most significant museums of Germany, France and North America."[43] He also secured patronage from rich individuals, such as Kommerzienrat Theodor Stützel, who funded excavations in Samos and the purchase of American fossils (including a *Titanotherium* skeleton), and German expatriates, like Otto Günther, one of the directors of the meat-extract factory in Fray Bentos, who sent fossils of Pampean mammals from Uruguay, and the doctor Karl Harberer who purchased fossil mammal remains from medicine markets in China, sold as "dragon teeth" and "bones."[44]

The size of the Munich collections was described in Maximilian Schlosser's 1912 guide: Five rooms were devoted to fossil mammals (more than any other subdivision of paleontology). Some displays showed southern German material, including a cave bear diorama made of specimens from Tischofer Cave in a family scene with "males standing upright, females in an ordinary position, and young sitting."[45] The other galleries demonstrated the international reach of the Bavarian collections. Visitors could compare two saber-toothed cats, the American *Smilodon* and the European *Machairodus*. There were fossil ungulates from North America, Europe, and a few rare specimens from China, as well as mastodons and South American mammals. The displays were vastly stocked, showing development, evolution, and taxonomy, in a wide geographic array. In Munich collecting, education, theorizing, and systemizing specimens were all linked.

Meanwhile across the Atlantic, universities mushroomed across the United States. As will be discussed in chapter 14, they were especially prominent in the western states, with the universities of Kansas, California at Berkeley, Nebraska, and Chicago developing strong paleontological programs. A further indicative example of US paleontology beyond the large museums is offered by William Berryman Scott at Princeton. After returning from Europe and being appointed professor of geology in 1883 (a post he held until retirement in 1930, after which he remained professor emeritus until 1947), Scott continued developing the program established by Guyot, teaching students, organizing research, and maintaining and expanding the university's paleontological collections at Nassau Hall. Scott was also a prolific writer, producing numerous scholarly works and an extensive autobiography.[46] The latter provides a rare, if highly mediated, account of the world of late nineteenth- and early twentieth-century paleontology, documenting his career while providing scholarly gossip and accounts of travel in Europe and South America.

Osborn and Scott maintained an alliance throughout their careers, growing from their common experience as students and interest in fossil mammals. However, Scott was from a less elevated background and therefore uneasy in the circles Osborn moved in. This can be seen in an anecdote from Scott's memoir, where he recalled being taken by Osborn to visit "Uncle Pierpont" to request $25,000 to fund illustrations for the publications of the Princeton expeditions to Patagonia. Scott recalls waiting in "trembling anxiety," and when Morgan agreed to the request, "in a tumult of astonishment and joy, I shot up out of my chair, as though propelled by a powerful spring." The meeting then ended with Morgan showing some "beautiful manuscripts" from his library to the group.[47] Scott's awkwardness in high society highlights how paleontological work depended on elite social connections. Scott also frequently noted financial difficulties and spoke of the "practical drawback to the paleontological method, namely its costliness," requiring "skilled workers to prepare the specimens, and great buildings in which to house them," as "distant regions must be examined and the whole world ransacked for material."[48] As well as a rare admission of the extractive nature of paleontology, this demonstrated the difficulties of work outside philanthropically backed collections.

Fieldwork was one area where university paleontologists could hold their own because they had a ready source of labor in the form of students, time for research during university holidays, and occasional grants from university regents for travel, equipment, and transportation. Many US universities conducted regular expeditions, visiting the classic sites in the badlands. These were rarely on the scale of the large museums, but were still able to gain much material. Scott continued the tradition of the Princeton geological expeditions, established by himself, Osborn, and Speir. Between 1882 and 1891 he spent seven summers leading student parties, primarily in Nebraska; some later expeditions in the 1890s were then led by John Bell Hatcher. While billed as scientific expeditions, these excursions had a strongly social and touristic element, with final-year students using them to see "the West." Student party members frequently wrote about their experiences. Stewart Patton, on reaching "Sam Smith's Stone" at Burnt Fork on the 1886 expedition, wrote to his mother that "scientifically considered the Expedition is a great success," having collected fossil rhinos, "enormous crocodiles," and "the skeleton of the ancestor of the horse—an animal about the size of a black and tan dog."[49] Patton's letter demonstrated an ideal outcome of the scientific expeditions, developing knowledge of the history of life, securing valuable fossils, and understanding the land.

However, other students valued different things. A journal by Cornelius Rea Agnew, Princeton Class of 1891, describing the Eighth Geological Expedition, mainly focused on sights of the West, with long descriptions of industrial pig-slaughtering in Chicago and visiting Native American reservations to purchase pipes and moccasins. Fossil digging was an arduous distraction, requiring early rising and hard work, as "everybody dug with a vim for about half an hour, and then the fossil fire began to dig out," with work only picking up when someone called "'Get a digging, get a digging,' . . . the sentence that hailed the approach of Scotty."[50] Hatcher was often unimpressed by these expeditions. He wrote to Scott at the beginning of the 1894 expedition that he wanted to take a 1,200-mile trip across the badlands, Black Hills, and Yellowstone.[51] However, by August, he wrote from Hermosa, South Dakota, how he was "not only disgusted but nearly crazy" by the speed of the work, delays in transportation, and the need to direct and escort the students. He wrote how in the past two months "we have packed only three boxes of fossils . . . it is certainly a burning shame to waste so much time and money and yet travel as a '*Scientific Expedition*'!"

He still said "the Expedition as a whole has been wonderfully successful, but in the future I shall ask to be excused from these tail end pleasure trips, they are a great bore."[52]

Hatcher was to stay at Princeton for another six years, but these points illustrate tensions around fossil work. The privileging of the American West for scientific dominance and adventure appealed to students taking part in fossil-gathering expeditions. However, it is also clear that fossils were not their priority but that paleontology more gave access to a constructed "western mythos." Tensions with Hatcher show fossil hunting itself becoming a hard-nosed business of extraction and productivity. Management of different goals, of touristic students and fossil hunters, allowed universities to maintain themselves within paleontological networks. However, this required a balancing of interests. University paleontology worked within the same world as the larger museums, but needed to be strategic to maintain positions.

Realizing Gaudry's Dream in the Paris Museum

While university departments were important players in paleontological work, a different trend was presented by older museums, which often had ambiguous relationships with the changes in museum work in the late nineteenth century. The two institutions examined in chapter 5—the British Museum of Natural History, and the Muséum d'Histoire Naturelle in Paris—provide important examples. The London museum, while old and authoritative (but not matching the great US museums in terms of funds), nevertheless became a major emblem of the new museum movement, often self-consciously moving against Owen's initial visions for the collection. William Henry Flower's 1893 Presidential Address to the Museum's Association in Britain on "Modern Museums" became one of the core texts of the movement, cited throughout the world.[53] While inherited structures meant the British Museum of Natural History could not renovate as much as Flower may have wanted, traditional centers of paleontological work in Europe could still renovate and defend their positions.

The institution that most clearly showed the mixture between the old and new, and the tensions of maintaining prestige in an old collection, was the Muséum d'Histoire Naturelle in Paris. By the 1880s Gaudry had won his conflict with the Department of Comparative Anatomy to control most of the fossils and established paleontology as one of the primary departments within the Muséum. However, the Paris paleontological department increasingly felt threatened by rivals in the United States and across the English Channel and the Rhine. Gaudry traveled extensively, defensively reporting on foreign collections to his assistant, Marcellin Boule. On visiting Munich, Gaudry was impressed by the dynamically posed cave bears, which were "infinitely better than our inanimate objects, all in the same position," although, he continued, "In general the mounts are very modest or even primitive."[54] Meanwhile on a trip to Britain Gaudry wrote to Boule about the development of the British Museum of Natural History and British commercial culture in general. He reported from the coastal resort of Eastbourne that "there are splendid hotels, shops like in London, hundreds of thousands of villas, each more charming than the next. We make an act of patriotism in working for the honor of French science, for we could never equal our neighbors in wealth, and we have no other assurance than being equal in spirit."[55] In the face of "splendid hotels" in Eastbourne, paleontology remained a science where French preeminence and "spirit" over materialism could be maintained.

Gaudry persisted in his dream of a more extensive gallery of paleontology showing the developing history of life. This was such a huge undertaking that a single department could not undertake it alone. Despite continuing tensions, the professors of comparative anatomy and paleontology put aside their differences to organize a new building to house their collections, along with those of anthropology. Ferdinand Dutert, designer of the Palace of Machines at the 1889 Universal Exposition, was hired to create a building following principles of expositionary architecture, with large glass windows, full electric wiring, and a hot-air heating system. An engineer named Seurat, noted in *Le Génie Civil* that this was a salutary contrast with the Muséum's "old buildings, poorly lit and insufficient from every point of view." The new galleries "harmonize the aspect of construction with conceptions suggested by the study of natural phenomena."[56] The building also reflected the balance of power between the departments, with the comparative anatomy displays on the ground floor, the paleontology exhibit on the first, and the anthropological galleries on the upper balconies, mainly displaying human skulls, anthropometric photographs, and prehistoric artifacts.

The new galleries opened in a ceremony on July 21, 1898, with invitations sent to the president of the Republic, various ministers and professors, and journalists from Paris's major newspapers. The Muséum's director, Alphonse Milne-Edwards, stated in his opening address that "like Comparative Anatomy, Paleontology is a French Science," and praised Gaudry for having "studied the development and chains of life across the first ages of the world . . . from the *Eozoön canadense* of primary times . . . to the large mammals which are contemporary with man."[57] The paleontological galleries were arranged as a walk through time, giving "proof of the unity of the animated world and its progressive development," showcasing the most important specimens and Gaudry's teleological theories. Visitors were to walk the seventy-five-meter hall, starting with the earliest invertebrates and fish, quickly being confronted by a *Pareiasaurus*, and then moving to a limited stock from the Age of Reptiles, described by Gaudry as "the reign of brute strength—its giants are stupid beings."[58]

The latter half of the gallery was a journey through the mammals. Gaudry wrote, "If, for Secondary fossils, we have much to envy in some foreign museums, this is not the case with the Tertiary fossils. We have an admirable series, taken mostly from French soil."[59] Cuvier's Eocene fossils from the Paris basin were seen first ("the most celebrated in our collections"), then *Megatherium*, the cast of *Dinoceras mirabile* from Marsh's collection, various early proboscideans, and as a culmination and "apogee of the animal world,"[60] the Elephant of Durfort at the back. Ideas of progress were further entrenched through artworks, most notably a series by Fernand Cormon in the building's classroom, depicting the animals of Pleistocene South America and Europe, before moving to a cycle showing human "progress" from prehistory to the ancient Gauls.[61] Simultaneously progressive, national, and universalizing, the Gallery of Paleontology was a bold statement of developmental theory. It was, to quote Louis Liard, director of higher education, "like a history, like a philosophy. It is in effect the history of animal creation, an interpretive history, rendered visible and tangible," and a true "natural history."[62]

The galleries were popular, receiving ten thousand and eleven thousand visitors, respectively, on the first two Sundays after their opening. Gaudry noted, "The public seems charmed and surprised. They pass reverently

▸ FIGURE 9.3. Part of the report on the Paris Gallery of Paleontology in *La Nature*. Marcellin Boule is depicted supervising the mounting of *Megatherium* in the lower image. Glangeaud, "Nouvelles galeries du Muséum," 308. Author's collection.

celle des autres services, forment une rangée continue sur le pourtour. On a usé du verre à profusion. Il laisse voir tous les détails d'un objet sans qu'on soit obligé de déplacer celui-ci et il offre un cachet qui n'est pas sans élégance. De grands échantillons montés sur pivots mobiles s'échelonnent au-dessus des armoires.

Fig. 1. — Vue d'une extrémité de la galerie de Paléontologie. On aperçoit à droite le *Megatherium*, à gauche le *Glyptodon*, au fond l'Éléphant de Durfort (4m,50 de hauteur) et le *Cervus megaceros*.

Les grands vertébrés font face à l'entrée et ce que l'on aperçoit d'abord c'est la large gueule du *Pareiasaurus* arqué sur ses pattes. Ce curieux Reptile qui possédait un œil au sommet de la tête et dont le palais était couvert de dents, comme celui des Poissons, a été trouvé dans des couches très anciennes du sud de l'Afrique.

A gauche on remarque un Ichthyosaure avec un fœtus, à droite, un Téléosaure admirable de conservation. Plus loin on voit des fémurs de Dinosauriens provenant d'Amérique et de Madagascar qui dénotent des animaux ayant de 12 à 15 mètres de long, tandis que les Mosasaures, ou Serpents de mer, qui vivaient dans les mers de l'Europe et dont il existe de remarquables spécimens dans la galerie, atteignaient une taille de 10 mètres.

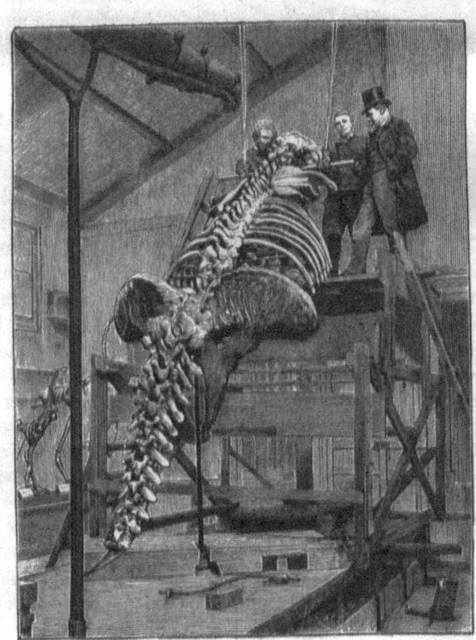

Fig. 2. — Montage du *Megatherium*.

En continuant à s'avancer vers le fond de la salle on rencontre un squelette complet de *Rhinoceros*, trouvé aux environs de Gannat, puis un superbe *Mastodon angustidens*, avec ses quatre défenses, provenant de Sansan (Gers). Dans les temps géologiques, vivaient en Grèce de nombreux troupeaux de Ruminants, de Cerfs, de Girafes, de *Rhinoceros* et d'*Hipparion*. Ce dernier animal, avec ses trois doigts à chaque patte, est l'ancêtre des chevaux. On a groupé cet ensemble de fossiles au milieu de la galerie. Plus en arrière, le *Megatherium*, ce curieux Edenté de l'Amérique du Sud, s'appuie sur un arbre brisé par la foudre. Il est flanqué du bizarre *Glyptodon*, autre Édenté, dont l'énorme carapace fait penser aux Tortues. Vers le fond de la salle et la dominant de toute sa hauteur se dresse le majestueux Éléphant de Durfort trouvé dans le Gard et qui atteignait une hauteur de $4^m,50$. Il faut signaler aussi le *Dinotherium*, ce Proboscidien dont la taille dépassait 5 mètres, le *Cervus megaceros*, qui possédait une ramure de 3 mètres et demi de large ; le *Dinoceras*, si étrangement cornu, avec ses

Fig. 3. — Vue de la galerie de Paléontologie prise en entrant ; au premier plan on aperçoit le *Pareiasaurus*, la gueule ouverte, arqué sur ses pattes.

Fig. 4. — Vue d'ensemble des deux galeries d'Anthropologie (balcon) et de Paléontologie ; le long des murs on voit les armoires vitrées, au milieu de la salle les grands Vertébrés et les meubles à tiroirs renfermant les Invertébrés.

canines en forme de poignard, les *Dincrnis*, ces oiseaux géants, sans ailes, de la Nouvelle-Zélande, ayant jusqu'à 3 mètres et demi de haut, etc., etc.

Voilà ce qu'on aperçoit au premier coup d'œil.

before those relics of beings which have been drawn from the stone where they have been for hundreds of thousands, even millions of years. People of very simple appearance become meditative before the fascinating mysteries of life hidden in the rocks, comparing the fossils of the successive ages of the world, they seek to understand what has happened since that distant day when the first puny creatures came to life, up until the day marked by the human creature."[63] The Muséum also promoted education, with specialist courses and a series of public Sunday Lectures, which frequently presented paleontological topics. The Department of Paleontology also branched into larger spectacles. At the 1900 Paris Exposition, Muséum workers set up a paleontological display in the Palais de l'optique, alongside exhibits of X-rays, kaleidoscopes, luminescent bacteria, and a hall of mirrors. The Salle Cuvier and Salle Gaudry showed the history of life across twenty galleries of magic lantern projections, beginning with the creation of the earth, showing combat between giant reptiles, and moving to the Age of Mammals, including reconstructions of Montmartre in the era of the gypsum *Dinotherium* and *Megatherium*, "the Auvergne on Fire" during the Miocene volcanic eruptions, the glacial epoch, and finally the appearance of humans. These aimed to "produce a satisfying effect through the brilliant illumination of incandescent lamps, tinted in diverse ways according to the subject they represent."[64] Paleontology featured at another attraction at the exposition, Le Monde souterrain, whose entrance was flanked by *Iguanodon* and *Megatherium* sculptures, leading to underground to displays on mining, fossils, and catacombs. The serial vision of the Jardin des Plantes was spread across the media, as were the long-running connections between fossil life and the industrial economy.

The opening of the Gallery of Paleontology was the culmination of Gaudry's career. To celebrate his jubilee in 1902 (marking fifty years since his first publications), "the Laboratory and Gallery of Paleontology were brilliantly decorated with hangings and flags, adorned with plants and flowers," in "a veritable Festival of Science."[65] Congratulatory telegrams were read from all over France, Europe, and North America. Edmond Perrier, then director of the Muséum, praised the gallery as a "magnificent palace, where all creatures of past ages are assembled before you, from the humble

◂ FIGURE 9.4. Further depictions of the 1898 Gallery of Paleontology in Paris. Glangeaud, "Nouvelles galeries du Muséum," 309. Author's collection.

Trilobite—King of Silurian times—to the gigantic mammoth which lived when man tamed the first horses," showing "this immense hymn of life."[66] Boule similarly discussed Gaudry's appreciation of the natural world, and how "previously, paleontology was a science of death, but with you it became the science of life!"[67] Gaudry's response narrated how paleontology was a profoundly French science, illustrating harmony across the epochs:

> Paleontology today is almost the opposite of what it was at its beginnings, when, in order to be established, it had to prove there had been creatures that were different from present animals. Now, instead of focusing on dissimilarities, it focuses above all on similarities, as it can be seen that, despite changes in appearance, the former world and the present world are one. Species are simply phases of development of types which, under the direction of the Divine Worker, follow their evolution across the ages. We discover chains from the days of the trilobites to the time when humanity appears. When saying chain, we say union; when saying union, we say love. The grand law which dominates life, is a law of love.[68]

The Gallery of Paleontology showed harmony and fellowship across the ages. And the collections of Paris, with their old traditions, new galleries, and prestigious scholars, aimed to be its center. Gaudry could use the gallery as a symbol of French dominance over paleontology while showing nature developing according to divine forces, uniting the whole of creation.

Gaudry retired two years after his jubilee, and Marcellin Boule was voted as successor by the Assembly of Professors. However, this succession was not as smooth as expected. Boule's named second, Charles Déperet, professor of geology at Lyon, was aggrieved, and wrote to the Muséum that there was broad agreement at "the evident superiority of my paleontological works and discoveries over those of my competitor" (something endorsed by Zittel and, apparently, Gaudry). Boule had been appointed for "reasons of sentimentality . . . and especially for the services rendered . . . to the collections of the Muséum."[69] This should not be regarded simply as a complaint by a sore loser, but as an illustration of important issues around the Jardin des Plantes, and large museums more generally. Building a museum career partly reflected scientific reputation, but also one's abilities to operate within the institution. In France this maintained authority within Paris, as those already enmeshed in the Muséum were most liable to be chosen for positions.

Boule was a canny operator in the complicated politics of the Jardin des Plantes and the international networks of paleontological science. He presented a possibly contradictory image, being a skilled network builder, but highly abrasive toward potential rivals.[70] He presented himself as a direct successor of Lamarck and Gaudry, seeing paleontology as a "historical science" in its own right and "not content to serve as an auxiliary to geology, or be some sort of complementary science to zoology."[71] However, his interests were distinct from those of his "masters," who respectively studied mainly invertebrates and Tertiary mammals. Boule focused on later periods. The British prehistorian Arthur Keith was later to call him "the indomitable reviewer and annotator of all that relates to the Pleistocene—its geology, its fauna, its stone cultures, its humanity."[72] Throughout the 1900s Boule wrote on the fossil fauna of cave sites across France and moved into what was termed "human paleontology," especially through studies of "the Old Man of La-Chapelle-aux-Saints," a well-preserved Neanderthal from southern France.[73]

Any decline in the Muséum's status was relative, and the scale of its collections meant it persistently maintained a strong position. However, there were still challenges. The paleontological gallery quickly became overcrowded. By 1907 (less than ten years after opening) the displays had expanded almost to breaking point, with Boule complaining, "The Gallery of Paleontology is now completely cluttered," and there was "a complete lack of storerooms, so the difficulty is not just displaying the rich collections . . . [but] protecting them."[74] This was exacerbated in 1908 when the twenty-five–meter-long cast of *Diplodocus carnegii* was donated to the Muséum. While a prestigious specimen, fitting it into the gallery was challenging, with the careful linear regularity needing to be modified (as well as the *Diplodocus*'s tail, which was curled round to fit in the allotted space).[75] Expansion was essential to building up a view of life and maintaining prestige. Limits on space worked against this.

Funding was also a problem. In 1900 Gaudry wrote to Milne-Edwards that "in comparison to London, New York, Washington and even Brussels and Munich, we have very weak resources in personnel and money."[76] Specimens and staff were expensive, and the department's budget needed to be supplemented by donations, gifts, and bequests. In 1905 the director of the Muséum wrote to the Ministry of Public Instruction, comparing the Jardin des Plantes to the British Museum of Natural History, which had considerably more resources and funding, despite its more limited remit.

Staffing in the paleontology division in London was much higher, with one curator, five assistants, three preparators, and a number of auxiliaries, while Paris had one professor, one assistant, two preparators, and one "garçon." The budgetary differences were even more embarrassing, with Paris having an annual budget of 3,960 francs as opposed to London's 43,750 (converted to francs and underlined as "15 times the budget of Paris"). Furthermore, in the conclusion it was noted that "the comparison with the American museums . . . would be even worse."[77] Traditional prestige allowed Paris to maintain its status, but this was despite troubled funding.

The years around 1900 therefore present a multifaceted image in paleontological research. Collections expanded dramatically and were reordered according to new methods and techniques around expositional architecture, industrial modes of work, and bureaucratic organization. While these trends were most marked in the "new museums," especially in the United States, they also spread elsewhere, including to universities and older collections attempting to maintain their positions. These other institutions often faced difficulties in resources compared to the philanthropic museums—and indeed a raw comparison of numbers shows their smaller scale. However, this should not indicate they were irrelevant. It was possible to follow other methods of gaining influence and status in paleontology, whether through taking advantage of educational networks or carefully calculated expansion and management of inherited prestige. Distinctive conformations of paleontological work formed in different places, and paleontology remained varied and heterogeneous.

CHAPTER 10

The Story of the Horse

Display, Evolution, and Paleontology

THE IMAGE IS INCREDIBLY FAMILIAR. A STEADY PROCESSION OF SKELETAL LEGS, MOVING from short limbs ending in five toes, through three, and then finely balanced on one, while the rest of the limbs become more elongated and flexible. Meanwhile a similar shift occurs in the teeth, starting short, relatively simple, and suited for soft vegetation, and transforming into a battery of high-ridged grinders, capable of working down the toughest vegetable matter. The rest of the body is almost an irrelevance, apart from changes in size, with a small doglike creature at the start growing into a huge galloping animal.

This is of course the image of the evolution of the horse, one of the most widespread scientific models of the last two hundred years. It rose to prominence around 1900 as paleontologists argued that their subject empirically demonstrated evolution and origins. It was then promoted and publicized by new visual technologies and media, in illustrated magazines, museum guidebooks, and arrangements of fossil specimens. For William Henry Flower "the anatomy and history of the horse" was "a test case of the value of the theory of evolution,"[1] "a view of creation which is the grandest, most sublime, and at the same time most reasonable, which has yet been presented to us."[2] William Diller Matthew spoke of how "the evolution of the Horse through the Tertiary period or Age of Mammals affords the best

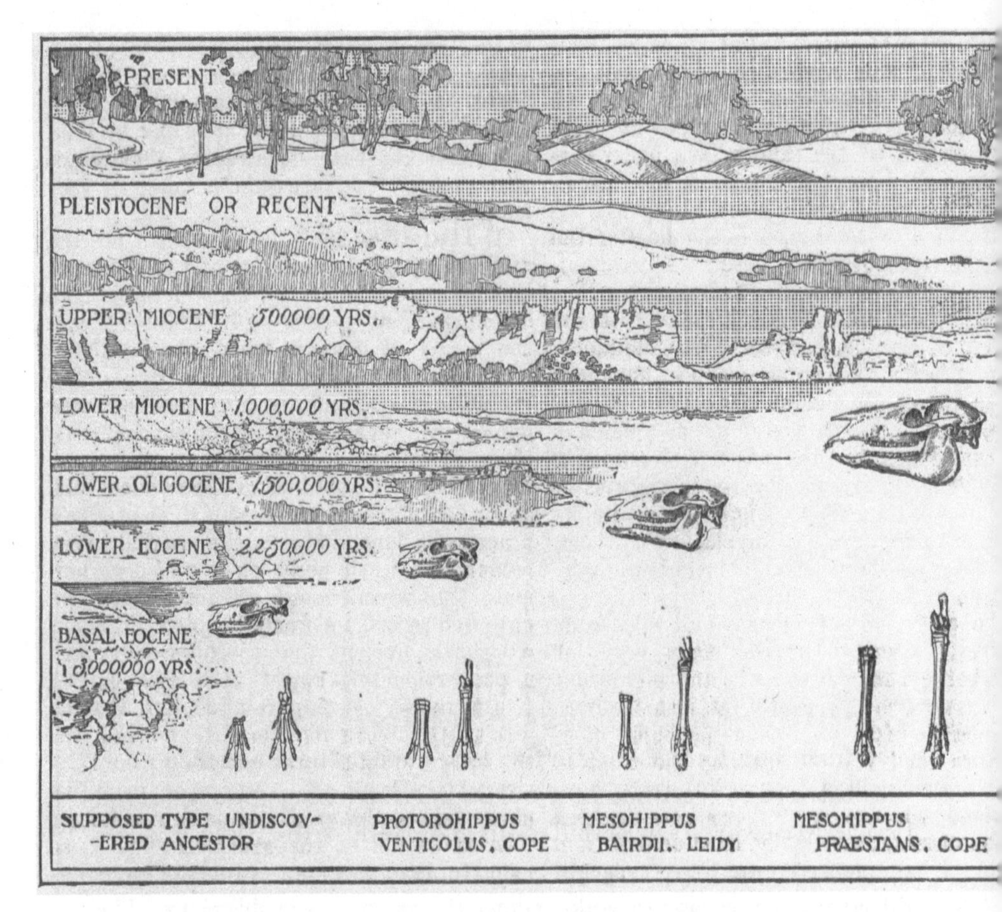

FIGURE 10.1. The evolution of the horse. Illustration by Bruce Horsfall. Frederic Lucas, "The Ancestry of the Horse: A Family Record that Reaches Back about Two Million Years," *McClure's Magazine*, October 1900, 514–15. Reproduced by kind permission of the Syndics of Cambridge University Library

known illustration in existence of the doctrine of evolution by means of natural selection and the adaptation of a race of animals to its environment."[3] In even grander terms, Henry Fairfield Osborn stated, "The horse is without doubt the noblest of our domesticated animals, and notwithstanding all the gaps which still interrupt our knowledge, no other animal presents such a complete and finely ordered ancestry; as compared with that of other quadrupeds his evolution may certainly be described as an *édition de luxe*."[4]

The model of horse evolution became a widespread trope in museum displays. However, its prevalence could obscure a great deal. Stephen Jay

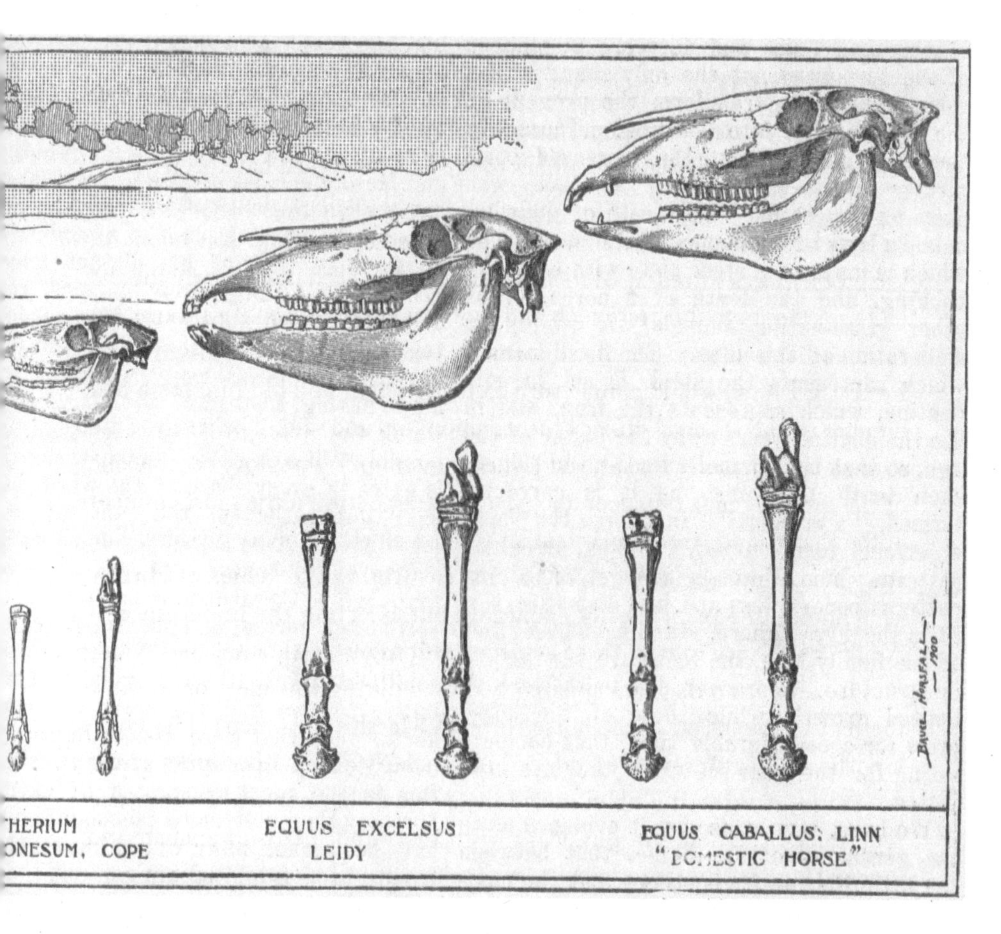

Gould highlighted how, throughout the twentieth century, the horse's early ancestors were persistently cited as the size of a "fox terrier," an almost meaningless comparator using a dog breed popular in the 1900s that had since gone out of fashion.[5] Even in the 1910s William Berryman Scott was almost apologizing for the prevalence of horse evolution, as "one cannot but hesitate to tell again this oft-told tale, which has been reiterated until it has become a hackneyed commonplace," although he nevertheless repeated it anyway: "On the other hand, no sketch, however slight, which purports to give an outline of the evidences of evolution, can in fairness omit all mention of this remarkable case."[6] The prevalence of the horse became as formulaic as it was widespread.

Even the way the evolution of the horse was consolidated became a trope around particular understandings of paleontology. The story went that in 1876, when Thomas Henry Huxley met Othniel Charles Marsh while lecturing in the United States, the two debated whether the horse originated in the New World or the Old. Huxley had used Vladimir Kovalevsky's and Albert Gaudry's ungulate researches to predict there must have been a five-toed ancestor "at some earlier period,"[7] but lacked fossils to demonstrate this. In the classic account of the meeting presented by Huxley's son Leonard, the problem was solved by Marsh's specimens: "At each inquiry, whether he had a specimen to illustrate such and such a point or exemplify a transition from earlier and less specialised forms to later and more specialised ones, Professor Marsh would simply turn to his assistant and bid him fetch box number so and so, until Huxley turned upon him and said, 'I believe you are a magician; whatever I want, you just conjure it up.'"[8] The story became one of the great legends of paleontology, a meeting between the most eminent of the British and American scientists to solve an evolutionary mystery. The story also alludes to a switch of batons, enshrining the genius of Huxley as deducing the principles of evolution, but asserting the importance of American science to turn these conjectures into material fact.

Huxley used a diagram apparently drawn during this meeting to show the regular evolution of the horse through Marsh's specimens. The Huxley-Marsh diagram was reproduced in scholarly and public works across the world, making it a ubiquitous image of development. Museums also reproduced the diagram physically with casts and fossils of ancient and modern equids. By 1914 almost every major natural history museum contained a display of horse evolution somewhere within its galleries, and the horse evolution series became as desirable a specimen as the remains of large beasts like *Megatherium*, *Dinotherium*, and *Dinoceras*. The promotion of the evolution of the horse was a managed affair where marketing was extremely important.

In many respects the iconic status of horses in scientific displays is unsurprising. Horses were ubiquitous in nineteenth- and early twentieth-century rural and urban environments, used for transportation, agriculture, industry, the military, and surveying. Horses were the most common large animal seen by most people and were deeply entwined into everyday activities, especially in large cities; in the 1900s London had 300,000 horses and New

▸ FIGURE 10.2. The Huxley-Marsh diagram of horse evolution. Huxley, *American Addresses*, 83. Author's collection.

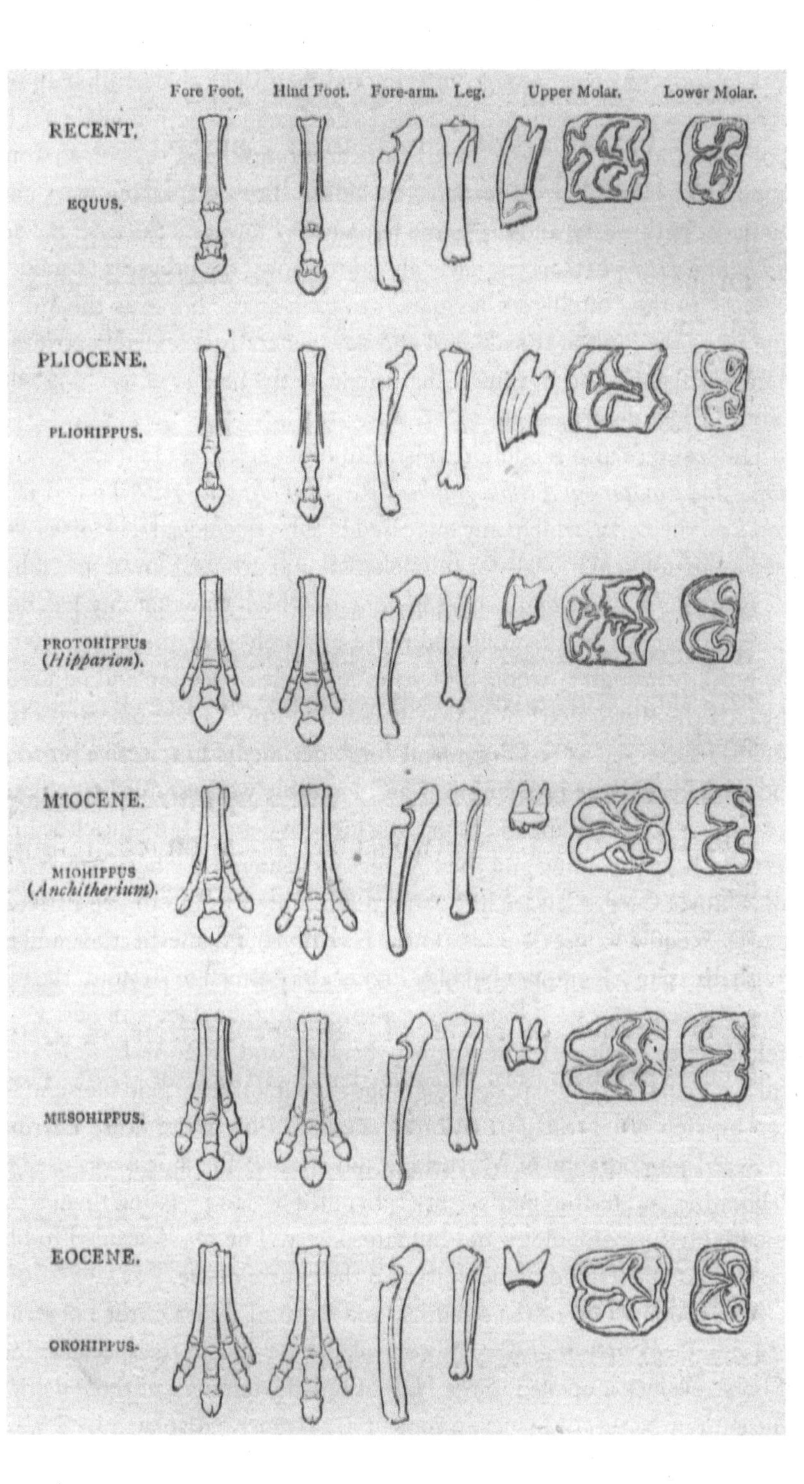
Fore Foot.
Hind Foot.
Fore-arm.
Leg.
Upper Molar.
Lower Molar.
RECENT.
EQUUS.
PLIOCENE.
PLIOHIPPUS.
PROTOHIPPUS
(Hipparion).
MIOCENE.
MIOHIPPUS
(Anchitherium).
MESOHIPPUS.
EOCENE.
OROHIPPUS.

York 150,000.[9] As paleontologists promoted their field, this familiar animal worked as a resonant focus. Richard Lydekker in Britain hoped that his "popular, and yet, I hope, scientifically accurate," book on horses would "appeal to a large circle of readers," including "breeders, racing men, antiquarians, naturalists, and big-game hunters."[10] Flower's *The Horse: A Study in Natural History* (1891) similarly aimed to show "the progress of modern science" to the "intelligent layman," "to look at the horse as the animal appears in the light of the modern and now generally accepted doctrines of Natural History, and in thus doing it may be the means of teaching what some of those doctrines are."[11]

Horses were also model organisms in other sciences. Huth's *Works on Horses and Equitation: A Bibliographical Record of Hippology* (1887) noted 3,800 works on the horse, which only increased in subsequent years.[12] Horses were used as anatomical models for physiologists and artists. Flower noted how "next to the body of man, there is none of which the anatomy has been more thoroughly worked out and more minutely described than that of the horse," through a "whole profession" of horse dissection and numerous "beautifully illustrated" books on horse anatomy.[13] The horse and other equids also served as test organisms for experiments in selective breeding and heredity. Horse breeding was an incredibly well-established pursuit, economically significant and impacting ideas of descent (and the close links between horse breeding and the eugenics movement have been widely noted).[14] James Cossar Ewart in Scotland attempted to interbreed different species of equid to test theories around heredity and domestication, and the Polish-Russian geographer Nikolay Przhevalsky aimed to "restore" the wild horses of central Asia. Horses were also important subjects in new visual technologies, such as Eadweard Muybridge's and Étienne-Jules Marey's widely circulated time-lapse photographs showing the horse in motion. The step-by-step serial transformation of the evolution of the horse mirrored the serial photography of Muybridge and Marey. The horse was used for elaborating and testing new scientific hypotheses, and relating them to the essential nature of biology and human society. The horse seemed to hold the keys to understanding heredity and the course of life.

The combination of the scientific and cultural values of the horse, and its use in ideologies of power and control, can be seen in the special exhibit of horse evolution opened in the Hall of Fossil Mammals at the American Museum of Natural History in 1908. This extensive display mixed fossils already in the museum's collections with new material acquired through the patronage of the politician, financier, and racehorse enthusiast William Collins Whitney (although Whitney died before the exhibit opened). The Whitney Fund paid for expeditions to the US West to locate ancestral horse fossils and to purchase skeletons of notable modern horses. Matthew wrote that "the American Museum collections of fossil horses are larger than those of all other museums put together,"[15] and were another indication of the institution's scale and reach.

The AMNH exhibit linked the past and present of the animal. It began with "complete skeletons of nine stages in the ancestry of the horse . . . showing every intermediate gradation from the earliest ancestor, no larger than a terrier dog, to the modern descendants." The skeletons, following the AMNH's display strategies, were combined with artistic reconstructions by Charles R. Knight (funded by J. P. Morgan) showing "the probable appearance of and natural surroundings of the animal during life."[16] The galleries moved on to show the horse under "the selection of man under domestication and breeding," as it became a companion, servant, and machine for humans. Prominent breeds of horse were displayed, including the skeleton of the racehorse Sysonby as the epitome of a modern horse, mounted in a running pose "made from direct observation and the instantaneous photographs of Muybridge, Hemment and Chubb."[17] There was also a particularly diminutive Shetland pony (purchased from Ewart) and a giant Ohio draft horse, comparing the variation of the animal under domestication. The display ended with a mount comparing the skeletons of a rearing horse with a human. The mount had a double purpose. The first was as an exercise in comparative anatomy, allowing the viewer to undertake (or, more likely, be shown through guidebooks) "comparison of the horse skeleton and the human skeleton, limb by limb and bone by bone." However, it had a second message, showing "the breaking and training of the horse by man. The rearing action expresses unwilling subjection, and the position of man—as if holding a bridle—of intelligent control."[18] The mount enshrined human supremacy, "showing the domination of man over this powerful animal through superior intelligence, in spite of relatively slight physical strength."[19] The horse simultaneously showed the power of nature to create a perfected organism, but also the power of humans, who dominated nature through their intelligence. The horse symbolized progressive evolution within the natural world, and the current control of that world by humanity.

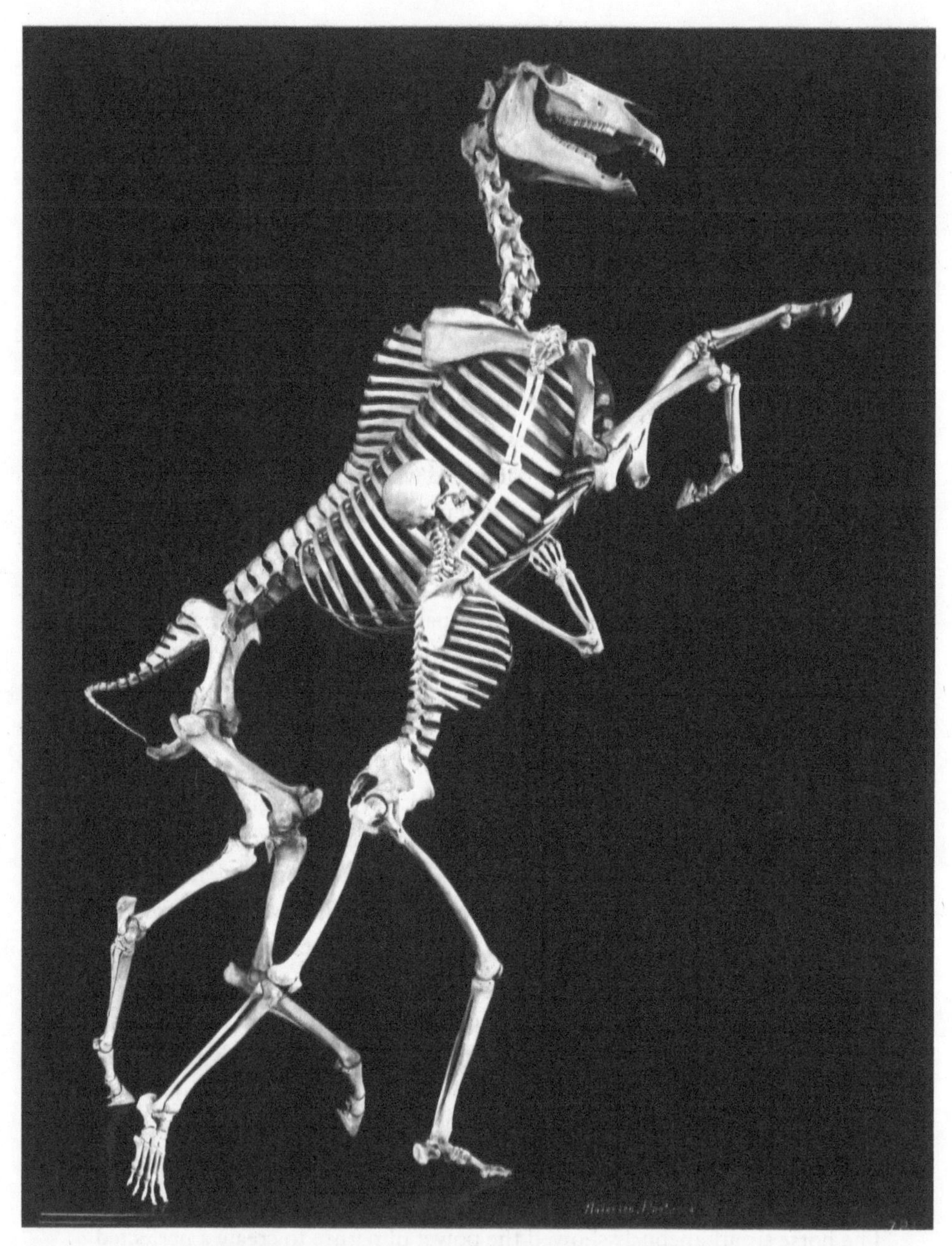

FIGURE 10.3. The culmination of the American Museum of Natural History horse galleries. Image #35320, American Museum of Natural History Library.

Evolution as a Linear Model

The serial model of horse evolution answered a major criticism presented by Darwin and others, that the incompleteness of the fossil record made it difficult for paleontology to contribute to debates on evolution. The evolution of the horse was now presented as a complete series, with no major gaps. It was also not limited to one organism, but could be used to understand many different creatures. Now that the evolution of the horse had been elaborated, paleontology required more support to resolve the lineages of other animals. The apparent "completeness" of the evolution of the horse could demand more status and authority to paleontology in order to "complete" the rest of the fossil record.

The evolution of the horse also spoke across multiple theories of development. The years around 1900 are often described as a time of great uncertainty in the life sciences, being the classic age of "the eclipse of Darwinism," a term coined by Julian Huxley but then examined by historians of science as a complex phenomenon, where evolution and heredity were debated between mutationists, Lamarckians, Mendelians, and others in the decades after the publication of *The Origin of Species*.[20] The horse could be used in almost all evolutionary models. It showed the gradual, incremental change implied by Darwinian evolution, but how its characteristics had shifted was open to debate. The constant assertions of the horse as a "perfected" creature and the seriality of the model also reinforced interpretations of evolution as steady improvement. Although not entirely confidently, as in the background was anxiety that not all creatures had followed this track to perfection, even if the horse had reached an "aristocratic" position.

While the image of horse evolution was flexible enough to be used by varied positions within the fraught late nineteenth-century landscape of evolutionary thinking, it was particularly significant for one model: "orthogenesis," or "line development,"[21] based on the idea that animals originated in a relatively plastic state, but were locked into particular courses early in their history, with anatomical features becoming more exaggerated as time progressed and core characteristics specialized into an eventual conclusion. Orthogenetic change was variably imagined. Sometimes it could lead to the overdevelopment of initially useful characteristics, which eventually doomed the organism. The giant deer was increasingly defined by a monstrous expansion of the antlers at the expense of all other bodily features.[22] The saber-toothed cat was also often understood by orthogenetically inclined writers as developing its teeth to such an extent that it became "practically impossible" for it to close its mouth, becoming a classic example of "variation going beyond its utility," creating "a hinderance even to the point of exterminating its possessor."[23] Orthogenetic change could be degenerative and lead to the extinction of the organism. However, the horse showed this was not always the case. For some creatures line development went toward perfection—as long as it was directed to the correct characteristics. As with horse evolution and heredity, orthogenetic ideas were not applicable just to animal lineages. They also had implications for human society, which needed similar management to ensure energies were devoted to progressive, rather than degenerative, ends. Those claiming authority and understanding of developmental processes—like Henry Fairfield Osborn, who in the 1930s would develop an even more elaborate theory of "aristogenesis"—could also claim authority over humanity.

The road which orthogenesis could take—whether in the animal or human world—was not certain. Anxieties around evolutionary development were thrown into relief by considerations of the horse's "relatives," both living and extinct. By the late nineteenth century the division of the ungulates into two groups—even-toed perissodactyls and odd-toed artiodactyls—was becoming conventional. The perissodactyls were particularly important in paleontology, comprising equids, tapirs, and rhinoceroses among the living animals and chalicotheres, titanotheres, and some South American forms among the extinct. The living perissodactyls were understood as creatures of extremes. The horse was the perfected organism, whose specialization led to constant improvement. However, the rhino and tapir showed different outcomes. The tapir was still regarded as an animal almost unchanged since the Eocene, retaining "primitive" characteristics. Rhinoceroses meanwhile were seen as having an elaborate and labyrinthine evolutionary history, but were now confined to only a few groups. Both the tapir and the rhino were prime candidates for extinction in the modern world: the tapir because of its limited range and primitive features, and the rhino because it was a target for hunters and had lost its vital energy, almost a prime example of "evolutionary senility." These valuations were accentuated by the great size and strange features of extinct perissodactyls, reinforcing the doctrine that overspecialization usually led to extinction. While the horse showed potential progress in evolution, other lineages gave warnings for future development.

Little *Eohippus*

Despite this developing interest, what exactly the earliest ancestor of the horse was, and even what it should be named, was contentious. Earlier works, including Flower's *Horse*, tended to refer to a creature named *Phenacodus* as the earliest ancestor[24] (very much following Cope's view), while Kovalevsky postulated *Palaeotherium* as the ancestor of all the ungulates. Attention also focused on other creatures. In 1841 Richard Owen named a fossil cranium "about the size of that of a Hare" found in southern English Eocene deposits as *Hyracotherium*.[25] While he described the animal as "probably intermediate in character between that of the Hog and the Hyrax,"[26] even giving it the name of "hyrax-beast," the creature was revised over succeeding decades. As more specimens were located, many scholars argued that the creature possessed equine features and should be included at the base of the horse line. This raised problems, however, as calling the earliest horse "the hyrax beast" was often regarded as ridiculous. As a result, a more impressive name for the ancestor was required, particularly as horse evolution became such a widespread model.

Especially in American contexts, the first ancestor was increasingly named *Eohippus*, or "the Dawn Horse," a name which cemented its horselike condition and germinal status, and connected with narratives around the "dawn period" of the Eocene. Richard Swann Lull wrote how "the first undoubted horselike animal appearing in the rocks of North America is a little creature not more than eleven inches high, known to science as *Eohippus*."[27] More important than the fragmentary fossils were evocative imaginings of *Eohippus* in its environment, the conventionalized tropical wetlands of the Eocene. Lull discussed how "during Eocene times North America was clad with forests in which grew both evergreen and deciduous trees distinctly modern in character. The moist climate gave rise to many streams and lakes, along the shores of which grew sedgy meadows that in turn gave rise to grassy plains. These were the conditions under which the horses made their first appearance, and the increasing development of grass lands gave the initial trend to their evolution."[28] The horse's earliest ancestors were small animals, threatened within a strange world, and their later development depended on environmental change. *Eohippus* presented a different view of the "wonders" of the past than the large dinosaurs and mammals that dominated many museums as centerpieces. It was interesting not for its character, but for its future.

FIGURE 10.4. "Little *Eohippus*," compared with the skull of a modern *Equus*. Image #28451, American Museum of Natural History Library.

Notably, public discussions of horse evolution often did not discuss the core lesson emphasized in scientific publications—that of a unidirectional, linear evolutionary drive being proved by fossils—but concentrated on the earliest ancestor. The middle stages of horse evolution were glossed over, and replaced with a sense of wonder at the change from a tiny Eocene animal to the familiar modern beast. Descriptions of *Eohippus* emphasized the creature's size, with extensive use of metaphors. As well as the "fox terrier" cliché discussed by Gould, *Eohippus* was presented as the size of a fox (the metaphor consistently used by Lydekker),[29] "not larger than the domestic cat,"[30] or was called a "Fuchspferdchen" (little horse-fox) by Wilhelm Bölsche.[31] Museum displays and popular illustrations directly contrasted *Eohippus* and *Equus*, highlighting the extent of the evolutionary change. Rather than a serial image of gradual transformation into perfection, public displays often took a different track, showing evolution driving a huge change in size, scale, power, and efficiency. The unassuming little *Eohippus* had, through the power of evolution, radically transformed.

The radical implications of horse evolution were not lost on contemporary commentators. While the elitist implications of Osborn calling the horse the "aristocrat" among animals and an "*édition de luxe*" of evolution are inescapable and tied to the patrician and racialized visions of development presented at the AMNH, horse evolution could be more politically flexible. The American feminist writer Charlotte Perkins Gilman, in her breakthrough 1890 poem "Similar Cases," opened with a verse on the little *Eohippus* "no bigger than a fox," which proclaimed:

> "I am going to be a horse!
> And on my middle finger-nails to run my earthly course!
> I'm going to have a flowing tail! I'm going to have a mane!
> I'm going to stand fourteen hands high on the psychozoic plain!"

The animal was then berated by the "the Loxolophodon," "lumpish old Dinoceras and Coryphodon so slow" (who were "the heavy aristocracy in days of long ago"), and told this was impossible:

> "You always were as small and mean as now we see,
> And that's conclusive evidence that you're always going to be."[32]

The rest of the poem, which discussed an early "anthropoidal ape" aiming at humanity and a Neolithic human moving toward civilization (and being similarly ridiculed by their conservative contemporaries) rammed home the point. Evolution showed tremendous changes could happen, despite the denials of elites—an argument with strong implications for contemporary Progressive causes. In this respect evolutionary ideas could be deployed in varied manners and used to rationalize both elitist hierarchies and radical social reform. The horse was again a highly flexible animal.

Reconstructions of *Eohippus* also varied over exactly how horselike it was thought to be, which had significant implications for theories of development. William Berryman Scott saw *Eohippus* as resolutely horselike, noting, "In spite of all these differences, there is something unquestionably horselike about *Eohippus*. You may not be able to say what it is. In the skull, in particular, you cannot put your finger on it and say 'That is *Equus*'; but the whole thing, the cut of its jib is that of a tiny horse. That is very interesting, showing how the stamp was put on the thing at the earliest stage and has never since been lost."[33] For Scott, evolutionary lineages were imprinted with definite characteristics at their origins. However, Richard Lydekker, unwilling to countenance these ideas, wrote that "the word horse must have some limitation; and the American practice of applying it to diminutive ancestral types of the *Equidae* no larger than foxes is one that is not to be commended."[34] Lydekker retained the term *Hyracotherium* and called them "small and almost fox-like creatures which in all probability frequented the swampy shores of lakes and marshes, and were little if any faster than badgers."[35] According to Lydekker, the first true horses were the three-toed *Hipparion*. These debates continued via reconstructions. Some gave *Eohippus* a highly horselike pose, while others showed a more generalized omnivorous animal with an almost doglike posture. Matthew wrote that "the proportions of the skull, the short neck and arched back and the limbs of moderate length, were very little horse-like; recalling, on the contrary, some modern carnivorous animals, especially the civets."[36]

The differing reconstructions of *Eohippus* implied different models of evolution. Were animals imprinted with essential characteristics early in time, which then persistently developed in an orthogenetic manner? Or were they plastic, with potentialities that were only slowly fixed? Reconstructions of *Eohippus* went to the heart of an important paleontological issue: Exactly how flexible were early creatures? As seen in earlier chapters, there was an idea in paleontological and evolutionary discussion that lineages tended to move from the generalized to the specialized. Along with this was a frequent anxiety over specialization, with the development of specialist characteristics often seeming to lead to misdirection or an inability to adapt to future changes. But when were these specialist characteristics and courses of development set? For Scott, organisms were locked into tracks early in their histories, which they then followed until their final development or extinction. For Lydekker and Matthew meanwhile, early creatures were more divergent, showing links between separated modern lineages. The roads of evolution, and the lessons drawn from them, were often contested, and whether patterns were nearly inevitable and predetermined, or whether there was scope for changes of direction or flexibility, was unclear.

Biomechanics and the Structure of Organisms

The evolution of the horse combined linear narratives and more holistic attempts to define the character of its first ancestor, *Eohippus.* It also linked paleontology, public interest, and other biological sciences. Throughout the nineteenth century paleontology had close links with—and often difficult demarcations from—comparative anatomy. Definitions of mammals and their various subdivisions were still based on comparative anatomical studies. Features based on reproduction, locomotion, digestion, or the nervous system continued to form the basis for typologies. Analysis remained focused on the teeth and limbs. In Lydekker's words: "Two factors have evidently been predominant in guiding this evolution, namely the necessity of collecting and assimilating food and of attaining a high degree of speed."[37] Animals and their evolution were defined in ways that recapitulated the models presented by Cuvier, concentrating on functional features most commonly preserved in fossils.

Defining the horse as a highly specialized creature marked by its history emphasized that animals should be understood paleontologically, characterized by a basic anatomy imprinted by their evolutionary history. Changing limb bones and digits told a story of increasing specialization and encouraged observers to think about the adaptability and changeability of organisms. While the reduction of toes was the most obvious change, this presaged the alteration of the limbs. The three upper limb bones were reduced and solidified, while the small bones which in humans and many other mammals made up the wrist, ankle, hand, foot, and digits were extended, expanded, reduced, and reworked into new structures, with the animal effectively walking on the nail of a single toe. Matthew opened his guide to the AMNH horse displays: "The Horse is distinguished from all other animals now living by the fact that he has but one toe on each foot . . . The hoof corresponds to the nail of a man or the claw of a dog or cat, and is broadened out to afford a firm, strong support on which the whole weight of the animal rests."[38] Many displays expanded this point, illustrating a human

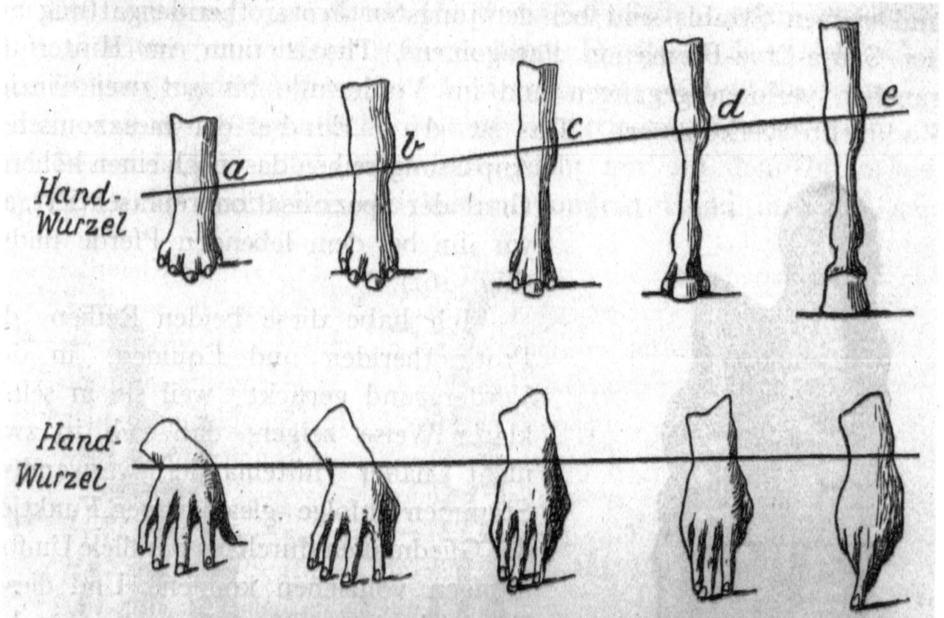

Fig. 163. Entstehung der monodaktylen Pferdehand (e) aus der pentadaktylen Urhuferhand (a) auf dem Wege Protohippus (b), Epihippus (c) und Mesohippus (d). Darunter eine menschliche Hand in den entsprechenden Stellungen. (Nach H. F. Osborn, 1910.)

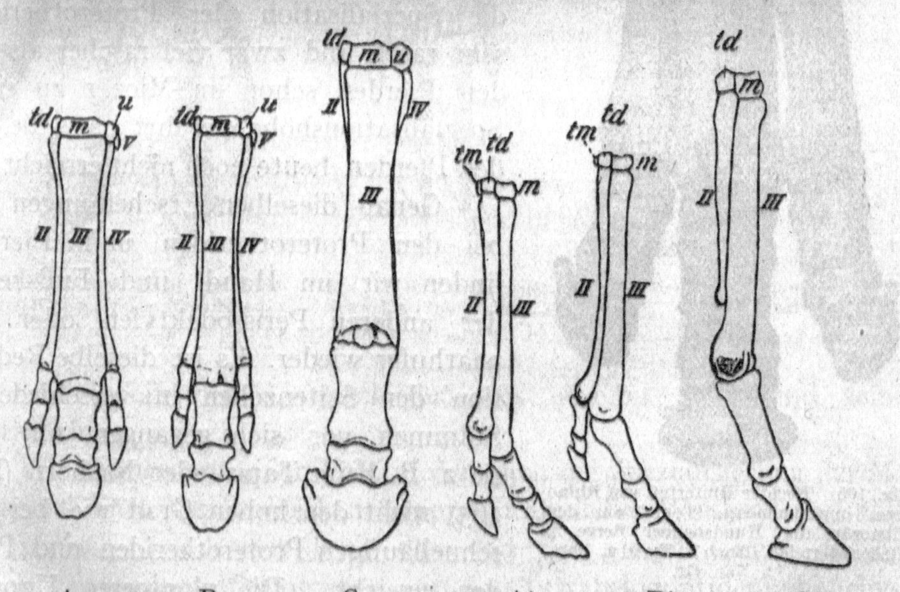

Fig. 164. Linke Hand von vorne (A, B, C) und innen (A', B', C') von Anchitherium (Miozän, A, A'), Hipparion (Unterpliozän, B, B') und Equus (C, C'). — Alle drei Gattungen eurasiatisch. (Nach A. Gaudry.)

hand slowly shifting from the splayed ancestral condition of *Eohippus* to modern *Equus*'s single digit— a gradual evolutionary change with a massive increase in force.

The horse's limb bones told a tale of ever-increasing strength and efficiency. An animal walking on its middle digit required tremendous muscular and anatomical development. The horse could be considered as a highly efficient running machine. This mirrored contemporary discourse using horses to link animals and machines, ranging from Muybridge's galloping photographs to the use of "horsepower" as a unit in transport and industry. Indeed, McShane and Tarr argue that the horse became regarded as a "living machine" at this time, and was treated in similar ways to human technology.[39] Paleontological presentations underlined this analogy: anatomical changes developed by evolution followed similar patterns to technological innovations, driving toward improvement and the reduction of superfluous characteristics. For Lull, the lengthening of bones meant "the pendulum-like motion of the limbs being all in one plane, the joints between the bones become pulley-like through the formation of interlocking tongues and grooves, which effectually limit any lateral motion."[40] Matthew made a further analogy on the horse's single "hard" toe, versus the "softer" feet of other animals: "A somewhat similar case is seen in the pneumatic tire of a bicycle; a 'soft' tire accommodates itself to a rough road and makes easier riding, but a 'hard' tire is faster, especially on a smooth road."[41] These links between the evolution of the horse and technological development also went in the other direction—technology was driven by similar processes to organic nature. It too evolved, sought new solutions, and worked through mechanical laws. Nature and human culture were again linked.

Paleontology did not simply render the horse a machine, though, but made it a creature imprinted by evolutionary history. The changes in the horse's limbs could explain how animals interacted with climatic and environmental change. As seen in chapter 7, scholars constructed a conventionalized history for the Tertiary, starting with the Eocene jungles, which became steadily drier as grasslands spread across the northern hemisphere, ending with harsh ice ages. The evolution of the horse moved with these climatic changes, beginning with a germinal, undifferentiated form suited

◀ FIGURE 10.5. A combination of Osborn's and Gaudry's visions of the evolution of the horse's limbs. Abel, *Grundzüge der Palaeobiologie der Wirbeltiere*, 235. Author's collection.

to sheltered forests to one evolved to gallop across open plains in search of food, escaping predators, and occupying land.

Climatic forces were also dramatized in the horse's teeth, which showed similar changes to the feet. Those of *Eohippus* were small, low, and lacked complex grinding surfaces, implying a diet of "succulent meadow grasses,"[42] fruit, and soft leaves, or even potentially an omnivorous diet. The modern *Equus* had a battery of high-crowned grinding teeth, renewed throughout the animal's life. This was a system for processing tough grasses. The transition between the horse's ancestors showed this change, with teeth becoming more complex, higher, and more ridged, as the lineage transformed from general forest dwellers to creatures of the plains. As Osborn stated (also arguing for North American origins for the lineage), "The American environment and the American horse evolved together—the teeth with the changes in the food, especially with the increasing spread of the grasses, the feet with the character of the turf, the speed with the development of speed among the wolves and foxes."[43]

Yet there were still uncertainties, especially in exactly how the horse's features had developed. As mentioned above, this period saw tremendous debate over heredity and evolutionary mechanisms. Argued strongly by Osborn, "the evolutionary law is simple: it is the development of the useful and the degeneration of the useless."[44] But how new traits were formed or passed on was unclear. It could be due to strict Darwinian pressures through random adaptive changes being preserved. It could potentially be due to sudden mutations, with limb bones and digits being reduced in certain individuals. Or it could have been Lamarckian processes, with running on increasingly open grasslands leading to anatomical changes through use and disuse, which were then passed on to later generations. Or changes could have been orthogenetic, with the early horse being stamped with an initial efficient gait, which was fortunately well suited to the Tertiary's changing environment. Or it could be due to a plan, with providential power constructing this perfected organism. This was the view presented by Lydekker, who concluded his book on the horse with the statement that "all these marvellous changes and adaptations are not due to any mere 'blind struggle for existence' or 'survival of the fittest,' but that they were directly designed and controlled by an Omniscient and Omnipotent Creator, is the settled and final opinion of the author of this volume."[45] These were very diverse views, which could all be accommodated through the project of paleontology by showing the course of evolution and demonstrating the pattern of development.

Models of horse evolution presented in specialist works were often more complex than the linear vision of many displays. For example, James W. Gidley, an assistant at the AMNH, conducted a general revision of all specimens in "the American Museum collection of Horses—from the Eocene to the Pleistocene inclusive—[which] now numbers several thousand specimens, including nearly fifty types and about as many casts of types."[46] Gidley placed the equids into four large subfamilies: an Eocene four-toed, short-crowned-tooth group; a mid-Tertiary three-toed and short-crowned-tooth group; a Miocene three-toed and high-crowned-tooth group; and finally the one-toed, high-crowned-tooth group from the Pliocene on. While this followed the sequence of progressive evolution, Gidley saw later representatives of each group as too specialized to evolve into new types. Instead, it was the earlier, less differentiated forms at the base of each group that were the ancestors of their successors. Evolution had a disjunctive rather than a continuous pattern.

Evolutionary change was often thought of as more "bushy" and "zig-zag"-like than linear. Even Osborn rendered the middle stages of horse evolution more complex by arguing for three distinct "races" of Miocene horses: the forest horse *Hypohippus*, a deerlike horse called *Neohipparion*, and *Protohippus*, the ancestor of the modern horse. This was an example of what Osborn termed "adaptive radiation," that early generalized creatures would diversify as they moved into different environments and adapt accordingly—something especially observable among mammals "because of their superior plasticity."[47] However, these theories of the complexity of the ancient animal past were often occluded in displays, which continued to show chains of development and linear progress. While the details may have been complex and bushy, the single line of development, and overall sense of improvement, was maintained—even if it could be more variable than the confident public statements might suggest.

Extinction and Domestication in a Changing World

H. G. Wells's "Story of the Stone Age" followed the adventures of a prehistoric human named Ugh-Lomi, living alongside other humans and Pleistocene beasts. Among these were ancient horses, who "lived apart" from the "cave men." This was described as a poor decision for the anthropomorphized horses; while "in those days man seemed a harmless thing enough," this was not to last. "No whisper of prophetic intelligence told the species of the terrible slavery that was to come, of the whip and spur and bearing-rein, the clumsy load and the slippery street, the insufficient food, and the knacker's yard, that was to replace the wide grass-land and the freedom of the earth."[48] The life of prehistoric horses was one of freedom and progress, while the modern domestic horse had a troubled existence, at odds with its proud evolutionary heritage. Valuations of the horse's perfection could aggravate a sense of unease, questioning the morality of how horses were used and controlled by modern humans—which seemed to go against their imagined history and wild state.

The story of the horse did not end with its evolution into the "aristocrat" of the animal world, perfected for the grasslands. The final chapter saw it shift to a new status—as a domesticated animal, living with "man" under "his" domination. The prevalence of domestic horses was a critical reason why horse evolution was such a widespread model. Yet the implications of domestication were uncertain. Control of the horse by humans could be understood as a great cross-species partnership, but also it meant this noble animal was now subservient, "enslaved," and largely extinct in nature. The current phase of the horse's history was a triumph for humanity, but could be read as a tragedy for the horse.

Debates over the domestication of the horse also connected with the extinction of wild equids in the modern world. Considerable attention focused on wild relatives of the horse, particularly zebra and onager, along with questions (as in Ewart's experiments) on whether these were actually different species or varieties of the same animal. The rarity of wild equids, particularly outside regions defined as "remote" like southern Africa and central Asia, meant they were understood as on the verge of disappearance. The quagga of southern Africa, driven to extinction in the 1880s, was frequently discussed.[49] The loss of the quagga was thought due to human action and "civilization," something which wild animals could not endure. Lurking in the background of discussions of the horse was the anxiety that even this perfect animal could disappear from the earth.

The Przhevalsky's horse of inner Asia, presented as a newly discovered wild species of horse in the late nineteenth century and then displayed in European and American zoos from the 1900s, demonstrates these concerns.

Many scholars saw them as escaped feral animals, rather than true wild horses. Those who accepted it as a "wild horse" had to acknowledge it was only a small, isolated, and remote population, which somehow had escaped human dominance and clung on as a relic of past ages. Ewart described that when Przhevalsky's horses were displayed in Britain in 1902, "the young colts looked ungainly and lifeless, the large head looked out of proportion to the undeveloped trunk, the coat was ragged, the gait awkward, the mane arched to one side as in hybrids, and the tail suggested a mule more than a horse. It was hence not wonderful that the Prjevalsky colts failed to realize one's conception of a wild horse."[50] They were drab and unimpressive compared to preexisting visions of the perfected wild horse promoted in paleontological settings.

Paleontological studies also raised concerns around the extinction of ancient horses. Much of the evolution of the horse seemed to have occurred in North America, with their fossils found up to the late Pleistocene. However, the American horse then suddenly disappeared, "in spite of the perfection of the adaptation."[51] This was disconcerting, as the standard mechanisms of extinction—inferiority, decline, and maladjustment—could not be invoked for this perfect creature. Matthew summarized the possibilities: "It may have been that they were unable to stand the cold of the winters, probably longer continued and much more severe during the Ice Age than now. It is very probable that man—the early tribes of prehistoric hunters—played a large part in extinguishing the race. The competition with the bison and the antelope, which had recently migrated to America—may have made it more difficult than formerly for the American Horse to get a living. Or, finally, some unknown disease or prolonged season of drought may have exterminated the race."[52] These factors—climate, competition with migrants like antelope and bison, human agency, or a mysterious disease—were all uncertain. Wilhelm Bölsche simply highlighted "the secret" of the American horse's extinction "in its last and most perfect development" as a mystery, showing how "*sic transit gloria mundi* is also the case in biogeography."[53] The horse's disappearance could be explained only through vague providential forces.

The fate of wild horses was tragic, and so potentially was that of modern domestic horses. The utility of the domestic horse was raised constantly in paleontological works. Osborn addressed the New York Farmers dinner at the Metropolitan Club that the horse "has been to man the most useful of all the domesticated animals, not barring the cow."[54] Horses moved across the world to become truly "cosmopolitan" animals. Human action also modified the horse's physical features through selective breeding to form a variety of types (as shown in the AMNH's displays of the horse under the "hand of man"). Humans could mold it to their needs and desires. However, this was not solely an expression of human dominance, as it still relied on the animal constructed by nature. The evolution of the horse showed—despite all modifications by humans—that the core aspects of the animal that made it such a supremely developed creature, namely its limbs and teeth, were forged by evolution, not humans.

While there was definite romanticization of particular types of domestic horse and writing on horses often showed great admiration and empathy, this coincided with assertions of how horses were mistreated or even "enslaved" by humans. Flower brought lessons from evolution to explain how modern treatment of horses was fundamentally cruel and against the animal's nature. He saw the horse's neck anatomy, evolved to hold its head up to look over the open plains, was cruelly restricted by halters and harnesses, raising the point that "of the numerous petty cruelties practiced by man upon the domestic animals in obedience to the dictates of fashion or custom, or out of mere thoughtlessness, the use of the bearing-rein as a regular part of the harness of a carriage or cart-horse is one of the least excusable."[55] Similarly, horseshoeing "has been too long left in the hands of ignorant mechanics," and should be adapted to the "state of nature[;] when the animal is free to choose the ground it runs over, the wear of the hoof is in exact proportion to its growth, and the organ always maintains itself in perfect condition."[56] Natural history needed to reform these injurious practices, just as humans themselves needed to avoid practices which "deform and injure our own feet by pointed shoes, and our own waists by tight lacing."[57] Proper treatment of the horse required a knowledge of its evolutionary history, which often went against modern styles.

The use of narratives from biology and evolution to question injurious current practices (both in humans and horses) raises the question of the evolution of the horse as a double-sided image. Ostensibly it provided a vision that evolution was based on progress, with horses being perfected into their current status as "aristocrats" of the animal world and the animal most useful to humans. However, underneath this overarching structure, the evolution of the horse was malleable and uncertain. The horse could be used for a range of social and political purposes, ranging from exclusionary eugenic visions and orthogenetic progress to reformist ideas of the inevitability of change and animal welfare. Scientific debates around the development of the horse were by no means certain either. How the horse had changed, and how its domestication and near extinction in modern times could be understood, gave an uneasy backdrop to confident linear models. Beneath the vision of progress remained doubt and uncertainty. However, this did not diminish the importance of the horse, but made understanding it, and claiming authority over its history, more important. As the horse's development became one of the most widespread images of evolution, it allowed paleontologists to assert authority over studying and valuing life—understanding development, heredity, welfare, economics, mechanics, environment, history, and morality.

CHAPTER 11

Ordering the Pampas and Patagonia

South America as a Zone of Innovation

UNDERSTANDINGS OF FOSSIL MAMMALS WERE IMPRINTED BY "NORTHERN" DYNAMICS. European and North American institutions defined Holarctic animals as superior and norms to compare with the rest of the world's life, in the past and the present. And yet the goals of paleontologists, of reconstructing the entire history of life and thereby understanding variation, development, progress, and hierarchy, provided a space for these centralizing emphases to be contested and in some cases overturned completely. The following chapters examine how research projects outside traditional core regions of paleontology reevaluated the history of life. In many respects, innovation and change came from outside the northern regions, destabilizing ideas of privileged centers.

South America was one of the classic areas for natural history, including research into fossil mammals. Remains of large mammals like great sloths, armored glyptodons, and other robust herbivores were excavated in caves and agricultural sites. European naturalists like Alexander von Humboldt, Alcide d'Orbigny, Charles Darwin, Alfred Russel Wallace, and Henry Walter Bates marked out South America as central for natural history and theorizing on development and variation (often defining the continent as distinct and peculiar).[1] Access to sites required interaction with South American government agencies and scholars and often work alongside commercial interests, especially the meat, hide, and guano trades. Natural history in South America was contentious, tied to both imperial and national dynamics.

Late nineteenth-century South America was seized by the same rush of collection building as other parts of the world. South American scholars and museums, often tied to reformist and liberal states, collected and discussed the continent's nature.[2] As state authority and control expanded, South American governments and elites also devoted great effort to knowing their own territories. This was partly economically based, seeking areas for settlement, mining, and livestock raising, but it also involved natural history, geography, anthropology, and ethnography, which became important nation-building fields, providing knowledge of the territory and its current and former inhabitants while asserting South American scientific institutions as equal to those of Europe and North America. Newly founded universities and museums aimed for international prestige and "progress" within the national community.

The dominance of biogeography as a research program increased the importance of South American natural history. All models agreed the South American Neotropical Region was extremely distinctive. Discussing the life of South America, Wallace wrote, "Richness combined with isolation is the predominant feature of Neotropical Zoology, and no other region can approach it in the number of its peculiar family and generic types."[3] Richard Lydekker stated that "next to Australia . . . South America possesses a greater number of peculiar types of animals than any other region," including "edentates, such as armadillos, ant-eaters, and sloths," "the camel-like animals known as guanacos and vicunas," and a "remarkable" abundance of rodents.[4] Deeper in geological history, before the faunas became "mixed" with North American types (through migrations across land bridges), "this peculiarity and distinctness of the South American fauna was even more intensified," with giant sloths, glyptodons, marsupial carnivores, and strange ungulates, whose fossils were unearthed in large numbers.[5]

Biogeographical ideas drew on older discourses of South America as a place of tropical wonder, strangeness, and degeneration. Many in the United States and Europe argued that the persistence of animals marked as "low" on the scale of life, most notably edentates and marsupials, showed the region's isolation from the major "theater of evolution" in Holarctica. However, this was not the only interpretation of South American natural history. Drives to acquire South American fossils could lead to alternative narratives. Dramatic creatures like *Megatherium* and glyptodons were important status symbols in museum collections, whatever values were placed on their position within the scale of life. In the 1900s increased attention focused on fossils from deeper strata in the far south of Patagonia. Many South American scholars used these animals to argue that the southern continent may have been the site of origin for many important groups. Even scholars who disagreed with these ideas (which was normal in the north) needed to explain how some southern animals, especially glyptodons and ground sloths, were not driven to extinction when the continents merged, but had moved north, colonizing much of North America. The extent of the collections in cities like São Paulo, Rio de Janeiro, La Plata, and Buenos Aires meant that South American scholars could gain significant authority. As South America became an arena for paleontological work, it was of global significance regardless of debates over the value of the organisms.

Conceiving "A Land of Skeletons" in Argentina

By far the most significant country for South American paleontology was Argentina. In the late nineteenth century the Republic of Argentina developed as an important economic power, something often occluded in "global" histories, which Matthew Brown has noted often neglect South American developments.[6] Argentina was the focal point of a variety of networks, especially around agriculture and livestock raising, and saw significant migration. The Argentinian state became more powerful and expansive, extending control into Patagonia and the northwestern interior, at the expense of Indigenous people—a process leading to political debate on the "Indian Question": whether Indigenous peoples could be integrated into the new state or if they were doomed in the face of "progress" (paralleling similar "extinction discourse" in the United States and British settler colonies).[7] Mapping, surveying, and ethnography gave scientific authority and fed directly into political and economic projects.

Despite the prominence of South American fossils like *Megatherium*, South American scholars faced difficulties in participating in international paleontological debates in the early nineteenth century. However, this shifted significantly in mid-century as new local institutions were established, often headed by migrant scholars from Europe. The Prussian naturalist

Hermann Burmeister, who escaped Germany following the revolutions of 1848, was appointed in 1862 to head the Museo Público in Buenos Aires (which gained the grander title of Museo Nacional in 1884).[8] While initially a varied collection of antiquarian, anthropological, and natural history artifacts, Burmeister reorganized the Museo Público to emphasize the fossil mammals of the Pampas formations. Indeed, the first issue of the museum's *Anales* was almost exclusively devoted to fossil mammals, and the second consisted of a long monograph on the glyptodons.[9] Burmeister controlled numerous institutions and promoted a catastrophist line. However, institution building and conflict in Argentinian paleontology reached a new level through the rivalry of two scholarly groupings, one centered around Francisco Moreno and the Museum of La Plata and the other around the Ameghino brothers.

Of these rivals, Francisco Moreno was the more established, although less studied in the secondary literature than the more colorful and overtly nationalistic Ameghinos. Yet Moreno was a significant figure, developing the La Plata Museum within the principles of the new museum movement. Taking part in and then directing expeditions across Argentina and Patagonia from the 1870s on, Moreno had interests across geography, cartography, anthropology, and ethnography.[10] In his view, scientific work was essential to understand and control the national territory, bind together the human and animal pasts, and assert the modern Argentinian republic as a progressive force over a territory with a deep, unsettling antiquity. He wrote how the territory of "the Argentine Republic is undoubtedly a vast necropolis of lost races," subjected to constant migrations of humans and animals, who competed across time and were layered in the landscape, as "some conquered and others were conquered, and were annihilated in our southern extremes."[11] This had resonances for the contemporary Indian Question. The current conquest of the southern cone by Argentina was the latest instance of the invasion of the territory by a new "race," and followed deep trends in the human and animal pasts. It also raised possibilities of mixture and incorporation, as well as death and expulsion. For Moreno, the Indian Question could potentially be resolved through bringing the Indigenous population into the new civilization. However, this needed to be carefully regulated and required anthropology to classify how far particular groups were "assimilable."

Moreno's museum was initially chartered in 1877 as the Anthropological and Archaeological Museum of Buenos Aires to host "the archaeological and anthropological curiosities which have been found on our still unexplored territories, vestiges of a lost past whose relics—if scientifically classified—would solve complex problems."[12] However, with the foundation of the city of La Plata as a capital for the province of Buenos Aires (with the city of Buenos Aires being the national capital), the museum was given a broader remit to include a whole range of natural antiquities. It initially had departments devoted to paleontology, anthropology, geology, zoology, and botany, and was intended as an institution of scientific research and public enlightenment. The first edition of its *Revista* featured a revised translation of William Henry Flower's 1889 address to the British Association for the Advancement of Science on museum organisation, along with a long reflection by Moreno on the layout of the British Museum of Natural History (while noting that "what was possible in London was not possible in La Plata").[13]

Despite Moreno's misgivings about comparisons with London, the La Plata Museum was a grand institution, constructed on the model and scale of the collections examined in chapter 9 as archetypes of the new museum movement. As well as being internationally oriented, it was also of major national importance, firmly inserting human prehistory into the world of the Pampas mammals. Jens Andermann has noted that it was one of the most expensive and largest buildings in the new city, and its displays promoted new ways of seeing time and sovereignty.[14] Entering through a neoclassical vestibule, the visitor saw painted scenes from South American prehistory depicting mastodons, *Smilodon*, prehistoric hunters, and early people dismembering a glyptodon—presaging the specimens and naturalizing the ancient Americans with the fossil beasts.[15] The visitor journeyed across three thousand five hundred meters of gallery space, whose circular displays were intended to "begin with mystery and end with man,"[16] first passing through an "outer ring" of minerals, paleontology, and modern natural history, and then inner and upper rooms of archaeology, human cultural artifacts and anthropology (including human remains and over five hundred skulls of Indigenous people). The paleontological galleries included a huge amount of material, with an entire room of ground sloth remains and another with a series of glyptodons. Moreno wrote that while Gaudry wished to culminate his galleries with "'a human figure, representing the artist and the poet' . . . at the Museum of La Plata, the galleries do not end—they run across one another in a grand central rotunda." It was the visitor, venturing round "the immensity of past times" and "vital forms of unremitting struggle," who occupied the culmination of the gallery.[17] The galleries reconstructed "the southern tree of life, its own tree and independent of the one born and raised in the opposite hemisphere,"[18] and demanded reflection on the viewer's place in this cyclical drama of conflict and migration. The gallery may have given the visitor a central position, but this was not necessarily a comfortable one.

While some critics saw the La Plata Museum as overly ostentatious, it was hailed as a great success in Argentinian publications and by foreign visitors. Moreno wrote to the minster of public works, that this "magnificent building" and "its collections have developed to be the envy of many of the great museums of the world."[19] Richard Lydekker waxed lyrically about how "the visitor will be absolutely lost in astonishment at the long array of perfect mounted skeletons of numbers of these creatures, while the unmounted skeletons and isolated bones displayed in the wall-cases will convince him that I am not exaggerating when I call Argentina a land of skeletons."[20] Henry Ward, the Rochester naturalist who had made a copy of the London *Megatherium* the centerpiece of his natural history trading company, made similar points: the La Plata Museum made him "so amazed . . . that my first visit seemed to me like a dream in which I was given over to savor the delights of fantastic visions. It was only after repeated visits that I was able to convince myself that this was indeed a reality."[21] The displays dragged paleontology out from Eurocentric perspectives, with Ward further noting, "How little do these gentlemen, whose travels are limited to the countries of Europe, know the inexhaustible sources of information in their chosen science which awaits them here, and which will one day astonish them!"[22] The museum was paralleled by Argentina itself, with Ward noting how the country was exceeding the United States in many respects, including the scale of migration and the number of telephones and banks.

Moreno's principal rival, Florentino Ameghino, followed a different approach to building a paleontological career, based on informal links and commercial culture rather than state patronage and articulations of the new museum movement.[23] Indeed, the prominence of Ameghino and his brothers in Argentina and internationally shows that the history of paleontology in the nineteenth century was definitely not an easy story of museums and professionals crowding out other actors; it was still possible to work through commercial culture and local connections (even if this did create challenges). Ameghino was born to Genoan parents in the early 1850s, and grew up in Lujàn, close to where the first *Megatherium* remains were excavated.

FIGURE II.1. The vestibule at the La Plata Museum. *Revista del Museo de La Plata 1* (1890–1891): plate 3. Reproduced by kind permission of the Syndics of Cambridge University Library.

Florentino Ameghino and his brothers Juan and Carlos developed an interest in fossil collecting, making early claims that the southern cone was a key center of evolution, which the more conservative Burmeister attempted to block from publication.

Florentino Ameghino has been regarded as a quintessentially nationalist scientist, building up Argentinian scientific institutions, promoting Argentinian fossils as central to understanding earth's history, and in the twentieth century being constructed as a national scientific hero in Argentina through the media and education system.[24] However, his early career took a migratory path, building links with Europe and North America. His interest in

FIGURE 11.2. The *Megatherium* displays in the La Plata Museum. *Revista del Museo de La Plata 1* (1890–1891): plate 5. Reproduced by kind permission of the Syndics of Cambridge University Library.

South American fossil mammals grew from his participation in the Argentinian pavilion at the 1878 Paris Exposition,[25] during which Paul Gervais introduced him to the collections at the Jardin des Plantes and highlighted the importance of South American paleontology. The exposition's South American fossils were then sold to another transatlantic visitor, Edward Drinker Cope, and were in turn eventually purchased by the American Museum of Natural History. On his return to Argentina, Ameghino served as vice director of the La Plata Museum from 1886 to 1887, although tension with Moreno meant relations broke down.

Inscribing Patagonian Remoteness

While South America was crucial for paleontological research since the late eighteenth century, most of the continent's fossil animals were from

Pleistocene formations. In Argentina the Pampean formations contained a range of dramatic creatures, including the megatherian ground sloths, glyptodons, and predators like saber-toothed cats. Santiago Roth, one of the scholars at the La Plata Museum, stated, "The Pampas must have been a paradise for mammals—here they could develop to a colossal size and diversity"[26] and the land was "indisputably very rich in fossil mammal remains and will never be exhausted" (although the destruction of fossils by weather and human carelessness was still a troubling issue requiring further work in the region).[27] The pampas layers were also relatively recent, and fairly easy to align with equivalents in Europe and North America. The shallow deposits and condition of the bones meant that they were uncontroversially dated to the Pliocene and Pleistocene. While there were debates over the taxonomy of the animals and their habits, when they had lived was not especially contentious.

Matters were different in Patagonia, in the far south. The region was annexed by Argentina and Chile in the late nineteenth century through brutal colonial wars against the Indigenous inhabitants. Control over Patagonia was another part of the "modernizing" drives of South American states, dominating a land conceptually marked as "wild" and "savage." However, while a network of coastal ports was constructed and the Argentinian state attempted to encourage migration, the Patagonian interior remained inaccessible. Distance and maritime routes made it a complex zone of mixture: Irina Podgorny notes that the journey to Patagonia from Buenos Aires was only a week less than the journey to New York,[28] and the dense network of British and US whaling and trade stations across the South Atlantic meant Patagonia was integrated into maritime networks, with towns like Río Gallegos having large populations of British settlers.

The Argentinian government organized several cartographic expeditions to Patagonia, several of which involved Moreno. Recounting one such visit in 1876–1877 in his *Viaje a la Patagonia austral*, Moreno described Patagonia as "sometimes terribly arid, at other times luxurious to the point it recalls the tropics, but always imposing, in its inhabitants, arid plateaus, immense volcanic mantles, high snow-capped mountains, volcanoes, lakes, rivers, streams and forests." Venturing there promoted sublime wonder and showed "what Patagonia can contribute to the prosperity of the Republic."[29] Patagonia was presented as dangerous, home to an Indigenous population depicted as wild and savage, and either in need of assimilation into the modern nation, or "fading away," with their extinction presaged in the history of the region, which had seen huge varieties of humans and animals die out. One incident, which was to be persistently recounted both by Moreno and almost all northern scholars who met him, was how he was captured by the Tehuelche. While he was quite quickly released, this bolstered narratives of a dangerous land where the people recapitulated ancient savagery (although of course, given the brutality of the colonial wars against the Tehuelche in these years, danger primarily came from the other side).

Patagonia was conceptually and economically significant beyond Argentina. It was classed as a remote region "at the end of the world," inspiring many European and North American travel accounts. Imaginings of the region were summed up in Florence Dixie's opening to *Across Patagonia*: "'Patagonia! Who would ever think of going to such a place?' 'Why, you will be eaten up by cannibals!' 'What on earth makes you choose such an outlandish part of the world to go to?' 'What can be the attraction?' 'Why, it is thousands of miles away, and no one has ever been there before, except Captain Musters, and one or two other adventurous madmen!'"[30] The answer for visiting Patagonia was, of course, "precisely because it was such an outlandish place and so far away," and to "be able to penetrate into vast wilds, virgin as yet to the foot of man."[31] Comments on the region's desolation and "uninhabited" state were common, although were usually followed by descriptions of meeting local guides and the Indigenous population. For European writers Patagonia was marked as a zone of utter primitiveness at the ends of the earth, with its human populations either disappearing or already vanished. Moreno's description of Patagonia as a "necropolis of lost races" fed into this: the modern Tehuelche and Mapuche were the latest of a long run of peoples who had died out in the land's history, and their modern "decline" was naturalized.

Patagonia was not just a land of the human dead. There were large quantities of animal fossils from geological strata far older than those of the Pampas—although exactly how old was unclear. The Patagonian strata potentially dated to the Eocene, the beginning of the Age of Mammals, or potentially (and more controversially) earlier. The fossils were also strange. Patagonian animals were given many names in the conflict between Moreno and the Ameghinos, from the predatory sparassodonts (tearing teeth) to the apparently herbivorous pyrotheres (fire beasts) and astrapotheres (lightning beasts). These had some similarities with northern carnivores and ungulates, but also had very distinct anatomical features, especially in the limbs and teeth. What they signified was unclear. Was ancient South America a separate continent where life had developed on lines parallel to but distinct from other regions? Or were these South American animals related to better-studied northern animals? The abundant fossils, and the importance of South America for biogeographic theorizing, meant that much attention turned to Patagonia.

Research in Patagonia was not—as was frequently presented in the narratives—movement into an unknown and mysterious region, but took advantage of government expansion, economic penetration, agricultural settlement, and maritime commerce. Much paleontological work in Patagonia was conducted by Argentinian scientists connected with the major museums in Buenos Aires and La Plata, often operating alongside cartographic projects—another instance of links between paleontology and government surveying. Moreno organized teams prospecting Patagonia to settle the disputed border with Chile, achieved through defining the underlying geology as much as the modern geography. The La Plata collections went partly to accumulate fossils, but also linked the human and animal pasts. Moreno was primarily interested in the anthropology of the modern and prehistoric "races" who, he argued, inhabited Patagonia alongside the extinct animals. Finds from Buenos Aires province and the south illustrated this, with Santiago Roth's discussion of the "skull of Pontimelo," human remains found inside a glyptodon carapace, indicating humans and extinct beasts had lived alongside one another.[32]

The most renowned paleontological collector in Patagonia was Florentino Ameghino's brother Carlos. After leaving the La Plata Museum, the Ameghino brothers established their own scientific networks. In 1882 they opened a bookshop, named the Librería del Glyptodon, as the center of their fossil collecting and exchange networks. This was a family firm where Juan managed the bookshop and financial arrangements, while Carlos was the field investigator, spending a huge amount of time in Patagonia. He traveled to the region fourteen times between 1887 and 1903 (Florentino only went once, on the final expedition). While the first of these trips was under the auspices of the La Plata Museum (in the brief period before the Ameghinos' falling out with Moreno), the majority of Carlos's journeys were undertaken independently, funded through the proceeds of the bookshop and the sale of fossils.

FIGURE 11.3. The study and storeroom in the Ameghinos' shop. William Diller Matthew, "Florentino Ameghino," *Popular Science Monthly*, March 1912, fig. 3. Reproduced by kind permission of the Syndics of Cambridge University Library.

The Ameghinos followed a commercial rather than a state-backed program. Carlos's journeys were small scale, consisting of himself and a team of workers. He regularly corresponded with Florentino.[33] Relations were expectedly close, given the family nature of the Ameghinos' outfit. They sent one another warm greetings, and Florentino expressed concern on the numerous occasions when difficulties in weather, transportation, or shipping delayed letters. One letter from June 1893, after Florentino had not heard from Carlos for several weeks, opened with an exclamatory "With what pleasure I finally received a letter from you, from which I can see that you are well!" followed by a long list of news.[34] Their correspondence usually did not go into detail on the management of the expeditions. The tense work of science at a distance, as shown between Marsh and his collectors, was unnecessary. Trust between family members and respect for Carlos's knowledge of the territory meant he could primarily get on with searching for fossils. Their discussions tended to focus on the number of specimens, the character and possible age of geological strata in particular localities, and news from Buenos Aires and La Plata from Florentino, or of other travelers in Patagonia from Carlos, who warily reported especially on the activities of potential competitors, like the Museum of La Plata and foreign expeditions.

Expeditions to Patagonia acquired large numbers of fossils, allowing Florentino to catalog the Patagonian fauna as unique. Carlos's long periods in Patagonia had further importance: studies of the geology and stratigraphy of the region reinforced the Ameghinos' more striking arguments, that the mammal fossils of Patagonia were far older than those of the rest of the world. If true, this indicated that mammals had themselves originated in the southern continents. Northern biogeographic theorizing defined South America as home to "strange" and "aberrant" creatures, especially compared to Holarctic animals. However, Argentinian scholars like Florentino Ameghino flipped these ideas. The features northern scientists presented as aberrant were actually ancestral, with the sparassodonts being the ancestors of modern carnivores, and the astrapotheres and pyrotheres being the earliest ungulates. Florentino and Carlos's studies of the Patagonian rock formations argued that the earliest mammal fossils in South America were not Tertiary, but dated back to the Cretaceous, with mammal fossils found alongside those of dinosaurs. The southern continents were where mammals first appeared, and were likely the center of dispersal for the whole group. Patagonia was not a strange region with curious fossils, but of supreme importance for global paleontology. The national dimensions of this research, making Argentinian science crucial within international networks, have been pointed out in a number of works.[35] It was also important for its biogeographic implications—the South was the true generative region.

While the Ameghinos' theories were regarded skeptically in Europe and North America, the strangeness and abundance of the fossils made relations with the Ameghinos useful for northern scholars. Similarly, difficulties in correlating strata meant that the Ameghinos' claims could not be easily dismissed, especially given their familiarity with the territory. Disagreements were often presented in generous terms. William Diller Matthew, in his obituary of Florentino, wrote how "heterodoxy is of the life of scientific doctrine, the surest indication of its vigor and progressiveness. Only in decadence will our theories degenerate into a 'body of geologic dogma.'"[36] Disputes were presented as invigorating for the field. Of course this commitment to scientific debate was not the whole story. Controversies over the dating of the Patagonian strata gave Florentino Ameghino an international profile. Meanwhile, for northern scholars correspondence with the Ameghinos was useful for gaining knowledge and potentially access to Patagonian sites. This made keeping any disagreement friendly an important strategy.

An especially fruitful relationship developed in the 1890s between the Ameghinos and Karl von Zittel in Munich. As seen in previous chapters, Munich was a crucial international center of paleontological work. Zittel followed news on South American fossils closely and had a long-running and warm correspondence with Florentino Ameghino. In 1892 he wrote how reading Ameghino's publications were "a completely unexpected revelation, which gives us the source and explanation of the southern fauna which is so different from that of Europe and North America,"[37] and proposed an exchange of duplicate fossils. Florentino's response provides interesting details on Carlos's work practices. He stated he did not have many duplicates for sale, as Patagonia "is so deserted and rugged, that transportation is very difficult and expensive . . . so I always recommend to my brother Carlos to try to take as much variety as possible, to always look for new objects, and not collect objects already known or represented by more beautiful examples."[38] Carlos had been living "an almost savage life in the solitudes of Patagonia" for over twenty months, organizing overland mule caravans "crossing rivers, sterile steppes and countries which are almost everywhere completely deserted." Florentino claimed the expeditions to Patagonia had cost nearly 150,000 francs, which almost "ruined myself, but I have the satisfaction of today of having one of the most beautiful collections of fossil mammals in the world, and far superior to those of the official establishments that have conspired to make me fail in my researches."[39] This dramatized Argentinian paleontology for an international audience, drawing attention to disputes between collectors and the sacrifices made for research.

Florentino eventually sold Zittel a large collection of fossils at a discounted price of 5,000 francs (this was still almost twice Munich's annual budget for fossil acquisitions). The discount was partly due to the Munich collection's limited funds, but also likely owed to Zittel's position. He specifically mentioned to Florentino that he was preparing new editions of his textbooks, which meant any Patagonian fossils sent to Munich would be featured, making them widely known. Another agreement was made in

1899. Zittel received a larger subvention from the Bavarian government, and sent 7,000 francs a year to assist Carlos's expeditions, on the condition that he acquire duplicate fossils for Munich.[40] An important alliance formed, as the Munich collection became a patron of Argentinian paleontology. The connection was beneficial for both sides: the Ameghinos received funds and a higher profile from the European institution, and Zittel received fossils from a region where collecting trips were difficult, and continued to build Munich's position as a center of education and publication.

The Ameghinos, through informal connections, could maintain a presence in Patagonia and build an important position within Argentina and internationally. These were part of the same process, with national and international connections being crucial for building status. The international links were not just about rivalry with foreign scholars with differing conclusions; indeed, as in the US case with Cope and Marsh, rivalry was much sharper within the national context. Allies from overseas were crucial for funds and status. Any foreign expeditions in Patagonia needed to interact with the Ameghinos, and they were skilled at leveraging foreign connections to promote their reputations and the importance of Patagonian fossils. This paid off tremendously in the 1900s when Florentino was appointed director of the Museo Nacional, which gave him a position equal to Moreno, as well as financial security.

International Collaboration and Competition in Patagonia

International interest consolidated around Patagonia, following a variety of links. Connections between the Argentinian and French scholarly communities were close. As already mentioned, Florentino Ameghino's initial interests in paleontology were inspired by his participation in the Paris 1878 Universal Exposition, and French was his main medium of international communication throughout his career. French scholars were also active in South America and, much like Gaudry's work at Pikermi, took advantage of informal influence. In the 1900s the aristocratic adventurer Georges de Créqui-Montfort led a series of expeditions in South America, primarily focusing on anthropological studies and archaeological excavations, but also locating Pleistocene fossils in Tarija in Bolivia (a region which had been known for fossil bones since the seventeenth century) and purchasing fossils from local notables. These specimens were donated to the Muséum d'Histoire Naturelle in Paris, and Boule eventually published a monograph on them in the 1920s.[41]

Another significant figure was André Tournouër, a French citizen who migrated to Argentina to raise cattle in Mendoza. His father, Raoul, had been active in Parisian scientific circles and was a friend of Gaudry.[42] According to Gaudry, "in the presence of the magnificent discoveries of Tertiary mammals made in Patagonia by scientists from the Argentine Republic, André Tournouër had the patriotic idea that our country should also take part,"[43] and collected material for the Muséum in Paris. Tournouër's published reports were brief, and it is difficult to piece together his experience and work. He traveled to Patagonia overland from Mendoza on three separate expeditions between 1898 and 1903, writing of his first trip that the area around Lake Colhué Huapí was "where I encountered the greatest difficulties . . . being absolutely deserted and bereft of vegetation, sixty kilometers from inhabited places."[44] With a team of locally hired workers and a large number of mules, he tended to ship material to France via Punta Arenas in Chile. While apparently dedicated to the promotion of French science, he needed to rely on Argentinian support, and Carlos introduced him to many sites. Gaudry wrote how Florentino Ameghino, "remembering the affectionate relations he had with us in his youth, when he worked at the Muséum of Paris, supplied information on the fossil deposits of Patagonia to Tournouër with the greatest liberality."[45] Meanwhile, in a letter to Carlos reporting on Tournouër, Florentino expressly said, "Try to be nice to him, if possible . . . this gentleman has already become quite skilled, and his friendship suits us, because the information he sends to the paleontologists of Europe and his publications are very favorable to us."[46] Florentino's interests in vertebrate paleontology had been inspired in Paris, and the prestige of French paleontology remained high. Collaborating with such an established institution was therefore an important strategy for the Ameghinos.

Tournouër could not understand the Patagonian mammals, writing of the *Pyrotherium*, "What a strange animal it is! Some mixture of an elephant and a *Dinoceras*?,"[47] and Gaudry agreed in a letter to Florentino Ameghino thanking him for supporting Tournouër: "Our old classifications are overturned" by these "astonishing creatures."[48] Unlike the Ameghinos, who were convinced that the fauna of Patagonia was ancestral to northern animals, Gaudry saw these lines as totally unrelated, but with similar adaptations and body plans. Work on South American material caused Gaudry to revise his older ideas of evolution. Since working at Pikermi, Gaudry had seen all organisms as linked by chains across the ages, and with evolutionary development moving on similar lines across the world. However, the South American creatures, and studies of "the march of evolution" in the Americas, and especially Patagonia, changed these theories.[49] Progress and development, even among the mammals, was not universal and could take different tracks and move at distinct rates. The ancient world was diverse and varied across geological regions, there being no single history of life.

After returning to France, Tournouër continued to reminisce on his South American experiences. He imaginatively reconstructed the deep history of "these desolate regions," thinking about how "if man had lived at that time in Patagonia, he would have contemplated a strange spectacle," when the pyrotheres and astropotheres "ploughed the ground" with their tusks "to look for their food," while "avoiding the immense burrows of Homalodontheridae and edentates, pursued by rare carnivores, which probably preyed only on the young." "The land[,] which would have been placed under his amazed eyes,"[50] was revived through the reveries of the scholar, transforming fossils and excavation sites into a vanished yet traceable world.

The final group of excavators in Patagonia came from the United States, a nation becoming simultaneously a leader in vertebrate paleontology and increasingly interventionist around the world. US involvement in South America was pronounced, with the expansion of American trade and geopolitical interests (especially around control of the Panama Canal). Relations with Argentina were less tense than with other countries. Given Argentina's significant economic growth and development, it was almost on equitable terms with the United States as a self-consciously "progressive" New World power (as shown in Henry Ward's statements on the growth of La Plata). There were also long-standing economic interests in Patagonia, particularly through the activities of American whalers and traders.

US commercial interests in South America were followed by paleontological work, and it was Princeton University that undertook the most extensive work in Patagonia. These Princeton expeditions were entrusted to John Bell Hatcher, who visited Patagonia three times from 1896 to 1899, with the final trip being a joint expedition with the AMNH, accompanied by Barnum Brown. Hatcher and his team traveled via steamship from New York first to Buenos Aires and then down the coast to Patagonia, to excavate fossils and (in some of the later expeditions) conduct ethnographic research for the Smithsonian.

Official sponsorship and honorary appointments from the US Bureau of Ethnology and Department of Agriculture, and letters of introduction from the Argentinian ambassador in Washington, DC, helped secure access.

The Princeton expeditions to Patagonia had two major aims. The first was to gain South American mammal fossils for the Princeton collections. The second was examining how the geological strata of Patagonia correlated with those of the northern hemisphere. This was a technical operation, involving careful analysis of sedimentation, cartography, and fossils. Working out the relative age of the Patagonian strata was critical to resolving the debate sparked by the Ameghinos, on whether South American fossil mammals were older than those of the northern hemisphere. Hatcher wrote that he hoped "the apparently conflicting observations and theories set forth by the Ameghinos would prove invalid, while the main facts would be found to harmonize with those already well established in the north."[51] The arguments of the north were intended to displace those from the south. However, they could never entirely do this. Northern institutions were dependent on southern scholars for access and comparative materials, and the geology itself remained difficult to assess.

The narrative of the expeditions written by Hatcher illustrates values around fieldwork. Hatcher went to Patagonia with his experience of working in North America in mind, arguing this gave him the skills to collect fossils—and endure the harsh environment. Before setting off on his first expedition, he was warned against conducting fieldwork in Patagonia in winter, to which he argued, "We had tented it for many years on the wind-swept plains of Wyoming, Montana, and the Dakotas, often with the thermometer far below zero, and had no uneasiness as to our ability to survive successfully whatever blizzards Patagonia might have in store for us."[52] Reading between the lines of the account though, Hatcher seems to have faced more difficulties than expected, frequently complaining about the cold and being injured on several occasions. At one point, he was advised by local people not to undertake a two-hundred-mile trip by horseback, which he disregarded as he was "accustomed in our own country to making trips of from five hundred to a thousand miles with one horse,"[53] and on the route back suffered a severe head injury after being thrown from his horse.

Hatcher and his team were expectedly dependent on local collaborators. Relations between Hatcher and the Ameghinos were cordial, despite their differing interpretations of South American chronology. Hatcher also gained a great deal of assistance from the governor of Santa Cruz, Edelmiro Máyer, who provided him with accommodation in his home, procured transportation, and gave letters of access. However, in the field Hatcher tended to work through the Anglophone settler community, especially for supplies and transportation. One British settler, named Herbert Stanley Felton, owned a large estancia at Killik Aike Norte, and gave Hatcher a field base and letters of introduction. Transportation was likewise negotiated opportunistically with passing ships traveling directly to the United States. "Striking a bargain" with merchant vessels was important for sending the material on and allowed Hatcher to circumvent "certain laws prohibiting the exportation of fossils from Argentine territory."[54] While expressed in terms of ingenuity, this indicates the underhanded nature of many expeditions, circumventing local laws on exporting fossils.

Hatcher wrote numerous reflections inscribing the landscape with deep geological antiquity. While staying with Felton, he was informed by a foreman of fossil teeth and bones near the mouth of the Gallegos River. Upon investigating, his "eye caught the reflection given off by the polished enamel of a tooth protruding from the surface" of a cliff, and on digging, "the discovery of more or less complete skulls and skeletons of other animals followed in rapid succession." The site contained the strange Patagonian fauna in a "vast cemetery, which for untold ages had served as nature's burial ground."[55] On another occasion he ventured onto high ground for a "commanding view" of the "broad and level plain," and reflected on its deep history:

> I could not resist the drawing of a mental picture of those comparatively recent times when the waters of the Atlantic washed the foot of the escarpment on the crest—on which I stood, and when the streams of molten lava were poured out over the surface of the great plain to the south. From this it was but another and comparatively short step backward to the time when, instead of a semi-arid region scantily clothed with vegetation just sufficient to sustain the present meagre fauna of these plains, this region supported a fauna rich in ungulates, sloths, armadillos, and giant flightless birds, all indicative of a mild climate and an abundant if not luxuriant vegetation. What a transformation had taken place since these animals had inhabited this region![56]

While the rising of the Andes made the territory dry and arid, the ancient past was knowable through engagement with fossils, landscape, and geology, making it knowable to those with consciousness of deep time. Sedimentary time moved back through a range of landscapes and faunas, which were linked and sequential but distinct. To understand this, empirical studies of the landscape and its fossils again required imagination and empathy.

In 1901 Scott visited Florentino Ameghino to study his collections and debate theories. He reflected in his memoirs how the two needed to speak in shared "bad French," as they drank tea and "discussed the problems of paleontology & squabbled with the most perfect amicability, for we hardly ever agreed about anything & yet we never lost our tempers & always kept the discussion on the purely objective plane."[57] While this might reflect politeness, it seems as if Florentino was well disposed when debating northern scholars who disagreed with his interpretations. Scott wrote to him after his visit about how "we differ in many of our opinions, but I trust that we shall always maintain sentiments of mutual respect and regard."[58] Florentino responded that his "feelings of friendship and sympathy" with Scott "cannot be diminished by differences in opinions, which are always useful, I would even say indispensable, for the advancement of science."[59]

In this case rivalry over theories was relatively easily diminished by commitment to the idea that friendly disagreement was key to science. Indeed, debates could be beneficial for both sides, raising Argentinian institutions to equal terms with northern ones and allowing northern expeditions to collect material from Argentina. Equally important was the conduct of paleontological excavation through complex negotiation and working through varied agencies and links. Particular lands, like Patagonia, were regarded as central for understanding ancient patterns of development. While scholars across the world disagreed on why the Patagonian fauna was significant, their valuations were similar—whether it was a land of the origins for all life, important for presenting a distinctly national fauna, or a land of strange developments that needed to be slotted into wider biogeographic diversity. These ideas bolstered the sense that Patagonian fossils were crucial for the history of life, making them (and those who controlled access to them) of great importance. These differing senses of value also bolstered the significance of research. Paleontology could be pursued through various manners, all of which constructed the unique "land of skeletons."

CHAPTER 12

Lands of the *Diprotodon*

Fossils and the Deep Past in Australia

IN 1909 *THE EVENING NEWS* OF SYDNEY PUBLISHED AN INTERVIEW WITH AN UNNAMED staff member at Sydney's Australian Museum, under the headline "The Great Wombat," describing bones found in a copper mine at Molong in New South Wales that "once formed the head of some very large animal." The interviewed museum officer identified the bones as "probably those of a big *Diprotodon*, or huge animal of the wombat species, now extinct." The reporter, significantly overestimating the size of "the dipro—the great animal here" as being "20 yards long," continued the questioning in a tone of mystery and fascination. The officer stated the creature "burrowed like the ordinary wombat," and "answered to the mammoth in bulk and was quite an Australian marsupial."[1] The animal was also deduced from fragmentary remains. The Australian Museum did not have a complete specimen of *Diprotodon*, and the rival South Australian Museum in Adelaide had only just pieced one together from a variety of bones. Uncovering this beast illustrated the significance of Australian fossils for global paleontology.

Throughout the nineteenth century scholars stamped Australia as a land of the past, not from studies of Australian fossils but from the imposition of concepts of antiquity onto modern Australian nature and humanity. The continent's botany and geological formations, diverse reptile and bird faunas, and uncertainties over monotremes and marsupials (and what they represented in evolutionary history and taxonomy) were tied to earlier periods of earth's history. Aboriginal peoples were also relegated by Western scientists as "prehistoric," and were conventionalized (along with the Indigenous people of Tierra del Fuego and the San of southern Africa) as the "lowest," most primitive people. For Ernst Haeckel, in crossing the Lombok Strait—the biogeographic boundary between the "Oriental" and "Australian" zones—"we step suddenly from the present into the Mesozoic age."[2]

Australia was regarded not just as an ancient land. The primordial and ancient qualities ascribed to Australian animals, landscapes, and Indigenous people contrasted with how the Australian colonies asserted themselves as youthful and driving progress and achievement, often in manners at odds with local environmental conditions—a set of tensions discussed by Libby Robin.[3] Economic exploitation, with dreams of expanding agricultural production, mapping the landscape, and discovering resources, was a central drive in these settler colonial projects. In arid regions water was particularly important, and searches for rivers and springs were major preoccupations.[4] Further ideas—gained from geographic speculation and Aboriginal knowledge—that the interior might contain water sources usable for agriculture were constant hopes. These drives interacted in uneasy ways with debates on Australian identity, which consolidated in the years around the creation of the Australian Federation (despite persistently strong local differences), but which also sought to incorporate aspects from the Indigenous past, whether this be human, geological, or biological, in a process discussed by Denis Byrne.[5]

These tensions meant that, in the words of Alison Bashford, "in Australia's past, then, 'natural' history, 'prehistory,' and modern history merge and perhaps collide, almost necessarily," and "in starker ways than in other 'national' histories."[6] Australia being conceptualized simultaneously as an isolated land of primordial antiquity and a laboratory of colonial modernity meant that Australia's nature and human populations were connected with ideas of salvage in the face of change and extinction. Aboriginal Australians were described as "falling back" before the expansion of colonial settlement and agriculture.[7] The native mammals were similarly threatened, under competition and predation from imported animals like cats, camels, rabbits, and cattle. The superiority of northern forms was taken as a given and underlay debates around "salvage." However, salvage projects also frequently valued the preciousness of the native fauna and desired to see it "rejuvenated."[8] Australian nature was also given other values, not necessarily marked by strangeness and atavism, but as a unique landscape and fauna that needed to be understood and could be bound into national and local identities.[9]

Deep time and the present were often linked in Australia, but the study of Australian fossils was fitful. While the Wellington Caves generated interest in the early nineteenth century, the following decades saw relatively little work on Australian fossils. While scholars interpreted modern Australian wildlife through the prism of the prehistoric, Australian vertebrate fossils were actually rare. On a material level the geological "oldness" of the continent, with most formations dating to the Mesozoic or Paleozoic, meant there were few sites from the Tertiary. Australian fossil sites were also difficult to work, often being many miles inland and far from water. They were also only sporadically located, sometimes identified during surveying expeditions, but more usually found as a by-product of expanding economic sectors, like pastoral agriculture, mining, and tourism, part of the expansion of settler colonial systems in Australia. Once Australian fossil mammals were excavated, they revised understandings of both the deep past and the present of the continent. Giant ancient marsupials were revived, showing a history of abundance and diversity. When aligned with Aboriginal stories of the land and colonial drives to expand science, agriculture, and settlement, the deep past also pointed toward the future.

Building Australia's Museums

Australia became an increasingly important place for scientific work (and especially natural history) over the nineteenth century. The growing cities of the Australian colonies established scientific societies and museums drawing on foreign examples, while aiming to create distinct Australian scientific institutions. Collecting and ordering Australian nature was a critical aim. This was partly for local education, showing Australia's urban population the products and inhabitants of their country, though it was also important on an international level. Australian collections were in prime positions to engage with debates on development and distribution, for which Australian animals were crucial.

This mixture of opportunity, comparison, and locality can be seen in the development of Australia's fossil collections. Here, relations with British collections were especially important, as many of the prime Australian fossils (such as those from the Wellington Caves) were then held in British museums. Additionally, Australian collections initially aimed to display the same "classic" specimens as European museums, and so relied on connections with British collections to secure these. One of the first large acquisitions of Sydney's Australian Museum was a cast of the London *Megatherium*, which was prominently displayed in its galleries.[10] In 1883–1884 Edward Pierson Ramsay, curator of the Australian Museum, traveled to Britain for the International Fisheries Exhibition, but also took the opportunity to observe "many of the types of Australian fauna of the old naturalists, and the important series of Australian and New Zealand fossils."[11] In the course of this, Ramsay arranged a major acquisition of "casts of gigantic fossil remains from the British Museum, including *Elephas ganesa*, *Mastodon andium*, *Toxodon platensis*, *Sivatherium giganteum*, [and the Australian] *Megalania prisca*, &c.,"[12] securing reproductions of South American, Indian, and Australian fossils. There were also close links among personnel. Edward Charles Stirling, who directed the South Australian Museum between 1889 and 1914, was born in Adelaide but was primarily educated in the United Kingdom. Meanwhile Bernard Henry Woodward, curator of the Western Australian Museum, was from a long-running British geological family, being the nephew of Henry Woodward, keeper of geology at the British Museum of Natural History.

The distances involved, and the complex scholarly relations between Britain and Australia, meant that conflicts over interpretation were common. A heated controversy raged over one of the skulls from the Wellington Caves, which had received little attention in the initial surveys but became a focus of argument in the 1850s. Named *Thylacoleo carnifex* (pouched-lion executioner) by Richard Owen in a conjectural reconstruction based on the skull and assumptions of ancient Australian ecology, the animal was understood as a ferocious marsupial predator. However, this interpretation was disputed by many scholars, notably the German-born director of the Australian Museum, Gerard Krefft, who argued the creature may have been herbivorous or an analogue of a giant rodent. The *Thylacoleo* dispute has been widely studied, both as an example of Owen's expanding professional status and conjectural methods and as an indicator of relations between colony and metropole in the history of science (or indeed whether such distinctions are a productive analytical framework).[13]

While historiographic attention has focused on the *Thylacoleo* dispute, a great deal—if not more—contemporary attention focused on the larger bones from the Wellington Caves, equivalents of which were found over much of the Australian continent. The large limb bones, which were initially assumed to be as the remains of an unknown Australian pachyderm, were reinterpreted by Owen as belonging to a huge marsupial herbivore and given the name *Diprotodon*. The existence of such a great animal in a continent conceptually marked as dry and arid, and inhabited by a fauna labeled "puny" and "primitive" by British naturalists, was significant. Such an animal would have required huge amounts of food, which hinted that ancient Australia must have been verdant enough to support these creatures.

However, *Diprotodon* was problematic. While relatively complete skulls, limb bones, and vertebrae were assigned to the animal, the creature's feet were difficult to locate, meaning its mode of locomotion could not be discerned. A few reconstructions were attempted, most notably a skeletal one by Owen that left the animal's unknown feet covered by vegetation. A more dramatic full-body reconstruction was attempted by the Brisbane-based scholar Frederick McCoy as an enormous hairy animal "almost exactly intermediate in general anatomical and zoological character between the living native bear and wombat," but over sixteen feet long and towering over a reconstructed dingo. McCoy thought this great beast "tore down with its immensely powerful arms the large branches of the gum trees, on the twigs and leaves of which it could then feed at leisure," much like *Megatherium* and *Mylodon*.[14] Images of the reconstruction were widely circulated and served as the inspiration for a life-size wooden-board model at the Melbourne Zoo.

The Diprotodons of Callabonna

A particularly significant paleontological project linked Adelaide's South Australian Museum with expansion into the interior and a range of international networks. Excavations at the site, with the contested name of Lake Mulligan or Lake Callabonna, built a new image for *Diprotodon* and assertive Australian science. *Diprotodon* became a native animal symbolizing the ancient continent, deeply rooted in its local context and globally significant due to international interest in marsupials. These elements were linked, and scholars based in Adelaide used these excavations to bolster their institution and work. However, they also drew on tensions and conflicts around Australian science. The materiality of the fossils and environment were self-consciously presented as difficult, while Aboriginal terms and knowledges were simultaneously valorized and subordinated.

The site had initially been named Lake Mullachon, apparently drawn from the Adnyamathanha word *malakanha* (a type of string bag), which was then transformed to Mulligan, with an assumption that the name was Irish.[15] The lake was a depression and claypan around an ancient river system. The land was used for stock raising in the late nineteenth century, highly dependent on Aboriginal people for labor and under constant threat of drought. Both Aboriginal people and European settlers noticed bones in the region, although the latter assumed they were the remains of horses and cattle. However, Frederick B. Ragless, a local station owner, described how the bones were reevaluated through discussions with an Aboriginal stockman named Jackie Nolan. When Ragless told Nolan about elephants, he responded he had seen elephant-like bones, or at least those much larger than a horse, in the country around the estate.[16] The two visited the bones, which were indeed huge, and caused a stir among the stockmen in nearby Callabonna station. One of these, John Meldrum, took some bones to Adelaide, where they were identified at the South Australian Museum as belonging to *Diprotodon*, indicating there was material to exploit in that area.

The South Australian Museum had a significant collection, although somewhat secondary to Australia's largest collection at the Australian Museum in Sydney. Its director, Edward Charles Stirling, was keen to expand the collections, and the site at Lake Mulligan provided an opportunity to do so. The South Australian Museum organized excavations in the region in 1893, first sending a team under Henry Hurst of the Queensland Geological Survey, which was soon replaced by one under Amadeus Zietz, assistant director at the museum. Both projects were beset by difficulties, and Hurst was replaced following Stirling's accusations that he was wasting money and damaging fossils, being "completely incompetent to conduct a work of such magnitude."[17]

Stirling's reports in *Nature* on the excavations emphasized the difficulty of the work. To reach Lake Mulligan, workers from Adelaide had to travel at least five days, four hundred miles by rail, and then up to another two hundred miles by road.[18] The team had a train of five camels loaned by the South Australian government and supervised by "two Afghans." All water and supplies needed to be procured from Callabonna station, six miles from

FIGURE 12.1. Excavating *Diprotodon* at Lake Callabonna. AA309/5/2/17/5, Stirling Collection, South Australian Museum Archives.

the dig site. Behind the scenes, matters were trickier. While the Raglesses were thanked for "their kindness and hospitality" in official reports,[19] the reality was more acrimonious. Stirling and Zietz accused Ragless of charging "extortionate" rates for transportation, storage, and water. There also seems to have been a long-running dispute involving Stirling, Meldrum, and the Raglesses over a rumored (or possibly promised) £1000 reward for the first complete *Diprotodon* specimen (Nolan's contribution was largely ignored).[20]

The environmental difficulties were presented as even more severe. The South Australian landscape was crucial for engaging with the deep past, but was also described as a threatening hinderance to the work. Stirling highlighted numerous difficulties, partly in terms of his teams overcoming hardships, partly to explain the ambiguous results of the project (which secured large numbers of fragmentary fossils), and partly to illustrate how excavation work—resting on camels, prospecting, and camping—was obstructed by the environment. The condition of the fossils was unpredictable, sometimes extremely soft, sometimes hard and brittle, and generally "saturated with what was practically a concentrated saline solution."[21] Sometimes when they were taken out of the ground and dried, the salt crystallized within the fossils, and they burst and crumbled. The old lakebed was also prone to flooding. Stirling wrote that when visiting the site in August 1893, a half-inch of rain was enough to make the ground "so boggy that further work on the field became for a time impossible," and the camels sank into the mud, so "it was occasionally necessary to remove their loads and dig them out of the glue-like mud in which they had sunk nearly to their bellies."[22] Meanwhile there were dust storms in October and November, and "innumerable flies were . . . a constant and maddening source of annoyance to man and beast, and so tortured the camels that the margin of their eyelids became quite raw." At another point the field site was beset by hundreds of rabbits, which ate boxes, gnawed fossils and equipment, and then died after either drinking saline water from the pan or being poisoned by the station owners. Eventually, "under such circumstances of heat, sand and effluvia, it is not surprising that the health of the party suffered eventually from ophthalmia and gastro-intestinal complaints," the team returned to Adelaide.[23] An interview with the just-returned Zietz for the *Adelaide Observer* in December reinforced this, noting "his health had suffered somewhat in consequence of the exposure, while his eyes had been considerably weakened through the sand."[24]

While much of the Lake Mulligan fossil material deteriorated shortly after excavation, and more crumbled to dust during transport, a large amount nevertheless survived. A correspondent for the *South Australian Register* visited Zeitz's workshop and was astounded to see

> the largest and most valuable collection of "dry bones" in Australia. They may be described as fossils from a wilderness and a wilderness of fossils. Here, there, and everywhere are bones, teeth and miscellaneous fragments. Some are on benches, others on boards or shelvings, classified, or put together, or awaiting classification. Here are skulls which have been embedded in clay mixed with cement, as yet unpacked saved as regards the upper moulds. There are huge thighbones and the pelvis of the gigantic diprotodon wrapped in bandages smothered with glue to prevent fracture during transport. To describe all these in detail would be a matter of difficulty, for in their immensity they are bewildering: but they all tend to create a profound impression regarding

> the giants of centuries ago, when, as Milton says, "by heavy beasts the ground was trod."[25]

The huge amount of material showed deep antiquity and required technical skill and patience to reconstruct. Fossils were judged to include giant wombats and kangaroos, as well as a huge flightless bird assumed to be equivalent to *Dinornis* or moa and eventually worked into a new Australian genus named *Genyornis*. The highlight was the *Diprotodon* remains. Zietz assumed that—when all the material was transported back to Adelaide—there would be at least one complete skeleton, which could be mounted as a centerpiece, cast for exchanges with foreign collections, and create new understandings of Australia's ancient environment.

Foreign museums followed the work avidly. In 1902 Stirling reported on a tour he had made of major US and European collections, observing their displays and negotiating exchanges of specimens. The Brussels Museum of Natural Sciences offered a cast of *Iguanodon bernissartensis* for casts of Australian fossils, and in New York "Professor Osborn . . . was quite prepared for a deal with us and appeared moreover to be liberally inclined," offering to swap *Diplodocus* limb bones for "the corresponding parts of *Diprotodon australis*."[26] *Diprotodon* was important scientific currency, worth as much as a great dinosaur.

Attention led to a rebranding of the land itself, with Stirling working with the state government to maintain the site as a fossil reserve while renaming it Lake Callabonna. This aimed to efface the Irish-sounding Mulligan with an assumed Aboriginal name, following the principle that "the native names of localities should, as far as possible, be retained"—despite noting that Callabonna "might erroneously suggest the possession of the scenic beauties of an Italian lake by an area which is not only waterless, but also almost unsurpassable for barrenness and utter desolation."[27] The renaming shows definitions of Australian-ness shifting to highlight Indigenous terms, in the same way Jackie Nolan's contribution was simultaneously mentioned but also effaced. Of course, that Mulligan most likely originally derived from a word in an Aboriginal language demonstrates that the romanticization of this "native" past by colonial scholars was often based on misunderstandings.[28]

The accumulation of *Diprotodon* bones in Adelaide also reversed authority between London and South Australia. A long-running correspondence between Stirling and the British Museum of Natural History in London saw British scholars attempting to gain material and leverage promises from Stirling. In 1898 the museum's director, E. Ray Lankester, wrote, "I am very anxious to see the skeleton of *Diprodoton* erected in the Hall here," and offered to prepare and mount two skeletons if the fossils were sent to London, so one could stay in London and the other be returned to South Australia.[29] Lankester's offer also contained a threat, that he had thus far "refrained from addressing the Government at Adelaide through the Colonial Office."[30] Lankester eventually did write to Lord Tennyson, governor general of South Australia, that "a word from you—expressing the desire to see South Australia thus represented at home, would help to decide Dr Stirling to despatch the bones to us."[31] Stirling was persistently apologetic, but explained that the diverse and fragmentary remains, lack of complete skeletons of individual animals, and difficulties transporting the fossils made Lankester's proposition impossible, highlighting how one skull "opened on arrival in Adelaide it consisted of nothing but a hopeless mass of dust and minute fragments."[32] Henry Woodward proceeded to give instructions, talking about how "Prof. W.B. Scott's experiences in exhuming and preserving fragile bones of Mammalia &c. in Wyoming & Dakota would have been of great service to you," referring to techniques of wrapping fossils in bandages held together by flour paste.[33]

Despite the fragility of the fossils, the South Australian Museum made considerable progress assembling a complete *Diprotodon*, of sorts. Zietz's initial assumption that the boxes from Callabonna would contain at least one complete *Diprotodon* proved incorrect. However, there were enough bones of similar size from different individuals to construct a composite skeleton and a complete image of the animal. Yet it had to be made up in plaster, as the bones themselves were too fragile to be mounted. Stirling wrote to the museum governors that Zietz "has met with greater success than he expected for he has been able with much expenditure of time and labour, to restore a fairly complete series of vertebrae and ribs belonging to one animal of which we have nearly all the other parts."[34] This meant Adelaide could construct its own *Diprotodon* without any assistance from London.

Building the cast was difficult, requiring much modeling and restoration from Zietz and other museum workers. The composite *Diprotodon* was finally unveiled in 1906. Stirling wrote another report in *Nature*, featuring an image of the skeleton and a reconstruction of the animal by artist Charles Howard Angas, depicted with "considerable resemblance to a gigantic wombat" standing in the modern "white expanse of one of those large salt-encrusted clay pans of . . . Lake Callabonna, where the bones were found."[35] The creature remained ghostlike in the modern desert, rather than the more abundant imagined landscape of ancient Australia. At ten feet long, the Adelaide reconstruction was rather smaller than McCoy's, although Stirling was at pains to point out the animal had probably grown bigger, as larger bones had been found. Locating the animal's foot bones enabled deductions of its lifestyle. The toe bones were unexpectedly thin, so it could not have been a digger like a wombat, and most likely lived on open plains. It was discussed in the press as strange, yet decidedly Australian. The *Sydney Mail* called it "a heavy, uncouth animal, carried on stumpy legs with peculiar feet, and possessing an out-of-date-looking head, diminutive ears, and short tail." It also referred to the recent transfer of *Diplodocus* casts around the world, that "it is a pity we have not a few wealthy persons interested in science, for our extinct animals are at least as interesting as the Dinosaurus that made Mr Carnegie's heart rejoice; and series of reproductions of them should be visible in our museums."[36]

The reference to the Carnegie *Diplodocus* was not unfounded, as *Diprotodon* became important scientific currency. The cast increased the South Australian Museum's prestige and funds. Rather than be exchanged, casts were sold for £50 (a considerable profit from the £17 5s manufacturing cost).[37] *Diprotodon* casts went to London, Cambridge, Paris, Pittsburgh, Brussels, La Plata, and New York, and the South Australian Museum gained *Diplodocus longus* limbs from Pittsburgh, original *Megatherium* and glyptodon skeletons from La Plata, and a horse evolution display from the American Museum of Natural History.

The Adelaide *Diprotodon* was inducted into the canon of fossil mammals, and the South Australian Museum became one of the most significant collections in Australia, potentially outshining its rival in Sydney (and indeed, the sale of a *Diprotodon* cast to Sydney was presented as a major achievement). This mode of "reversing the tyranny of distance" through securing valuable local specimens for international exchange has been discussed by Ruth Barton in relation to Julius Haast at the Christchurch Museum in New Zealand, who exchanged moa bones to build up a large collection and reputation.[38] Stirling had Haast as a model, writing how in "Ethnology and Paleontology the South Australian Museum might be made to acquire, in reference to Australia, a reputation such as the Christchurch Museum

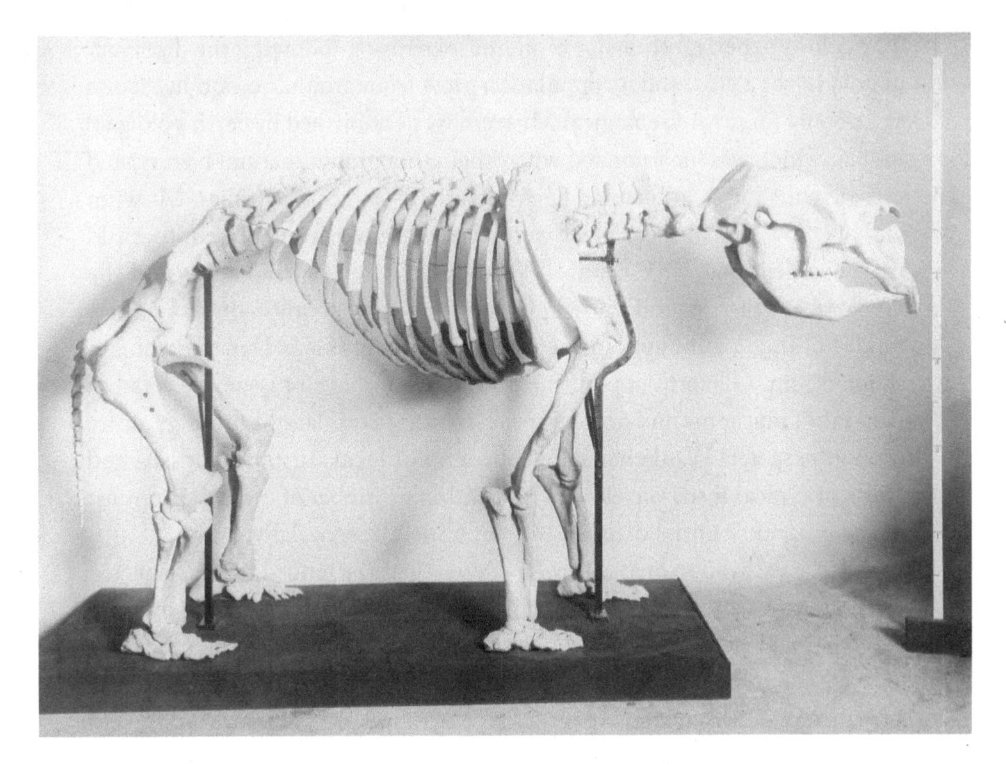

FIGURE 12.2. The *Diprotodon* cast. AA309/5/2/17/5, Stirling Collection, South Australian Museum Archives.

has acquired in relation to New Zealand."[39] However, unlike Haast, who supplied foreign museums with actual moa bones, the *Diprotodon* was a composite cast, if a technically accomplished one. Casts therefore continued to have a significant value within scientific networks, especially for rare specimens. Quite extreme artifice, building a complete animal from copies of fragmentary bones of many individuals, was not a barrier to authoritative knowledge. With increasing public displays and desires among collections to show the entire history of life, reproductions became even more significant.

The Mammoth Cave of Western Australia

Another paleontological project occurred on the other side of the continent, in Western Australia. The colony's capital Perth, over 1,300 miles from Adelaide, developed as an important municipal center and Indian Ocean port.

The colony experienced major economic expansion following the discovery of gold in the 1880s, and its population grew from around 30,000 in 1880 to 324,000 by 1914.[40] A Geological Museum was established in Perth's old gaol in 1891, which was incorporated with other ethnography, natural history, and artistic collections to form an agglomerated Western Australian Museum and Art Gallery in 1897. Staff were also recruited from Britain to build up the collections, with Bernard Woodward hired as head curator of the Geological Museum and Ludwig Glauert, a geologist originally from Sheffield, holding a joint appointment with the state Mines Department and the museum. Glauert's appointment raises an important issue: in Western Australia, museums and geology were a strong corollary of mining, a core economic sector. While having a collection of local Australian fossils and "casts of typical fossil vertebrates," including a number of mammals,[41] most of the museum's initial displays were of minerals, especially gold.

Yet the excavation of fossils in Western Australia tended to be connected to economic sectors other than mining. *Diprotodon* remains were found by geological surveying parties, although these were usually fragmentary and "so much weathered that it was valueless except as a record."[42] Pastoral agriculture tended to be more productive. Around 1908 Bernard Woodward corresponded with sheep farmers who found bones around the Balladonia station. Initially these had been ignored, and the *West Australian* quoted Woodward stating that they were "so common in that district some years ago, that the settlers considered them of too little value to be worth preserving."[43] Woodward nevertheless cultivated relationships with the stockmen. One farmer, John Sharp, wrote to Woodward in 1908 that a limestone deposit "contained long bones shoulder Blade, ribs, scull, teeth & they were sent to Melbourne about 20 years ago we was advised they were of no value as they were to be found in a good many places of that State."[44] He then promised to keep any new specimens for Western Australian institutions.

The Western Australian Museum fostered relations with another farmer, William Ponton, who sent teeth and bones. Bernard Woodward initially wanted to sponsor Ponton to collect material, but he found this difficult. Hiring men for excavations was expensive, and there was only a small window of opportunity during the wetter winter season to excavate fossils, which coincided with shearing time. As at Lake Callabonna, environmental issues posed problems. Ponton wrote that another stockholder had found fossil bones while excavating a dam, although "this was two year ago and unfortunately he left them exposed & his camels got at them." Ponton also could not visit the museum in Perth, as due to "want of rain and dingoes in the paddocks I have not been able to leave the station."[45] These exchanges, bringing farmers and prospectors into contact with elite scientists in municipal centers, promoted knowledge of the colony, but were also difficult to manage.

The museum sent Glauert to excavate, with guidance from Ponton. Glauert returned to Perth with fragmentary fossils of *Diprotodon*, wombats, some kangaroos (including a large specimen named *Macropus magister*), and what seemed to be *Thylacoleo* and a Tasmanian devil. This range of animals allowed Glauert to weigh in on the still live controversy over the habits of *Thylacoleo*. Gnaw marks and disarticulated specimens indicated the bones were "the refuse after a meal partaken of by the Marsupial Lion," which in Glauert's view fed "after the fashion adopted by the hyena, acting as a scavenger and feeding upon the dead and dying . . . forming its lair in caves, and dragging thither the more or less mutilated bodies of its victims."[46] The site was an Australian equivalent of William Buckland's hyena dens nearly a century before, which could show the behavior of ancient animals and supply their bones.

The most extensive site in Western Australia was connected with another expanding economic sector: tourism. About two hundred miles southwest of Perth, a series of limestone caverns with dramatic stalactite formations were explored by geographers and prospectors, and documented by Tim Connelly, the caretaker of the site. The Western Australian government set up a Cave Board to promote tourism to the region. By 1904 visitor numbers reached 2,463, there was a licensed cave house, and "a number of wild deer were let loose on the cave reserves in the South-West as an experiment," so that "as time goes on the presence of these animals will lend an added pleasure to visitors driving through the reserves on the way to the caves."[47]

As the caves were transformed, fossils were found. In 1904 Connelly made a cutting to improve the paths around the caves and unearthed some bones and teeth. These were sent to Perth, and Ernest Albert Le Souef, director of the recently established city zoo and trustee of the Western Australian Museum, visited the "Mammoth Cave" and encouraged Bernard Woodward to organize excavations. Glauert surveyed the cave in early 1909, continuing until 1915. He wrote of bones of large and small kangaroos, a great echidna three times the size of the largest living species, which was

FIGURE 12.3. Ludwig Glauert in 1914, with fossils from Mammoth Cave. State Library of Western Australia: Image #013891PD.

duly named *Zaglossus hacketti* (Hackett's giant echidna), "in honor of Sir J. Winthrop Hackett, KCMG, etc., the President of the Board of Trustees, as a slight acknowledgement of his generous support which alone rendered the exploration of these caves possible."[48] Glauert's "most thrilling experience" was finding what first seemed to be "the skeleton of a 'dingo' lying in a pool, and to discover upon careful examination that the luckless creature, that had perished so far from the light of day, was not a dingo at all, but a Tasmanian wolf." The thylacine, then rare and found only in Tasmania, was not yet known from Western Australia, leading to a sense of loss and antiquity. "How many years had those bones lain there undisturbed? It is impossible to give an answer."[49]

The press was also fascinated by the fossils from Mammoth Cave. The *West Australian* reported that "from these few scattered remains the scientist is able to resurrect the past and formulate their theories of immeasurable importance which the continent and its animal inhabitants have undergone during the flight of centuries."[50] Perth's *Sunday Times* noted that Glauert had collected over ten thousand bones from the "animal cemetery" by 1914.[51] Successive pieces in the *Western Mail* noted the greater size of the fossil animals compared to their modern counterparts, and one report showed a photograph of Glauert amid a huge number of bone-filled boxes and trays.[52] Another article described how Western Australia had its own distinctive fossil life:

> To those who have heard and read of the extinct monsters of the other continents—the mammoth and mastodon of Europe, the gigantic bird-reptiles and dinosaurs of North America, the great sloths and armadillos of South America, the sivatherium of Asia, and the huge moas of New Zealand, it will not come as a surprise to learn that Australia, too, has its "dragons of the prime."
>
> For nearly a century the huge diprotodon—a kangaroo with a head as large as an elephant's—was known to have existed in the southern continent ages ago, for traces of it have been found in all the states, including Western Australia.[53]

While garbling scholarly views of *Diprotodon*, the report marked Western Australia as important for the history of life, as its special biogeographic position provided local wonders that could stack up against the better-known fossil animals of the north.

Commentary on Western Australia's fossils also had a darker tinge. The naturalist José Guillermo Hay wrote a letter to the *West Australian* situating the finds within contemporary discourses of the "extinction" of Australian people and animals. Drawing attention to the thylacine fossils at Mammoth Cave, he wrote that the site must have dated to the period before Tasmania was cut off. Hay argued for the wide range of the Tasmanian devil and thylacine across Australia at this time, and that the ancestors of "the Tasmanian natives" were also present. He added that "this race (an inferior one) occupied the mainland before the severances by the sea of Van Diemen's Land, and before the advent of the present aborigines,"[54] and highlighted the common trope that the Indigenous Tasmanians were driven to extinction in the nineteenth century. The fossils were therefore used as evidence of "inferiority" and "extinction" of Indigenous fauna and people, who were linked. Hay continued that the "noble quest" to understand Australia's deep history should receive more funding, and "in Canada and the United States there would be individuals tumbling over one another to find funds for far less profitable results."[55] Local pride required the study of loss and extinction.

Transformations in the Landscape

The 1890s and 1900s increased fossil work across Australia, drawing on museums and the settler economy. The traditional center of much scientific work in the Australian colonies, Sydney, benefited but also slightly diminished in relative prominence. The Australian Museum displayed its cast of the South Australian *Diprotodon* in its Osteological Hall, but—not to be outdone—unveiled a new prize exhibit in 1912: "a very interesting restoration of *Palorchestes azael*, Owen, an extinct gigantic kangaroo."[56] This was another lost marsupial reconstructed in a dialogue between Australia and Britain. A fragmentary cranium sent to Owen by Ludwig Becker was imagined as a giant kangaroo, to be named *Palorchestes* (ancient dancer) in 1873. The Australian Museum followed this assessment and extrapolated a model "from computed measurements of the known bones of the above extinct form, compared with all the details of an 'Old Man' Kangaroo. It affords an idea of the size to which marsupials grew in prehistoric days."[57] The reconstructed animal was nine feet seven inches tall and covered in the skins of red kangaroos; it was exhibited in the entrance hall, opposite a lion diorama and beneath a sperm whale skeleton. It became a major attraction, and was eventually replaced with a further "more realistic restoration" in 1945.[58] However, this was destroyed a few years later, when further studies interpreted the animal as not kangaroo-like at all, but quadrupedal like a *Diprotodon*, potentially with a trunk like "a marsupial-tapir."[59]

Paleontology reevaluated Australian life. Attention focused not so much on the "strangeness" and "primitiveness" of Australian mammals, but on their diversity. Many Australian marsupials seemed to mirror northern types in their anatomy and lifestyles, indicating larger evolutionary trends. Scott wrote that while marsupials were "of more primitive structure and

▸ FIGURE 12.4. The *Palorchestes* model at the Australian Museum. "Kangaroos of Former Days," *Daily Telegraph* (Sydney), August 27, 1912. State Library of New South Wales: BN 215.

greater antiquity" than other mammals, "the order is one of very great variety in size, form, appearance, diet and habits, and mimics several of the higher orders in quite remarkable fashion."[60] In Brussels, Louis Dollo noted anatomical similarities among marsupials to argue that they all diverged from a single tree-dwelling ancestor, showing how evolution could lead to massive diversification from a single stem.[61] The Australian fauna also showed Osborn's principle of adaptive radiation, that animals would diverge from early germinal forms to fill a range of available lifestyles. This seemed to have happened in Australia, with the rhino-like *Diprotodon*, giant kangaroos, and large predators like the marsupial lion. Australia became as much a land of abundance and diversity as of "backwardness."

Yet the great extinct marsupials reinforced theories that modern Australian animals were doomed. What had caused the extinction of the larger marsupials was something of a mystery. Some scholars placed responsibility on the Aboriginal Australians. In the 1909 interview on the "Giant Wombat," the museum officer stated that "the diprotodon was probably exterminated by the blacks. Being a large and clumsy animal, he would fall an easy prey, and you know, a spear put through your inside is generally enough."[62] However, other causes were also invoked. Assumptions of marsupial inferiority meant that the most culturally laden of Australian placentals, the dingo, could have outcompeted the larger predators (this connected with controversies over whether dingoes had arrived with the Aboriginal Australians or earlier). Arthur Henry Lucas and W. H. Dudley Le Souef's *Animals of Australia* asserted that the dingo had more likely "reached the continent without the aid of man, and by Pliocene times," and while "at first he had to compete with the marsupial ground carnivores . . . these he soon ousted, endowed as he was with far superior strength and intelligence."[63]

A more widely posited cause was that climatic change had turned the land into desert. The old landscape of giant marsupials was imaginatively reconstructed as a wetland environment in several works. A 1929 article by Glauert stated that the geology of Australia and other southern continents (including Antarctica) showed this ancient change: "At one time, early in the 'Age of Mammals,' much of the south of our State was under a shallow sea. When the land rose again the climate was pleasant, and most of the country, if not all, enjoyed a bountiful rainfall which favoured the growth of rich and luxuriant vegetation." So "with such a wealth of food, it is not in the least surprising to discover that Australia had roaming over its surface a most wonderful variety of animals," including *Diprotodon*, giant kangaroo, and "many other quaint and huge beasts." Diminishing rainfall caused the decline of these animals and explained why their fossils were usually found in ancient lake beds: "During the rainy seasons the pools and watercourses were visited by the thirsty creatures; but when these dried up the animals had to turn to the lakes for water. Here thousands and thousands of mammals, birds and reptiles came to a miserable end in the treacherous boggy margins, where they were slowly but surely engulfed in the sticky mud."[64] The fossil sites not only contained bones, but evidence of mass death and shifting ancient climates.

The idea that Aboriginal Australians interacted with ancient marsupials was prevalent, going beyond assertions that their ancestors were responsible for the animals' extinction. Aboriginal Australians were crucial for identifying fossils and sites, as shown by the discovery of the Wellington Caves and Jackie Nolan's location of the Mulligan site. However, this direct involvement of Aboriginal Australians in excavation projects was often sidelined in official reports. Instead, more abstract valuations of Aboriginal knowledge were instrumentalized, with traditions cited as gateways into the deep past. In ethnographic collections of Aboriginal traditions, undertaken in those years through salvage projects, water creatures and cave spirits were often cast not as myths but as ancestral memories of interactions with extinct creatures. Aboriginal knowledge was tied to the past, and the expertise of Western scholars was required to decipher and extract material of scientific value. Aboriginal Australians were rendered as exotic bearers of "ancestral" or "racial" memory, connected with the deep past but not the present. Their value was tied to the idea that they were lost in the modern world.

Possibly the most elaborate combination of Aboriginal knowledge and deep-time scholarship was presented by the Scottish geologist John Walter Gregory. His 1906 book *Dead Heart of Australia* synthesized material to argue that the Australian interior could be returned to its ancient lush state. Gregory highlighted the large marsupial fossils and Aboriginal stories of the bunyip and *kadimakara*, opening the work with a poetic account linking geology and myth, and that "according to the traditions of some Australian aborigines, the deserts of Central Australia were once fertile, well-watered plains. Instead of the present brazen sky, the heavens were covered by a vault of clouds, so dense that it appeared solid." This meant "the rich soil of the country, watered by abundant rain, supported a luxuriant vegetation, which spread from the lake-shores and the river-banks far out across the plains," and "in this roof of vegetation dwelt the strange monsters known as the 'Kadimakara.'" These legends were a "shadowy reminiscence of the geographical conditions which existed in some distant ancestral home of the aborigines, or of those which prevailed in Central Australia, at some remote period."[65] Traditions and fossils built a new history for the land.

Much of *Dead Heart of Australia* followed Gregory's travels to the interior, reflecting on the landscape's past, present, and future. While the land was discussed in terms of conflict with the environment, Gregory also expressed optimism at the potentials of colonialism. He did not find any significant *Diprotodon* fossils, but noted a history of water around the depression of Lake Eyre, imagining this as once a huge inland sea; "unfortunately for the economic prosperity of Central Australia, this condition of affairs did not endure." When Lake Eyre was cut off from the ocean and "the lake shrank in area, less and less rain fell upon its shores . . . the giant marsupials died of hunger and thirst; hot winds swept across the dusty plains, and the once fertile basin of Lake Eyre was blasted into desert."[66] But despite this tormented history, the land held keys to its future. Gregory wrote, "Given but water, that country would be as fertile as a garden; and if it remained as free from mosquitoes, malaria, and flies as it was during our visit, it would be an Eden. . . . Various schemes for the improvement of Central Australia have been mooted; and the most daring of them is one, that proposes to secure for Central Australia a fairer share of the good gifts of the universe, by flooding the basin of Lake Eyre from the sea."[67] This was a gigantic undertaking, but geology and paleontology showed it was feasible. Yet restoration would not restore the giant marsupials, or their land to Aboriginal Australians. The future belonged to white settlement, crops, and cattle, not Aboriginal people and *Diprotodon*. They were part of a past that had set up the potentials of the land, but were superseded in this new era.

Gregory's instrumentalization of Aboriginal knowledge was not the only way the Australian past was made relevant to the present. In a report for the *West Australian* in February 1910, Daisy M. Bates, an ethnographer with interests in Aboriginal affairs, wrote that paleontology gave reasons to listen more closely to Aboriginal stories of the landscape and greater urgency for cultural "salvage." Australian cave sites were largely undisturbed owing to Aboriginal traditions of these being "the 'homes' of the giant or 'spirit' kangaroo" or "the abode of 'man-eating dogs' (probably the thylacoleo or thylacinus)," again suggesting that Indigenous spiritual beliefs recalled ancient fauna. While Bates believed that "only a small portion of what was apparently a rich mythology has been secured," this made salvage more imperative.[68] Bates sought to deploy a settler version of Aboriginal knowledge to understand the landscape and how humans lived alongside the great beasts. There was also an unsettling undertone: Aboriginal Australians were relics similar to *Diprotodon* and other lost animals. Their stories were fragmentary and broken and provided only a glimpse of their original culture. Australia was again marked as an ancient land, which underwent tremendous transformations, in the present and in deep time. Loss and potential were again simultaneously presented.

CHAPTER 13

Africa as the Source of Life

Elephants, Imperialism, and Internationalism in Egypt

IN HIS 1905 WORK *EXTINCT ANIMALS*, E. RAY LANKESTER, DIRECTOR OF THE BRITISH Museum of Natural History, described "the origin of the elephant's trunk." This was not the story told by Rudyard Kipling, whom Lankester called his "friend," in the recently published *Just So Stories*, where the young elephant's nose was seized and stretched by a crocodile, but reflected "slow change in long ages of time from other and more simple animals." Lankester described how the British scholar Charles William Andrews had recently brought fossils from the Fayum in Egypt, after a journey into "an absolute desert waterless region, establishing a staff of camels which daily brought up water as far as three days march into the sandy wilderness."[1] The fossils, dating from the Eocene, showed a "comparatively normal pig-like" beast named *Moeritherium*, which developed into a more elephantine creature, *Palaeomastodon*. When placed at the base of the lineage, they showed how "the wonderful elephant, with his upright face, his dependent trunk, and his huge spreading tusks, has been gradually, step by step, produced,"[2] in a similarly linear story to the evolution of the horse.

Lankester stated that this solved a major "mystery" within paleontological science—the evolution of the elephants. As already discussed, living elephants had a special cachet in popular and scientific culture throughout

FIGURE 13.1. The "old" African fauna. From Bölsche, *Tierwanderungen*, fig. 10. Author's collection.

the nineteenth century. They were also increasingly integrated into global and colonial economies, being used in southern Asia as beasts of burden, while elephant populations in Africa were destroyed for the ivory trade.[3] Elephants of the past were also important. Fossil proboscideans were central paleontological specimens, with the Elephant of Durfort, the Ilford Mammoth, *Dinotherium*, frozen mammoths of Siberia, and fossil elephants of Siwalik and the North American plains all being emblems of paleontological science. However, these specimens all dated from the Miocene on.

FIGURE 13.2. The "later migrating" African fauna. Bölsche, *Tierwanderungen*, fig. 18. Author's collection.

Even the earliest, like the *Dinotherium* and mastodons, were considered too "specialized" to be originary. Solving the mystery of elephant origins was therefore a crucial problem.

The Fayum beasts solved even more expansive biogeographic mysteries, "filling in the map" of ancient creatures and points of origin. In his 1897 *Geographical History of Mammals*, Lydekker noted that "nothing is known of the mammalian Tertiary paleontology of the Ethiopian region,"[4] arguing the whole of Africa had not been paleontologically studied. But by 1914 the German popular science writer Wilhelm Bölsche organized African mammals into two groups: an early "indigenous" fauna of elephants, hyraxes, and manatees; and newer migrants from Asia, including giraffes, antelopes, big cats, rhinos, and equids.[5] This was not a migration that saw the pouring of dynamic northern forms into the ancient "Dark Continent." The southern forms were important and powerful animals, and Africa was a generative region in itself. This characterization of African animals depended on fossils from the Fayum, with one Egyptian region standing in for the whole continent.

Colonial tropes and biogeographic theorizing overlay more practical issues. While Lankester presented solving the mysteries of elephant evolution as the work of a single British scholar, the research rested on Egyptian, colonial, and international networks, which played out in fossil work in Egypt in the 1900s as European and American scholars attempted to gain access and specimens. Paleontology in Egypt was framed as solving grand "mysteries" around the history of life. However, the conduct of work and gathering of material was messy, opportunistic, and not always successful. Working with local collaborators and external patrons was crucial, and the expansion of paleontology in Egypt showed tensions in the field, as much as grand results.

British Colonialism and Fossil Work in the Fayum

For much of the nineteenth century Egypt was ruled by a series of modernizing khedives such as Mehmet Ali and Ismail, who promoted economic development and state reform, often connected with European commerce and scientific activity. The construction of the Suez Canal between 1859 and 1869 in partnership with the French, and the expansion of cotton plantations were important instances. Another development was the establishment of scientific institutions and surveying. In 1873 Ismail patronized an expedition to the Western Desert under the German scholar Gerhard Rohlfs. The expedition consisted of a group of German scholars (including Karl von Zittel), over a hundred Egyptian participants, and a large camel train—although it was largely unsuccessful, with the team almost dying of thirst after running out of water. Another participant in the Rohlfs expedition, Georg August von Schweinfurth, later collected whale fossils in the Fayum basin and in 1875 became the first head of the "Khedivial Society of Geography," modeled on the geographic societies of London and Paris and aiming to expand Egyptian power and knowledge in the Red Sea, Sudan, and Western Desert.[6] The society consisted primarily of Europeans (and its working language was French) and shows the partnership of Egyptians and Western scholars in this earlier period.

In 1882, Egypt was occupied by Britain, partly due to the khedive's default on debts owed to Britain and France and geopolitical desires for control over the Suez Canal. After this Egypt was in a complex position, still technically under Ottoman suzerainty but in practice controlled by British colonial authorities, with most local government conducted by Egyptians—with significant involvement of French, German, Italian, American, and other external powers. The British authorities devoted much attention to knowing the country. The Egyptian Geological Survey was established in 1896, headed by Henry George Lyons, a military officer with personal interests in geology, fossils, and antiquities, who later became a prominent figure in the Royal Society and director of the London Science Museum (demonstrating the links between geology, the military, colonialism, and metropolitan museums). Lyons organized the Geological Survey in line with British military and economic interests. Its main tasks were surveying for resources, and organizing cadastral and topographic surveys for land and agricultural taxation.[7] These aimed to chart a country which was socially and politically unfamiliar to its new rulers, and also shifting in geography, as the movements of the Nile regularly altered the landscape. Typifying colonial attitudes and orientalizing stereotypes, the London *Times* reported on Lyons's work: "By hard and steady work order is gradually being evolved out of something nearly resembling chaos."[8] Yet this control was more aspirational than real.

Western scholarship in Egypt also followed interest around the Pharaonic past. Archaeological excavations and the transportation of ancient Egyptian artifacts had been major Western scholarly preoccupations since the late eighteenth century.[9] Excavations and smaller-scale artifact trading were important parts of the Egyptian scholarly and commercial landscape. As Donald Reid has demonstrated, Egyptology in this period was a contested field, being reappraised and used by scholars from Egypt and European imperial powers.[10] Egyptologists sought to excavate an ancient past which had resonances to biblical and classical antiquity, yet also drew on new techniques in excavation, labor management, and prospecting.

Colonial "modernization" and excavations of the Egyptian past combined in the Fayum in the 1900s. Like most paleontological projects, this grew from a mixture of economic expansion and chance encounters. The Fayum was a basin sixty miles southwest of Cairo around a small freshwater lake, the Birket-el-Qurun, and was the remnant of a much larger body of water, known as Lake Moeris in Ptolemaic times. The southeast of the Fayum was suitable for agriculture, but the rest of the basin was largely desertified. The Fayum had been surveyed by archaeologists and was known for the naturalist "Fayum portraits" from grave sites. During the British period the area was marked for economic expansion, with a railway to Cairo built in the late 1890s to transport cotton and sugar. The Fayum was a priority for surveying by the Geological Survey of Egypt, which tasked Hugh Beadnell with the project.[11] Beadnell's works, resting on Egyptian observation and labor, emphasized supplying water to increase agricultural production, and connected the region's stratigraphy with the Egyptian past, highlighting the numerous tree and whale fossils and the extent of settlement in ancient times.

Paleontological work in the Fayum developed through a shaky alliance with the British Museum of Natural History. In the 1890's Charles William Andrews, an assistant at the museum, developed a health condition (most likely tuberculosis) and followed medical advice to spend the winter in a hot arid climate. After traveling to Egypt by the British Museum's arrangement, Andrews accompanied Beadnell on his surveys, primarily to collect desert mammals and birds for the museum's zoological collections. However, Andrews became increasingly drawn to the fossils around the oasis. Beadnell's stratigraphic analysis dated these to the Eocene, the originary period of the Age of Mammals—particularly significant from a biogeographic perspective, as Eocene fossils from Africa were almost unknown. The coalescence of expertise meant the survey party, and their Egyptian laborers, collected fossils rather than trapped desert creatures.

Andrews wrote several reports for the *Geological Magazine* in 1901, describing fossils of turtles, crocodiles, and a "considerabl[e] quantity" of fish."[12] Most attention focused on large mammal fossils. "One of the most important specimens" was an elephant-like jawbone, named *Palaeomastodon beadnelli*. There was also "a great quantity of remains of an ungulate about the size of a large tapir,"[13] named *Moeritherium lyonsi* after Lake Moeris and honoring the expedition's sponsor at the Geological Survey. Further reports described large hyrax-like fossils, and a hastily published note by Beadnell (under the auspices of the Egyptian Public Works Ministry) described the skeleton of a "a large, heavily built, ungulate, about the size of a rhinoceros,"

to be named *Arsinoitherium zitteli*—the genus after the Ptolemaic Queen Arsinoë II (after whom the Fayum was named in Ptolemaic times), and the species "in honour of the eminent geologist, who may be regarded as the pioneer of geology in Egypt."[14] Field photographs showed the animal's enormous horns while presenting its desert context.

The collaboration between Beadnell, trained in geology and colonial engineering, and Andrews, a natural historian, mirrored that between Hugh Falconer and Proby Cautley nearly seventy years before. Excavation in Egypt also depended on colonized people for labor, utilizing teams of Egyptian workers. However, there were important differences. While Falconer and Cautley engaged with Indian religion, Andrews and Beadnell conceptualized a subordinated antiquity. The fossils and region were naturalized as decidedly "Egyptian," but this was the Egypt of Ptolemaic and Pharaonic times, rather than the modern country. The current inhabitants were largely effaced (even ignored in expedition photographs). Any references to modern Egypt used antiquity to inspire colonial development. Beadnell wrote how Lake Moeris had been "used as a regulator of excessively high and low Nile floods," but, according to the archaeologist Flinders Petrie, had declined "under the Persians or Ptolemies," and "continually and gradually sunk to its modern dimensions."[15] The implications were that Lake Moeris could potentially fulfill this purpose again, if restored by an enlightened colonial administration. The landscape was understood in terms of decline from earlier, grander periods.

Andrews's private writings showed aggrieved frustration at almost every aspect of the project. His correspondence with the British Museum of Natural History described difficulties of the environment, excavations, and interaction with Egyptian workers and the Geological Survey. Beadnell and Andrews worked separately in areas marked off for them (which Andrews suspected reserved the best localities for the Geological Survey). Work on fossils also divided the British Museum of Natural History and the Egyptian Geological Survey. Fossils were sent to London to be prepared and described, but then returned to Egypt. This seems based on Egyptological practice, where the partage system gave Egyptian institutions the first choice of artifacts (even if this was often circumvented or ignored by

◂ FIGURE 13.3. Skull of *Arsinoitherium*, depicted in the desert. Beadnell, *Preliminary Note on* Arsinoitherium zitteli, plate 6. © The British Library Board: 7208.ddd.

well-resourced Western collectors). London therefore made do with casts and lower-quality duplicates, rather than the most complete fossils and type specimens. These arrangements could lead to conflict. In 1903 Andrews wrote that he had received "a furious letter from Lyons," who heard that Arthur Smith Woodward, keeper of the Department of Geology, had referred to the fossils sent to the British Museum of Natural History as "rubbish." Lyons "says he will take care the BM is humbled with no more rubbish which means I take it that no more duplicates or anything will come to us, at least until he changes his mind."[16] While the matter was smoothed over, relations remained tense.

The Fayum material spurred the Egyptian Geological Survey to develop its own museum in Cairo to store and display the fossils (and material from other sites). The British Museum of Natural History sent one of its preparators, Frank Barlow, son of the mason formatore Caleb Barlow, to assist the initial arrangement of the museum, build cabinets and cases, and accompany Andrews on collecting trips. Barlow reported on sorting the fossils, making new ironwork for mounting, and "secur[ing] a young man from the Technical Schools with a view to teaching him as much as possible of my work. He is an Egyptian but speaks English well and has been through the course of five years instruction in these splendidly equipped schools."[17] The Cairo museum became an important conduit for knowledge of fossil work, with practical skills coming from London, but with material maintained in the country of origin and with Egyptians trained in aspects of the work (unfortunately the identity of the Egyptian preparator is impossible to trace). The Geological Museum was eventually opened in 1904 in a grand neoclassical building designed by Marcel Dourgnon, with "the place of honour" in the paleontological gallery "occupied by the head of the huge herbivore *Arsinoitherium zitteli*,"[18] and with *Palaeomastodon* and *Moeritherium* remains in wall cases. These were unique native fossils, showing the deep past of the territory—and the importance of the Geological Survey (and British colonialism in Egypt).

The fossils created a stir in the British Museum of Natural History, where Andrews worked on a monograph on the Fayum vertebrates. The excavations were presented as uncovering several unknown forms. "By far the most striking" was the *Arsinoitherium*, described as "somewhat like a large and heavily built rhinoceros." Enough remains of the animal were found to reconstruct a complete composite skeleton. The *Arsinoitherium* was deduced as representing a now entirely extinct lineage—much like the North American Dinocerata. Lankester, in an article in *The Sphere*, attempted to understand its lifestyle, writing how "the use of horns is in almost all cases either for defence against animals of prey or for those fierce contests between the males of a species in which the victor becomes lord of a number of females. Possibly both purposes were served by the horns of Arsinöitherium."[19] The beast had disappeared, but could be understood through modern analogues.

While *Arsinoitherium* was a centerpiece, the most important specimens were connected with elephants. Andrews wrote that "the primitive members of the Proboscidea are perhaps of greater scientific interest, because they help to fill, at least to a large extent, one of the most important gaps in our knowledge of the extinct Mammalia." Andrews's monograph was generally dry and technical, but contained flashes where the Fayum fossils were transformed into animals interacting with a changing environment. *Moeritherium* was "about the size of a Tapir, which it must have greatly resembled in general appearance," and from its teeth and humerus was "probably an amphibious, shore, or swamp living animal." The later *Palaeomastodon* was suited to a drying climate, "more adapted to terrestrial life," and "must have resembled a small rather long-necked Elephant."[20] Andrews's text followed conventions in paleontology, mixing technical comparison of fossils and anatomical structures with creative inferences. His publications also featured physical reconstructions of the major Fayum animals by Alice Woodward. These imagined the *Moeritherium*, *Palaeomastodon*, and *Arsinoitherium* following the habits described by Andrews and modeled (albeit in a stylized, art nouveau manner) on modern animal analogues.

The material from the Fayum was classified in multiple ways as the fossils were rendered into scientific objects and used to reconstruct the animals as they may have appeared in life. Andrews was tasked with rearranging the Proboscidean fossils at the British Museum of Natural History, producing a new *Guide to the Elephants*, bringing the Fayum remains into dialogue with the elephant specimens acquired over the preceding century. Readers were told "the British Museum possesses the most extensive series of Proboscidean remains to be found anywhere," and visitors could take the book "and can to some extent check the accuracy of the various statements for themselves."[21] The guide emphasized the spectacular nature of elephants, and the newly "complete" knowledge of their evolution. Elephants

FIGURE 13.4. Reconstruction of *Moeritherium* by Alice Woodward. Andrews, *Guide to the Elephants*, fig. 8. Author's collection.

▸ FIGURE 13.5. The serial evolution of the elephants. Andrews, *Guide to the Elephants*, fig. 5. Author's collection.

were "perhaps the most remarkable" of living mammals, due to their large size and a trunk "which is at once a sensitive organ of touch and a most efficient means of grasping objects." Additionally, their teeth "[reach] a degree of complication not to be found in any other animals."[22] Elephant evolution from the tapir-like *Moeritherium* to the modern elephant showed the regular development of these characteristics. The animal's body size and brain volume increased, and the skull altered to accommodate the proboscis. The teeth also transformed: the incisors elongated into tusks, the molars acquired complex grinding ridges, and the rest atrophied and eventually disappeared. The elephant became another example of progressive evolution on a linear track.

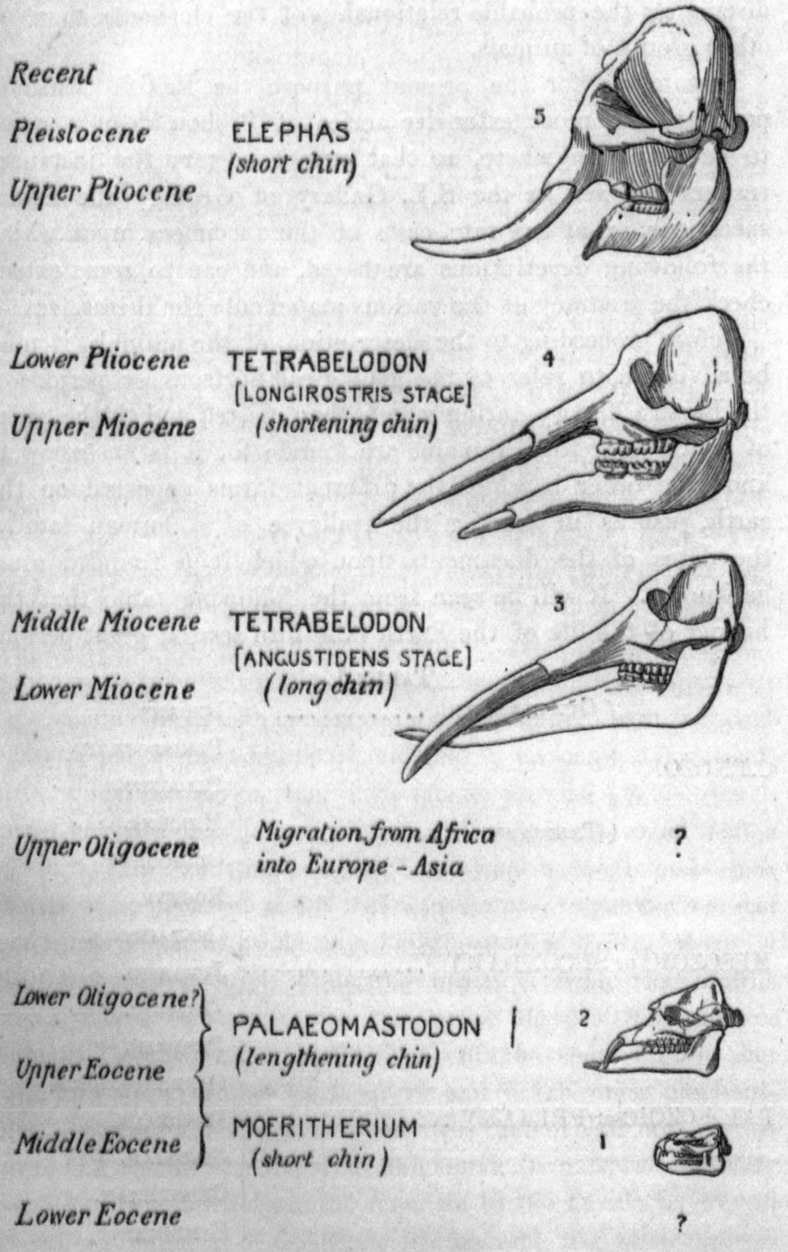

Recent

Pleistocene — ELEPHAS (*short chin*) — 5

Upper Pliocene

Lower Pliocene — TETRABELODON [LONGIROSTRIS STAGE] (*shortening chin*) — 4

Upper Miocene

Middle Miocene — TETRABELODON [ANGUSTIDENS STAGE] (*long chin*) — 3

Lower Miocene

Upper Oligocene — *Migration from Africa into Europe - Asia* — ?

Lower Oligocene? }
Upper Eocene } PALAEOMASTODON (*lengthening chin*) — 2

Middle Eocene } MOERITHERIUM (*short chin*) — 1

Lower Eocene — ?

Diagram showing some stages in the gradual increase in size, and alteration in form of the skull and mandible, occurring in the Proboscidea from the Eocene to the present day.

Fraught Opportunism in German and French Involvement in Egyptian Paleontology

Britain was the dominant colonial power in Egypt, but other countries had interests there too. This international rivalry was played up in Andrews's reports. During the dispute with Lyons, he wrote to Arthur Smith Woodward that he hoped the matter would be resolved soon "before Schlosser or some other German gets hold of the things."[23] These comments had some basis in fact, as German scholars had long-standing links in Egypt and worked informally through German-speaking expatriates. A particularly important go-between was Richard Markgraf,[24] a German speaker from northern Bohemia who initially worked as a bricklayer before joining a touring musical troupe. However, his health deteriorated while the troupe was in Egypt, and—following an idea similar to Andrews's that dry arid climates had restorative properties—Markgraf stayed in the country and spent a great deal of time in the desert. He developed a sideline as an antiquities dealer and returned to Cairo for additional income as a hotel piano player.

Markgraf worked with southern German institutions, allowing them to work through informal connections. German paleontologists recognized Markgraf's important (although idiosyncratic) role. In an obituary, Ernst Stromer von Reichenbach of the Munich collections wrote that "Markgraf was only a poor fossil collector and small-time trader in natural history, but he still deserves to be honored in a scientific journal."[25] His entry into paleontology occurred through quite serendipitous circumstances. In 1897 Eberhard Fraas of the Stuttgart Naturalienkabinett visited Egypt almost by accident. He intended to visit the formations in German East Africa where large bones had been reported (which were to eventually be exploited at Tendaguru), but the outbreak of the Maji Maji uprising made visiting the region impossible. Stuck in Cairo, he was put in touch with Markgraf by the Württemberg Consul and went with him into the Western Desert to prospect for fossils. Stromer wrote how Fraas, "who knew how to attract lay-people to collect fossils," taught Markgraf excavation and preparation techniques.[26] The extent of Markgraf's later fossil collecting implies that he developed considerable expertise. Markgraf also became a pointman for negotiating and dealing with Egyptian workers and for organizing camels and provisions for travel into the desert—something which most foreign scholars seem to have been unable to manage (from their stated annoyance at having to deal with Egyptian people).

Fraas maintained a close relationship with Markgraf, paying him to collect fossils for Stuttgart. In 1906 he wrote to his superiors that "the Royal Naturalienkabinett has received some outstanding paleontological finds from the Tertiary strata of Egypt in the last few years, and it can probably be said that in some cases these surpass everything that has up until now been found in European museums."[27] Later that year Fraas went on a second trip to Egypt, traveling with Markgraf to the Western Desert. While spending only eighteen days collecting fossils, the expedition returned with significant material. The visit also enabled Fraas to appeal to public audiences. He wrote an account of his trip to "Egypt—the ancient wonderland of Herodotus" for the popular-science journal *Kosmos*. Rather than express annoyance and frustration, as was often the case for paleontologists in Egypt, Fraas described the wonders of the human and animal past and the romance of the desert. He took the reader back in time, although he avoided the country's Islamic history by beginning with the Pharaonic period, moving backward through the "predynastic" period elaborated by scholars like Petrie, before reaching eras that "in Europe would be called the Diluvial or Ice Age, although this expression does not match Egypt, since we have no glaciations there, but at most an increased amount of rain and more moderate climate." Finally Fraas reached the early Tertiary, "long before the first appearance of humans and where we recognize the first development of the mammals." He then imagined *Arsinoitherium* as "a herbivore the size and probably habits of the hippopotamus, which assembled in the swampy lowlands of the primordial Nile," living with the "primordial elephants [*Urelefanten*]," *Palaeomastodon*, and *Moeritherium*.[28]

As well as reflecting on ancient beasts, the desert provided wonder and excitement, although the expedition required considerable preparation, as "a journey in the Libyan desert is quite a different affair to one in our Swabian Alps." Fraas praised his "loyal collector and comrade Markgraf," describing him as "a simple, modest man, who has roamed the desert regions for years with tireless zeal," and "traveling with him was a pleasure and joy, because I was relieved of all worries about the route, the camels, the Bedouins, and—above all—food and thirst—and so was able to devote myself entirely to geological studies."[29] The ostentatiously bourgeois Fraas's reliance on his assistant was continued through the main hardship he mentioned, that their water was carried in petroleum cans (which could also be used as boxes for packing fossils), so his soup and morning coffee always tasted of oil.[30] These difficulties however faded with the majesty of the location: "The first night camp in the desert! What poetry lies in pitching one's tent in the infinite field of gravel and sand, far from any human place."[31] The work of the fossil hunter, if carefully supported by go-betweens and Egyptian workers, was a romantic endeavor, allowing the scholar to commune with the landscape.

Other projects emanated from Munich, where Ernst Stromer von Reichenbach developed numerous researches in North Africa over the 1900s, also working with Markgraf. Stromer's excavations are primarily remembered for uncovering the dinosaur *Spinosaurus*.[32] Yet most of his work retraced the sites of other German scholars in Egypt, searching for fossil whales. He published accounts in German periodicals and delivered talks with "numerous magic lantern-slides," which "were received with great applause."[33] In his reports and field photographs, Stromer presented himself as almost the archetypal colonial explorer, retrieving treasures for German science from the desert.

Unlike Fraas's romantic account, Stromer was more matter-of-fact, describing fossils and geology in great detail—and complaining extensively about dealings with Egyptians. On one expedition he wrote, "Our people, contrary to their terms of the contract, had not taken enough camel food with them." Buying additional supplies from fishermen was impossible "due to their too high demands," and so "our camels had to content themselves with tamarisks and reeds growing at the lake."[34] In another report from 1904, he complained about the "wind and weather" and "the spirited but unreliable and greedy natives,"[35] again disparaging the local people. Such complaints were common in European accounts, and are often the only times Egyptians appear in these sources at all. However, reading between the lines (and connecting with approaches to the history of Egyptology developed by Christina Riggs, Alice Stevenson, and Stephen Quirke, which highlight the essential work of Egyptians in archaeological excavations),[36] we can see several things. Western travelers like Stromer were completely dependent on Egyptians for extracting and transporting fossils and purchasing supplies. The frustration and annoyance expressed by Stromer, Andrews, and others demonstrate that behind the rhetoric of colonial mastery (and exploitation of Egyptian workers in many instances), Egyptians could protect their interests and take advantage of the fossil hunters.

Max Schlosser, one of the paleontological specialists in Munich, systematized their Egyptian fossils in a 1911 monograph. While Andrews's monograph had focused on large and robust mammals, Schlosser focused on smaller animals, whose fossils were numerous but barely mentioned by earlier researchers. Particularly noteworthy were the hyraxes, which currently existed only as a "poorly-shaped and inconspicuous type of mammal," but had developed in "an astonishing wealth of forms as soon as they appear" in the Eocene, "some of which were quite large.[37] This was not simply interesting for the history of this little-studied group, but for reconstructing the ancient landscape and fauna. The diverse early hyraxes had relatively simple teeth suitable for aquatic vegetation, but no obvious anatomical adaptations for swimming. This meant "these hyracoids inhabited swampy forests, where they did not need any noteworthy locomotive ability, because there was abundant food for them." In a paleontological morality tale, Schlosser described how they were placed under great pressure as the climate dried, exacerbated by the migration of new predators from Asia like *Machairodus* and early rhinoceroses like *Aceratherium*, "which like all rhinocerotids, must have been a belligerent fellow. A single charge of such an animal could well have been enough to throw a defenceless hyracoid to the ground with broken limbs, where it would then perish hopelessly."[38] Climatic and biogeographic change created a narrative of decline and fall for the hyraxes. Nature was again defined by migration and invasion.

While German scholars developed several projects in Egypt, French work was more fitful—something potentially surprising, given the long history of French interest in Egypt and France's colonial position throughout Africa. While French influence in Egypt declined somewhat with the establishment of British colonial authority, it nevertheless persisted. The Egyptian Antiquities Service remained dominated by French archaeologists, and its head, Gaston Maspero, reserved important archaeological sites for French teams while coming to a modus vivendi with British excavators and colonial authorities.[39] The Egyptian Museum, inaugurated in 1902, was designed in a Beaux-Arts neoclassical style by a French architect, and had more plaques honoring French Egyptologists than any other nationality (followed by Britons, Germans and Italians).[40] And the French language maintained a high level of prestige within Egyptian society.

Andrews's researches were discussed in the French scientific press and in the Muséum d'Histoire Naturelle. One scholar, Claude Gaillard, wrote that in the Geological Museum in Cairo he saw "beautiful jaws of *Palaeomastodon*," and "several superb skulls of *Arsinoitherium*, *Moeritherium*, etc.," and "I could not hide my annoyance at the thought that all these precious documents were taking the road to London."[41] The prominence of the Fayum fossils in international circles and French aims to maintain the Paris Muséum as the world's key institution for the study of fossils drove desires to secure similar "precious documents" for France.

Paris scholars attempted to work with a local collaborator, René Fourtau, a French railway engineer employed by the Egyptian government who had previously sent them echinoderms and other fossils. Fourtau reported British and German excavation projects to Marcellin Boule, attempting to gain sponsorship for similar work. He wrote that "the Geological Survey of Egypt has placed guardians everywhere and excavations are prohibited"[42]—to the extent that Andrews himself was forbidden from excavating in the Fayum in 1903. Boule in turn wrote to the Egyptian Ministry of Public Works that "the English paleontologists" had found interesting fossils, and it "would be very important that the Gallery of Paleontology in the Muséum of Paris, which is aiming to present the general tableau of evolution of the animated world in the course of geological ages, might have some representatives of those curious creatures which have been received from the soil of Egypt."[43] This seems to have been successful. Gaston Maspero responded to Boule that "Captain Lyons made no objection to French work in the Fayum,"[44] and met with Fourtau directly to discuss the project, promising "to delay Stromer's final authorization"[45] to collect in the region, thereby giving the French party a head start. Long-running French influence in Egyptology played in the Muséum's favor.

However, difficulties soon arose. Boule attempted to gain funds from Edmond de Rothschild, one of the Muséum's major patrons, but there was doubt about Fourtau's ability. It was decided that the project either needed to be entrusted to a scientist from the Muséum, or Fourtau would need to come to Paris, take Boule's courses in vertebrate paleontology, and then begin work in Egypt. On hearing this, Fourtau wrote a long letter on how the Muséum "must not delay," as rivals were securing material: "Mr. Beadnell is working to deplete the deposit as much as he can," Stromer was working in the region, and other French museums were asking Fourtau if he would collect for them. This meant that "if M. de Rothschild does not make up his mind . . . I will accept the offers of the city of Lyon," and "there will not be much left for the Muséum."[46] This convinced Boule and Rothschild to forward 3,000 francs to Fourtau, with a telegram instructing him to "start excavations immediately."[47]

In total, Fourtau was in the Fayum for approximately two weeks. He claimed he was delayed for ten days while waiting for water containers (for which he again blamed "Arab workers").[48] However, he also "diverted a Bedouin who has served Beadnell for two years, and led me directly to the right places for 1fr25 per day."[49] Pleased with his work, Fourtau sent a box of fossils to the Muséum, describing the "inexhaustible" deposits that were likely to contain *Arsinoitherium* remains, if money kept coming for excavations. This limited excursion however was not what Boule had hoped for, and he reacted angrily, initially going two months without sending any response (which inspired a worried letter from Fourtau on whether the fossils had even arrived).[50] Eventually, Boule wrote that he initially thought Fourtau's letter was "a mistake," and did not expect "your first campaign would be so short, after having spoken about it with such urgency."[51] Fourtau wrote two missives, explaining that the region was difficult to work in during the summer due to malaria, which meant "the Bedouins abandon this country with their herds for other pastures."[52] He had plans for more research, but required more funds due to the difficulty of the terrain: "I will just repeat one more time, that it needs a lot of money. Here we are not in Sansan or Pikermi, but in the desert in the middle of sand dunes and three and a half days from any inhabited place."[53]

Following this, Boule simply demanded the unspent funds be returned. It is unclear from the archival records whether this happened, but Fourtau's association with the Muséum ended (although he did later work with the Geological Museum in Cairo, studying both fossil echinoderms and mammals). Boule then went through official channels to gain Fayum material. Andrews visited Paris in 1904, informing Boule that "the Museum in Cairo has more *Arsinoitherium* skulls."[54] Boule secured some material from Lyons, including partial skulls of *Arsinoitherium* and *Palaeomastodon*, which arrived at the Muséum in 1907.[55] However, these were not the fine, display-quality fossils hoped for and were relegated to the storerooms, not being studied until 2008.[56]

The contrasting fates of the German and French projects in the Fayum indicate the importance of informal connections within paleontology. German scholars were able to secure large collections of fossils through commercial links and Markgraf's abilities negotiating with Egyptians. Personal connections were crucial, rather than the more formal colonial programs of the British scholars. The failed French case—with a relatively inexperienced go-between in the person of Fourtau—saw collaborations break down despite the intervention of influential supporters like Maspero. Fourtau's case also shows the difficulties of managing work at a distance. Whether it failed because of a misunderstanding on the project's scope, or whether Fourtau took advantage of the Muséum's patronage, working through intermediaries was difficult, and there could be many breakages along the way.

Money, Prophecy, and Biogeography in the AMNH Fayum Expeditions

The Fayum fossils also attracted attention in North America, especially in the American Museum of Natural History. The AMNH had grown into a dominant US institution and expanded internationally, taking advantage of American commercial and diplomatic power (in much the same way as US archaeological and art institutions also expanded during this period). Osborn and the scientists at the AMNH followed Egyptian paleontological work closely, corresponding with researchers in London and Germany. Osborn wrote to Fraas, "I envy you this most interesting trip and have in mind a similar trip of my own some of these days; perhaps we can go together."[57] Fraas however was occupied with work in the dinosaur fields at Tendaguru (and becoming increasingly ill), but still advised Osborn to visit Fayum and recommended Markgraf as a field-worker. Osborn also wrote to Andrews, praising his monograph: "Now that you have published this volume I feel free to visit the Fayum . . . I would not have thought of collecting there while I knew you were working on the scientific description of the materials, but now that your memoir is complete I feel entirely free to do so."[58] Osborn desired to acquire fossils within a new paleontological region. However, there was also the recognition of priority—and not repeating the conflicts and duplication of the Cope and Marsh period. This strategy seems to have been successful, with Andrews recommending sites and workmen, although noting that the most productive localities were exhausted.

The rather grandly titled Fayum Expeditions of the American Museum of Natural History were organized through a subvention of $3,471 from Morris K. Jessup, the museum's president.[59] The team was supported by letters from President Roosevelt (a friend of Osborn) and Charles Walcott, addressed to Lord Cromer and Lyons. The Egyptian Geological Survey also offered assistance. Walter Granger, the lead AMNH paleontologist, would later state how "every facility of the Department was extended to the party, the Geological Museum was offered as headquarters, camping equipage was freely loaned, the services of an experienced native was offered, gratis, the construction and shipment of packing cases was attended to, and valuable assistance given in the organization of the caravan and working party of natives."[60] The party consisted of Walter Granger and George Olson from the AMNH, Hartley Ferrar of the Geological Survey, and fluctuating numbers of Egyptian workers, averaging around twenty-five (Osborn accompanied the expedition, but went on a number of "surveys" visiting prominent Egyptological sites, rather than busying himself with excavations). The group also located Markgraf and hired him to collect material. Illustrating hierarchies in labor, Markgraf's wages were $282.80, while all the Egyptian workers together were paid only $606.73.

The American party generally did not complain about difficulties in transportation and climate, having experience working in deserts in the United States, although Granger did complain extensively about the condition of the fossils. Most sites had already been prospected, and "in point of preservation, the bones are poor. Petrification has not taken place."[61] Granger also noted the superiority of "the American method of pasting bones" to preserve them over Markgraf's use of carpenter's glue, noting that had he used the AMNH's techniques "in his previous collecting for German institutions he would have saved several fine skulls from destruction in transportation."[62] The exchange of techniques and arguments for the superiority of American paleontology were important parts of the process.

The AMNH scholars relied on Egyptian excavators, although they had difficult relations with them. Granger was extremely disparaging of the Egyptians, writing an account of "Native workmen," complaining that "native labor is all that is procurable in Egypt. At its best it is none too good and at the worst is most exasperating." The first team of workers were accused of being "careless and inclined to shirk, and not susceptible of training," and were discharged after two weeks. A new team was hired from the village of Quft, who were trained "under Mr Quibell and other excavators along the Nile." Granger wrote that "the most efficient native workers . . . were two aged men from Tamia who had been several seasons at the bone pits under Beadnell and Andrews, and who evinced considerable understanding as to the occurrence of the fossils and the importance of care in uncovering them, at which work they showed surprising skill."[63] Osborn was more appreciative, although he understood Egyptian workers in the same patronizing terms of colonial improvability, writing, "The twelve workmen from Sakkara entered in to the search with great care and intelligence, and although they had never hunted for bones before showed an enthusiastic reverence for a delicate object which was delightful to watch as expressed in the display of smiling rows of ivory white teeth at a word of encouragement." This may have been inspired by his observation of archaeological excavations by the Boston Museum of Fine Arts at Giza and the Metropolitan Museum at Lisht, where he wrote, "The natives are believed to be direct descendants of the builders of the pyramids of five thousand years ago and the tools and carrying baskets have remained unchanged."[64] In order to be appreciated, Egyptians needed to be described in the language of antiquity.

At the end of the season Osborn wrote to Andrews that he had "a delightful trip to Egypt,"[65] but the paleontological results were mediocre. Twenty-seven cases of around five hundred fossils were collected, but were not their priorities—high-quality skulls of iconic Fayum animals suitable for exhibition. The two *Arsinoitherium* skulls had missing horns, there was a complete though crushed *Moeritherium* skull, some broken and fragmentary *Palaeomastodon* skulls, and a "nearly perfect" skull of a predator named *Apterodon*. More material was simply noted as "numerous fragmentary jaws and skeletal bones," indicating the lack of importance given to non-cranial material.[66] Echoing Granger, Osborn wrote that "the region has been so thoroughly prospected since 1901 that the chances of easily securing fine surface prospects are very remote."[67]

Despite these problems, in 1907 Osborn wrote an article, "Hunting the Ancestral Elephant," for *Century Magazine*, a popular monthly he frequently used to galvanize public interest in the AMNH's work. Osborn wrote, "In the year 1900 I ventured a prophecy which placed the original home of the elephants and of several other great groups in Africa," and was "convinced of the probability that Africa in early geological times was a great centre of independent evolution."[68] The piece did not just praise Osborn's own prophetic abilities, but cited Andrews's and Beadnell's work as "epoch-making, marking a turning-point in our knowledge of the history of the earth, and arousing such wide-spread interest that for the time north Africa becomes the storm-centre of paleontology." Fossil work in Africa was also presented in terms of biogeographic exchange. Highlighting the proboscidean fossils in America, Osborn wrote, "Now that many of the great extinct animals in the American Museum paleontological collections had proved to be of remote African origin, what a temptation to secure some of the diminutive ancestors, to place beside their American descendants, the majestic mammoths and mastodons!" The *Arsinoitherium* would "rival its four-horned American contemporary *Uintatherium*, named after our own Uinta Mountains!" In the context of arguments that camels had originally evolved in North America, Osborn wrote, "How it pleased the fancy to take a caravan of camels, animals which were the gift of our Western American plains; to bring forth the remains of the elephants, which were the gift of Africa to all the other continents!"[69] Paleontology showed links across the world, connecting ancient nature, biogeographic migration, and the development of life.

Osborn's article was accompanied by Charles R. Knight's illustrations depicting the Fayum creatures in violent, social, and developmental scenes. They included "the strange *Arsinoitherium* and a primitive carnivore in combat, *Prozeuglodon* swimming near the shores of an ancient sea, *Eosiren*, an ancestral manatee, *Palaeomastodon*, *Moeritherium*, and *Megalohyrax*."[70] Knight's images were a key selling point of the article, and Osborn commissioned him to send "a set of colored restorations to the Egyptian Geological Museum executed from the pictures you have already made, in return for their courtesy to our expedition."[71] Knight's work could be exchanged in lieu of fossils.

Two of Knight's images stood out. The first dramatized conflict in ancient Fayum, showing *Arsinoitherium* fending off carnivorous hyaenodonts. Taking inspiration from conventions in contemporary natural history illustration, Knight showed an active, lively animal, living in social groups and behaving like a rhinoceros or buffalo. He also depicted the most significant implications of the Fayum fossils—the evolution of the elephant. A full-page triptych showed the elephants attaining their development, starting with *Moeritherium* emerging from the watery swamps, moving to *Palaeomastodon* entering drier land, and reaching an apogee with the largest known elephant, the Columbian Mammoth. The images showed progressive transformation, where the full potentialities of the organism became apparent. They also suggested environmental change was key to progress, with the creature adapting to the Eocene marshes of Africa and to Pleistocene North America. This was a tale of expansion and environmental change, with movement to the open plains of the New World being key to the animal's superiority, and with elephants reaching their fullest development in North America. The elephants "were superbly equipped for foreign travel and conquest . . . as secure as the Romans, with their phalanxes and their legions, these quadrupeds marched over one or other of the land bridges as soon as they were formed" and "in the whole history of creation no other animal, with the single exception of the horse, accomplished such feats of travel."[72] Biogeography showed deep imperial histories, driven by animals in the prehuman past.

The AMNH maintained interests in the Fayum, although it worked at a distance rather than through expeditions. Osborn attempted to enlist Markgraf to secure "more perfect examples of the skulls of these important animals for exhibition in our halls." He promised an advance of $100, and would pay $200 for "a really fine, large complete" skull of either *Arsinoitherium* or *Moeritherium*, or $150 for *Palaeomastodon*, noting "these are much larger amounts than I have ever offered before for similar specimens, but I observed while in the Fayum that the fossils are becoming very rare and it will be by no means easy to procure such specimens."[73] After some reluctance, Markgraf attempted to find these skulls, but warned how difficult it would be, as "whole skulls are very seldom found and nearly always are incidentally stumbled upon on the surface, which is pretty well exhausted now."[74] Although the AMNH promised to fund Markgraf at a generous rate, he does not seem to have devoted much time to the search. While he did send bones of small animals and a further broken *Arsinoitherium* skull, the AMNH could not succeed, despite its wealth.

Paleontological work in Egypt was extensive, and works like Lankester's *Extinct Animals* and Osborn's "Hunting the Ancestral Elephant" publicized the research. As well as enhancing a sense of colonial adventure and scientific exploration, these expeditions reevaluated Africa's place in natural history, presenting it as a dynamic center of origins. Osborn concluded his piece: "Africa, far from being a continent parasitic upon Europe, was proved to be a partly dependent, but chiefly independent centre of a highly varied life."[75] This stood out in contrast to some earlier presentations, which saw the continent as a passive dumping ground for animals forged in more "progressive" regions, which then degenerated in the tropical environment.

▸ FIGURE 13.6. Charles R. Knight's illustration of ancient life in the Fayum. Osborn, "Hunting the Ancestral Elephant," 821. Author's collection.

CHAS. R. KNIGHT
1907

Drawn by Charles R. Knight

EVOLUTION OF THE ELEPHANT. (DRAWN TO THE SAME SCALE)

3 The Mammoth (*Elephas Columbi*) of the State of Indiana. (Based upon a specimen in the American Museum of Natural History.) 2. The *Palæomastodon* of Eocene Libya. (Based upon the restoration by Dr. Andrews.) 1. The *Mœritherium* of Eocene Libya. (Based upon a skull figured by Dr. Andrews.)

It revived older tropes of Africa being "the home of animals," discussed by Sandra Swart,[76] but not in the sense of it being a pristine wilderness. African animals in the ancient past, known from the fossils of the Fayum, played a significant role in earth's history.

The significance of this research was also framed by imperial ideas. The serial models of elephant evolution illustrated that development moved on particular tracks, with the ever-privileged elephants being perfected forms. Elephants were not just "perfect" and dominant in their anatomy, but also in their history. Their origins in Africa and spread throughout the world, showed they were dynamic conquering animals, which "invaded" all the continents except Australia and Antarctica. Wilhelm Bölsche imagined this global march of the elephants from Africa across northern Eurasia and into the Americas: "It must have been a great sight in earth history: the first appearance of the numerous herds of elephants, young and old together, their giant drooping ears, their constantly moving trunks, their bright trumpet calls, moving to a whole new landscape: Africa on a world conquest. It is impossible to not think of the later historical events of the Carthaginian Hannibal, who took African elephants over the Alps."[77] Biogeography constructed a deep history as dramatic as human conquests. Colonial projects in the Fayum bolstered ideas that the history of animal life paralleled human history, with similar invasions, migrations, and expansions. However, in this instance it unsettled older typologies that saw the north as the home of progress and dynamism. Africa, the source of the elephant lineage, was progressive and generative. Biogeography could be an imperial science, but in doing so could also center modern colonized regions.

◂ FIGURE 13.7. Knight's illustration of the evolution of the elephant, through *Moeritherium* and *Palaeomastodon* to the Columbian Mammoth. Osborn, "Hunting the Ancestral Elephant," 834. Author's collection.

CHAPTER 14

New Communities

The Expansion of Paleontology in the Western United States

THE PREVIOUS CHAPTERS EXAMINED HOW PALEONTOLOGICAL PROJECTS IN SEVERAL territories—Patagonia, Australia, and Egypt—built new collections, and new conceptions of life. However, the expansion of paleontology did not just involve work in regions under-researched by paleontologists in the previous decades. It also reconfigured territories which had long been sites of paleontological exploitation. Changing economies, patterns of settlement, and new educational institutions altered the balance of power, allowing local fossils to become valuable resources across scholarly networks. This especially occurred in the western states of the United States. As seen in chapter 6, the North American West was exploited for fossils since the early nineteenth century (following the track of US annexation, transportation building, and economic exploitation). Fossils became crucial for understanding the character and potential of the territory—a point strongly made by Daniel Zizzamia.[1] In the 1900s these dynamics continued, but also shifted, as the West was not merely exploited by outside collectors, but became a place where new institutions were built, claiming authority over the paleontological past.

The growth of local paleontological work in the US West grew from the wider transformation of the region, as it became increasingly conceptually and economically significant. Jeremy Vetter has examined the tight connections between settlement, railroad and telegraph networks, and field sciences in the West in these years.[2] With the culmination of the wars against the Plains nations and massive expansion of agriculture and mining, the western states became increasingly densely settled. Some western cities, like Chicago, San Francisco, and Los Angeles, became hugely significant, rivaling older East Coast cities. Indeed, Chicago's Field Museum matched East Coast collections, especially in dinosaur research.[3] State identities were also built through expositions, regional symbols, and prestige buildings. Of course these ideologies of progress and expansion rested on the expulsion of Native Americans and negation of their rights over their land.

Western states could build strong positions in vertebrate paleontology. Geology was essential for mining and industry, and geological education expanded tremendously in the university systems of the West, and was frequently linked with state and commercial surveying for coal, oil, and other resources. Universities therefore offered ways to build paleontological programs, and university careers were often preferable to subordinate positions in large museums: Samuel Williston, who was professor of geology at the University of Kansas before moving to the University of Chicago, reflected that while Pittsburgh's Carnegie Museum had wanted to hire him as a curator, "I have so long had the academic freedom that I do not feel like being subjected to the direction of any one as to my daily going and coming." He did however note that if he had been offered the position of director of the museum, he "would very willingly accept."[4]

While western institutions never matched the funding of the AMNH and Carnegie Museum, they show there were other ways of maintaining influence in paleontology. Proximity was important, enabling access to sites at short notice and working with local communities. Western paleontology gained a personal character based around cultivating relationships with people who owned the land containing fossils. It grew within the world of universities, local patrons, geological surveys, farmers, and settlers. The connections of western paleontology with universities and field research also colored its agendas. Rather than construct the entire history of life or move to the display of large dinosaur fossils, western researchers usually reconstructed the fauna of their local territory as ecological assemblages, in some respects refiguring work on sites like Sansan, Pikermi, or the Siwaliks. Much attention focused on fossil mammals, because of the values placed on them, but also because Tertiary sites contained abundant fossils that were easier and cheaper to work than the bulky and often fragmentary remains of dinosaurs (which were increasingly displayed in larger museums in what has been termed the "Second Jurassic Dinosaur Rush").[5] Links with biology became especially close. Just as field biology in the 1900s often became based on systematic surveying and examinations of ecological communities, the ancient past was also understood in terms of structured localities and assemblages of extinct animals.[6]

Paleontology was conducted across the western states. As well as Williston in Kansas, paleontological programs, private collections and museums developed in Colorado, Oklahoma, Texas, and other states. This chapter focuses on two particularly significant examples of western paleontology, which developed new institutions, engaged proactively with established centers on the East Coast, and imagined the deep past of the territory. The first was in Nebraska, where Erwin Hinckley Barbour and the Cook family linked local society and long-standing, although often fraught, relations with larger museums to transform Nebraska into not just an important region for fossil extraction (as it was during the Leidy, Cope, and Marsh period), but a place where fossil research was conducted. The second was in California, where major institutions—centered around the University of California at Berkeley and Rancho La Brea in Los Angeles—built local paleontology to a huge scale in a short period of time, taking advantage of economic and university growth, patronage, and productive fossil sites. These examples illustrate how western paleontology drew in a range of actors and collaborators, building new centers of authority. Local landscapes of fossil work were constructed, and prehistoric faunas used to imagine the deep past of the locality and assert scholarly importance in the present and future.

E. H. Barbour and the Building of University Paleontology in Nebraska

Erwin Hinckley Barbour gives an example of successful career and institution building in the western states, bringing together established paleontological networks, collaborations across local society, and sponsorship by local patrons and state institutions. Barbour was born in Ohio in 1856 and trained under Marsh at the Peabody Museum. However, like many under Marsh's authority, he fell out with his master (his attack on Marsh's preparation methods has already been seen). In 1891 he moved to Nebraska to be appointed simultaneously professor of geology and zoology at the University of Nebraska, state geologist, and director of the university's museum—varied responsibilities linking universities, economics, and state politics. His writings were full of a desire for expansion. In 1894 he wrote to George Baur at Clark University that he was "exceedingly well pleased with my present outlook" in Lincoln, "a city of 60,000 and will double that in the next ten years. There are 800 students now, with assurance of 1000 next year. The college is a state institution and has immense grants of land that are very valuable"[7] (land that had recently been seized from Indigenous people). Lincoln was a prime place for the opportunistic expansion of paleontological work.

Barbour's staff was relatively small, and his sister, Carrie Barbour, played an important role. After initial education as a wood and ceramics worker, she joined the University of Nebraska art department a year after her brother's arrival. Carrie was quickly seconded to the university museum, working as fossil preparator, teacher, and field-worker. While Carrie produced only a few short publications herself, she was hugely significant within the department, becoming a member of the Nebraska Academy of Sciences and being appointed assistant curator of paleontology in 1895, remaining in the post until the 1930s. In the largely male world of paleontological collections, this was distinctive and shows that western museums often gave women more opportunities to build careers than their East Coast counterparts.

Nebraska paleontologists focused on fossil mammals, but moved beyond the badlands fauna debated by Cope and Marsh. Instead, they tended to examine Miocene formations, which included mammoths, a diverse range of ungulates, saber-toothed cats, and strange forms like the entelodonts (huge, apparently carnivorous piglike animals) and the oreodons discussed by Leidy. These fossils showed the abundance of nature, the resources of the state, and potentially how productive the lands were in the geological past. They also allowed Barbour to intervene in national debates. In an early attempt to build his reputation, he published several articles on spiral structures in Miocene deposits, known as "Daemonelix" or "Devil's Spirals." Arguing against interpretations that these were the burrows of an ancient rodent, Barbour used his emplacement in Nebraska to argue that they were relics of ancient plants. While Barbour's interpretation was rejected over the following decades (especially as fossils of a rodent, *Paleocastor*, were found in the spirals), this was nevertheless an instance where Barbour used his "peripheral" position to debate with more established institutions.

Barbour's emplacement allowed him to take advantage of finds made by farmers, prospectors, and other Nebraska citizens. His correspondence is full of attempts to secure the bones of mammoths and mastodons, which were emblematic local fossils but also fragile, requiring trained excavation if they were not to be destroyed. One was found in the cellar of a house in Lincoln, with the tusks protruding through the basement wall. While a team from the museum extracted some of the remains, more extensive work would have caused the building to collapse and work had to be abandoned (several years later when the owners moved out, Barbour recommended that the university purchase the house to unearth the rest of the mammoth).[8] Another mammoth was found near the farm of R. E. Barney, who noticed prairie dogs bringing up bones and teeth, having constructed their burrows around a skeleton. Barbour urged Barney to leave it alone, as any untrained excavation "will only add to the damage which the prairie dogs have already done," and the fossils will "fall to bits unless treated with various chemicals beforehand."[9] Barbour's attempts to obtain this specimen show strong people-managing skills. He went to the site and cultivated a relationship with Barney by taking a photo "of the elephant tusk with your family in the background."[10] However, there were less scrupulous elements: he wrote a "confidential letter" to a local lawyer to keep the completeness of the skeleton secret, as "if we write to Mr Barney and tell him the facts, ten chances to one he will at once completely exaggerate its value and there will be no such thing as securing it for the State University, the place where all such things should go. . . . What we want is to secure this before some of the eastern colleges hear of it."[11] The desire to possess fossils for the locality could lead to underhanded dealings and plying social influence.

Barbour worked extensively in economic consulting around minerals, which connected him with leading businesspeople. Nebraska elites could never provide the resources of Andrew Carnegie and John Pierpont Morgan, but still gave crucial support. Barbour's most important supporter was Charles Henry Morrill, who settled in the West after serving in the Civil War and, following an unsuccessful career as a farmer, became a store owner, then a banker, and then a land developer. He also claimed to have met Marsh in the 1870s, who told him about the local fossils.[12] Morrill made a fortune purchasing and reselling grants of land seized from the Lakota

FIGURE 14.1. The Barney family, alongside the mammoth remains excavated on their property. Image #320101–00312, University of Nebraska-Lincoln, Archives & Special Collections.

(estimated at "one hundred thousand acres . . . to improve and develop new wild sections of the West"),[13] again illustrating how frontier politics and paleontological institutions often grew from the dispossession of Indigenous people.

From 1892 Morrill sponsored the Morrill Geological Expeditions of the University of Nebraska for $1,000 per year. To secure funds Barbour appealed to Morrill's civic-mindedness and personal pride. An unnamed letter from someone in Barbour's department (possibly Carrie) to Charles Schuchert at Yale noted how Morrill "is a man who has come through the poverty of a New England hillside to a position of considerable affluence. His College training having been limited to a few terms in a Connecticut preparatory school has put him in sympathy the more deeply perhaps with educational affairs and it appeals to his pride to contribute anything to the cause of education and science."[14] At one point when it seemed as if Morrill was going to stop funding, Barbour advised the university's chancellor on how to maintain patronage: "Mr Morrill's weakest and most vulnerable spot is the pride he takes in the relation which he has had to educational affairs." Strategically securing Morrill's support was essential as "with Yale, Harvard, Princeton, Carnegie, American Museum of Natural History, and Field Museum all in the field . . . competition is rather active. They are taking remarkable things out of the State."[15]

Collaboration and Competition over the Fossils of Agate Springs

Nebraska paleontology was conducted not just through universities and museums. Local people whose lands contained fossils were also important collaborators. As with the notables at Fort Bridger who had invited Cope and Marsh, people close to productive fossil sites could use them to build status. However, this could be a tense process, requiring delicate negotiations with both local and East Coast paleontological institutions—relations that were sometimes productive, but sometimes broke into conflict.

The example of James Cook shows a self-conscious confluence of "western" tropes with fossil work.[16] Cook was from Illinois and in the 1870s went to Texas to take part in the cattle drives, during which he claimed to have brokered the collaboration between Marsh and Red Cloud.[17] After this Cook worked as a guide, escorting parties of East Coast and European sportsmen in the West, which gave him a large amount of money and an ability to deal with elites. After a time in New Mexico (where he also informed Cope's collectors of fossils), he married the daughter of a Nebraska ranch owner and began purchasing land that eventually became the Agate Springs Ranch, a large landholding turned over to cattle raising in Sioux County, close to the Pine Ridge reservation (the ranch was also a source of employment for the reservation's residents, and Cook developed interests in collecting Indigenous artifacts).

As well as being prime, well-watered grazing land along the Niobrara River, Agate Springs was close to deposits of Miocene fossils. As a landowner with a public reputation, Cook was enmeshed in the state's social networks, inviting Barbour in 1891 to visit Agate and see the fossils. In 1904 Olaf Peterson, conducting work for the Carnegie Museum in Nebraska, visited Agate Springs. According to later reminiscences by James's son, Harold J. Cook, when shown the site "Peterson was greatly excited," and rather than depart as planned, "he waved his arms wildly when we approached the ranch. 'Put the team int the barn,' he shouted, 'we aren't going anywhere!'"[18] Following this, the Carnegie Museum conducted numerous excavations, staking one productive site as Carnegie Hill. There were nevertheless tensions, especially emanating from the museum's director, William Holland, who attempted to control the work from a distance and secure as many fossils as possible. This raised tension with James Cook, who felt the "old bones" were plentiful enough to be shared by museums around the country, generating prestige inside and beyond the state.

Harold Cook, became increasingly interested in paleontology, writing in his memoirs how he "worked with Peterson every spare moment I could get away from the ranch work; and, probably, many times when I should have been riding," and purchased a copy of Flower's *Osteology of the Mammalia* on his advice.[19] This provided an opportunity for Barbour to secure an interest at Agate Springs and develop a close relationship with Harold. A letter exchange in 1905 cemented this alliance, with Harold reporting on the Carnegie Museum's work and that he had "found an immense field, so large that they could not work it out in years, so that there is plenty of material for other parties to work with."[20] Barbour, aiming to secure the site for the university, asked him to "keep it for us without fail and do not allow any one to poach upon it. What you say is exactly true, every one has been robbing the state of its very best things and the state itself has scarcely anything."[21] Harold responded that he was also keen to reserve material for Nebraska institutions, and went to "secure these beds for the state at once, as I can tag them while riding."[22]

The relationship between Harold Cook and Barbour developed over the following years. One of the first fossils Cook sent Barbour, a four-horned antelope, was named *Syndyoceras cooki*, with Barbour honoring the collaboration in the nomenclature by naming the animal after Cook. Barbour frequently visited Agate, gave Harold advice in paleontology (with their correspondence becoming increasingly technical), and engaged in social activities—hunting ducks and exchanging flowers and gardening tips. Harold eventually married Barbour's daughter Eleanor in 1910 (a major event in Nebraska society, attended by such local luminaries as William Jennings Bryan). While the Carnegie teams worked, Cook reserved productive sites for Barbour, and by 1913 Barbour estimated that Cook had directly supplied him with $20,000 worth of fossils.[23]

While the Carnegie Museum and University of Nebraska had long-standing interests at Agate, the American Museum of Natural History also became involved, but adopted a different strategy, relying on personal engagements with the Cook family. Henry Fairfield Osborn's uncle, John M. Adams, president of the Nebraska Cattle Association, asked James Cook whether the AMNH would be able to conduct excavations. Cook responded that he "would be glad to have him come to my home at any time and look at the fossil beds," but that the Carnegie Museum teams might take issue with this, as "these 'Bone hunters' like the old time cow men claim to have 'range rights' and the first one to work in a quarry claims the earth about that spot for a certain number of miles or leagues I don't know which."[24] He emphasized that Elmer Riggs from Chicago's Field Museum had already been warned off by Holland (although smaller teams from Kansas, Yale, and Amherst College were also active). When Osborn asked Holland about this directly, the idea was met with displeasure, with Holland stating that "Mr Cook is in the habit of inviting the world to come and dig," with the passive-aggressive addition that "I cannot oppose to your wish to go there if you decide so to do, but if I were to undertake to begin digging at the Bone Cabin Quarry I do not think you would like it."[25] This reference to one of the AMNH's prize sites seems to have been enough to stop Osborn's plans to work at Agate.

While the AMNH backed off from Agate, Osborn adopted a longer game by personally patronizing Harold Cook. This was partly mediated through Barbour, who maintained good relations with Osborn, and through sending museum scientists like William Diller Matthew to Nebraska to prospect other sites with Cook. This culminated in Osborn inviting Cook to work at the AMNH in 1908, in a paid position to fit around his studies and ranch duties:

> You will get a splendid training as a paleontologist which I believe you cannot get so well anywhere else in the world as in our museum, which is so rich in mounted and exhibition material and so well supplied with books and with a staff of earnest, intelligent and most kindly men, all of whom are working together for a common end, the advancement of paleontology and the bringing together of a great collection. I am willing to make this engagement very elastic so that you can give such time to your ranch work as your father requires and your duty as a son demands.[26]

The extract is worth quoting in full, showing Osborn's view of paleontology as a collaborative endeavor mobilizing men and resources. Osborn paternalistically emphasized duty and respect, allowing Cook to follow paleontological interests while acknowledging his other responsibilities. This was part of the AMNH's strategy, fostering local collectors who could channel fossils to New York (much like the relationship with Markgraf).

In Pittsburgh work continued on Agate fossils, but relations between the Carnegie Museum and field site became tense. In April 1908 William Utterback, one of Peterson's key field-workers, wrote to the museum's secretary that "I have decided to give up the fossil business for all time," as Holland's control was unbearable and "perhaps in the future when Dr Holland gets a collector to work as hard and faithful in the interests of the Museum as I have endeavored to do, he will not allow any one person to use him for his own personal glorification." Peterson had forced "old Bill out to do the hard dirty work," and he "should be more than pleased to give Dr. Holland just a few parting words." Utterback was also irate with James Cook, noting "the life of a collector is hell, without being subject to restrictions from such bombastic windbags as J.H. Cook."[27] Much as with Marsh's attempts to control his field-workers, fossil digging was both difficult labor and a source of pride, and required respect and fair renumeration.

Holland visited Agate in late 1907, and James Cook wrote, "He is a totally different man than what I supposed him to be. He came 'right down to earth' in his visit to us, and we all as a family enjoyed his stay."[28] However, this did not alleviate the long-term situation. Holland was intent on maintaining the Carnegie Museum's dominance—although this seems to have further soured relations with the Cooks. When it became apparent that Carnegie Hill was just outside the boundaries of the ranch, James Cook became concerned that Holland or another party might attempt to formally claim ownership of the site. The provisions of the recent Kinkaid Act meant it was fairly straightforward for James Cook to extend ownership over Carnegie Hill, and these tensions over land seem to have broken off relations with the Carnegie Museum.[29] Holland's 1909 report as museum director noted, "We have reached the conclusion that it is more to the profit of the Museum to abandon work there," making an excuse that "the animal which is most frequently encountered in this quarry is a species of small rhinoceros of which we already possess portions of some two hundred and fifty skeletons, and it does not appear desirable to add to their number."[30]

At the Carnegie Museum, preparation continued, especially on its prize specimen, *Moropus*, a large chalicothere (a clawed herbivore).[31] The mounted skeleton of *Moropus* was unveiled in 1911, accompanied by a monograph cowritten by Holland and Peterson,[32] gaining international interest. Extensive European researches on chalicotheres was a major reason for focusing on *Moropus*, keeping up Holland's relations with European scholars built through the presentation of *Diplodocus* casts.[33] While corresponding with Barbour to secure *Moropus* casts from him, Holland boasted "from the Jardin des Plantes in Paris I have received as a gift casts of all the material representing the chalicotheres which they possess, including the type specimens of Cuvier, Lartet, Filhol, and Gaudry."[34] The implication was that if the great collection of Paris had sent casts, Nebraska could hardly refuse.

Reactions at Agate to Holland's monograph were considerably less favorable than its international reception. Harold Cook was not sent a copy until he asked Holland personally, recalling "when I did get it, I could see reasons why he was not particularly anxious to send a copy."[35] Holland's account of the Cooks' management of the work was taken as incendiary. One footnote referred to the various collectors at Agate with the expression "where the carcass is there will the eagles be gathered together."[36] Harold regarded this as a slur against other collectors and against his family's agency. It was "a positive insult . . . The parties who worked here came at our invitation, and were not buzzards, or intruders."[37] More egregious was an accusation that the Cooks initially thought the fossils were the remains of "Indian horses" rather than extinct animals and destroyed many fossils while attempting to extract them, before trained museum teams arrived. Jeremy Vetter argues it is impossible to verify this story, but private correspondence demonstrates Harold Cook was particularly stung by these insinuations.[38] Given his developing scientific education, this was a public slight on his status.

Harold wrote an irate letter to Osborn (also copied to Barbour), stating that Peterson and Holland had been "persistently unfair, not frank, and had been underhand in their dealings with us, while we were furnishing them with a house to live in, pasturage for their horses, storage for specimens and outfits and going out of our way at every turn to do them favours," and "we are sick and tired of such bickering and a very little will prompt us now to close the quarries altogether to those outside of the state."[39] Harold wrote to Barbour that "I have an abhorrence of squabbles, but a person's honor demands that they do not stand for more than a certain amount of prodding."[40] Codes of hospitality were broken, and honor was key. The Cooks were so incensed they requested the Paleontological Society issue a censure motion, planned to take the matter to Andrew Carnegie, and referred to a bill about to go to the Nebraska Legislature forbidding the export of fossils from the state. As a result, Osborn requested a detailed report from Matthew on the matter, aiming to defuse tensions and cement the AMNH–Cook relationship. Osborn responded to Harold: "I entered the paleontological field in 1879 when everyone was at war with everyone else and no paleontologist had a good word to say about any other, and I became thoroughly sick of the controversial spirit, because I saw how damaging it was to the progress of science; so I have persistently thrown my influence on the side of peace and have allowed a great many things to go by good naturedly rather than take affront and stir up trouble."[41] The Cope–Marsh struggle cast a long shadow, making US paleontologists rhetorically committed to collaboration and demarcation. Osborn wrote that a law forbidding the export of fossils "would be a great step backward,"[42] citing similar laws in Alberta as diminishing paleontological work, and that "France is the only country in the world which has stopped collecting in one of its colonies, namely, Madagascar, and the result is that work there has ceased entirely."[43] Despite likening Nebraska to an African colony of a European power, Osborn's letter allayed tensions. Harold wrote that "father and I both feel that your advice . . . is decidedly good, and the most sensible course to follow. But as you know, it is not always easy to 'smile and take it,' when you do not deserve it."[44] The Cooks and Barbour also reassured Matthew they would now oppose the bill against fossil exports and "do not think there will be any difficulty in knocking it in the head."[45]

Work at Agate continued in succeeding decades, if declining in scale. Jeremey Vetter has discussed how Harold Cook maintained a strong position in western paleontological circles, although he was cut off from a national profile.[46] The mixed fate of Harold Cook's career was not matched by the University of Nebraska, where Morrill continued to fund excavations and Barbour expanded his collections, culminating with the opening Morrill Hall in 1927, centered on a gallery of mounted fossil elephants. The AMNH also maintained Nebraska institutions within its network of collaborators, setting up a display of Agate Springs fossils in the 1910s. Osborn wrote that new *Moropus* skeletons from Agate showed "an altogether different animal from that mounted by our friend Holland, who knows more about theology perhaps than about *Moropus*."[47] Personal connections and bolstering actors away from traditional centers could aid in building collections and new understandings of the past.

Tar, Accumulation, and Sensation in the Los Angeles Pleistocene

While the northern plains were a classic region for paleontological work, a different part of North America loomed large in the paleontological imagination in the 1900s: California. Paleontology here showed similar trends to Nebraska, of local institutions, often connected with universities, gaining support from local patrons, exploiting productive local sites (linked with natural resources), and building independent programs of research to comprehend the state's fossil past. However, there are differences, most notably the scale of Californian paleontology, which quickly rivalled the East Coast collections, and the lack of competition from eastern scholars. Californian paleontology became a powerful and autonomous force.

As in much of the world, paleontology in California mixed with the wider study of life, and two local trends were particularly significant. First were valuations of California as a land of natural wonders in need of preservation. While initially heavily exploited, the state's redwood forests and marine life were central in germinating conservationist discourse, including the establishment of Yosemite National Park and the work of the Sierra Club.[48] Equally significant was California's reputation as being a paradise for economic expansion. The gold rushes of mid-century were merely one instance, followed by intense surveying and prospecting. While scholars from eastern states initially investigated parts of California, the distances involved meant there was considerable space for California institutions. So rather than fight for autonomy in a region already under the influence of eastern collectors, as in Nebraska, Californian paleontology was far more independent.

Given the importance of mining, the state had a long history of geological education, concentrated in the San Francisco Bay area. The prime builder of the University of California at Berkeley's geology program was Joseph LeConte, who arrived in 1869 as one of several southern professors leaving the former Confederacy, owing to what he described as "the chaos of emancipation and reconstruction," as "the catastrophe of the war and the resulting emancipation of course swept away everything I owned as property," namely a slaveholding family estate.[49] LeConte shows a confluence of environmental policy, university expansion, and stark racial politics. He wrote how hierarchy was necessary within humanity and nature, and that slavery (as practiced in the antebellum South) was an inevitable relationship deriving from racial differences. (He also wrote "the Negro" was "plastic, docile, imitative," and suitable for subordination, while "the American Indian" was "highly specialized and rigid," meaning their "extermination is unavoidable.")[50] LeConte was politically and socially active in other areas, being a prime figure in the early California conservation movement and aiming to reconcile geological and evolutionary ideas with Christian doctrines, regarding them as manifestations of a singular process of natural development. So while the geology department at Berkeley was frequently directed to the exploitation of mineral resources, LeConte's influence meant there was space for paleontology, in a form that linked racial and environmental politics.

Paleontological work at Berkeley was later directed by LeConte's former student, John C. Merriam, who returned to Berkeley in 1894 after completing his studies in Munich. Merriam expanded paleontological education and gained a sponsor in Annie Alexander,[51] a further instance of patronage in paleontology operating in the western states. In a confluence of colonial and commercial relationships, Alexander's father made a fortune acquiring a stranglehold on the Hawaiian sugar trade before being killed in an accident while on tour in East Africa. She had a strong interest in natural history, and her considerable wealth allowed her to follow these interests in a manner unusual for women of the period, who were usually excluded unless they carved careers as popular writers, gained status through family connections (like Carrie Barbour or Alice Woodward), or undertook low-level or voluntary work within museums.

In a story told by both participants, Alexander began auditing Merriam's courses in 1900–1901, which she later recalled "inspired me with the idea of financing a fossil expedition."[52] More privately, she wrote to her friend Martha Beckwith that "I have not missed a lecture. I like it more and more, the study of our old, old world and the creatures to whom it belonged in the ages past, just as much as it does to us today. Perhaps the study is all the more interesting because it is incomplete, there is so much yet to find out."[53] Once again, fascination with the deep past drew from connections with the modern world and desires to solve scientific mysteries. Alexander undertook field trips with the university and donated to support paleontology. This was part of her wider patronage at Berkeley, also funding the Museum of Vertebrate Zoology.[54] Using Alexander's funds, the Berkeley Department of Geology expanded tremendously. In 1912 Merriam wrote, "No other university in the country has as many persons engaged in paleontologic investigations as we have in the University of California."[55]

While the old center of paleontological work in California was in the north, in Southern California science, commerce, and resource extraction coalesced around a productive site just outside Los Angeles—the tar pits at Rancho La Brea. La Brea was exploited and marketed to become one of the world's foremost paleontological sites, reconfiguring ideas of ancient North America and providing fossils for national and international exchange. The exploitation of the tar pits also rested on older usages. Oil and tar had been known in the region for thousands of years, being used by Indigenous people and settlers for waterproofing baskets, roofs, and boats. In the late nineteenth century the Los Angeles oil industry boomed, first through demands for asphalt for roads and then for petroleum-based engines and generators. Oil wells were dug across the region in a drive for profits, causing considerable environmental damage, as oil and tar spilled into the city streets.[56]

One opportunistic individual who purchased oil-rich land was Henry Hancock, who acquired Rancho La Brea in the 1860s and employed Chinese workers to excavate the asphalt.[57] Bones were well known from the site, and it was called *la huesomenta* (the bone-yard) by local people,[58] although the remains were assumed to be of modern animals. In 1875 one set of remains was identified by the geologist and spiritualist William Denton as belonging to a saber-toothed cat. However, this was not followed up on. Past knowledge was effaced in most accounts, which tied the site's exploitation to the oil industry. Samuel Clover noted that in 1901 "workmen, in drilling a water well near the Hancock ranchhouse, brought to the surface a number of bones larger than anything of the kind seen before." The petroleum geologist William Warren Orcutt "recognized one of the huge bones as the fossil of an animal of a former era and he at once concluded that hidden in the pits must be a wealth of material of the greatest scientific value."[59] After securing permission from the Hancocks, Orcutt spent several years excavating remains.

According to a later account by his wife, Orcutt initially worked "alone, because he feared that the precious fossils might be broken by careless handling." However, with the discovery of a *Smilodon* skull, "the first entire fossil skull of this terror of the pre-historic animal kingdom,"[60] he pitched an excavation project to Stanford, his old university. As Stanford did not have a paleontology program, they recommended the Berkeley department because it was the largest paleontological unit in California. Merriam visited La Brea in 1905 and organized a series of excavations. The Berkeley team initially consisted of five men, but increased to sixteen during particularly busy periods. The work was described as hard, involving "asphalt and clay varying from the consistency of muck to the hardness of a cemented asphalt pavement, but the larger part of the work was in disentangling the intricately matted bones." Nevertheless, a huge amount was excavated. By 1913 Merriam noted that twenty tons of La Brea material (at least 250,000 individual bones) "occupies all of the floor space available in our store room."[61] Once cleaned, the bones could often be assembled into complete skeletons of camels, ground sloths, bison, lions, wolves, and saber-toothed cats. Merriam reported to Alexander with some plausibility: "The results of this work give the University of California unquestionably the best collection of Pleistocene vertebrates known in America."[62]

La Brea was important for its extent, but even more so for imaginings of its origins. In 1908 Merriam wrote an article for *Sunset Magazine*, "Death Trap of the Ages," describing the site and explaining its ancient history.[63] Subtitled "Tragedies of Aeons Ago: . . . A Pathetic Page from the Prehistoric Past," *Sunset*'s editor described the account as being as gripping as a "three-volume novel":

> The sticky pool of water and tar has been a Death Trap of the Ages. Here for centuries, evidently, the enormous ground-sloth and other clumsily moving

> creatures of his kind came for water, only to be held relentlessly; herds of bison and horses were entombed, extinct forms with whose bones mingle those of the mammoth and the camel. To this helpless prey, snared for them in this bird-lime bed, came the lords of that era, the huge saber-tooth tiger and the monster wolf, the largest of the dog family. Trapped in their turn, they, too, fed the black maw of the asphalt pool and the death trap baited itself anew.
>
> To-day the records of that ancient death drama are disclosed.[64]

In this prehistoric world the landscape and environment themselves were perhaps the greatest danger, with the sticky tar, so useful to modern humans, trapping the animals. The number of fossils—while an immense trove for science—indicated a site of mass death.

Like the Cook family, the Hancocks initially felt the amount of fossil material meant the site could be worked by several groups. However, unlike the Cooks, they invited only California teams. Whether this was due to local pride or because East Coast collections did not have significant infrastructures in California is unclear. One of the most proactive teams was not a museum or university, but a local school. James Zacchaeus Gilbert joined the Los Angeles County High School as a science teacher after studying with Williston in Kansas, and took his pupils to La Brea to excavate the pits. Specimens were prepared and mounted around the school, gaining press attention. The *Los Angeles Examiner* in 1912 noted that the pupils had already mounted a *Smilodon* skeleton and were now working on a ground sloth. As well as being educational, this enabled the public to commune with the ancient creatures, as "speaking further of this prehistoric monster, Prof Gilbert sank into the remotest realms of science and paleontology," saying the animal "belonged to the class of Edentates and became extinct in the Pleistocene age, which, he supposed, meant that it lived a considerable number of years before America was discovered."[65] The excavations at La Brea gained considerable attention, allowing opportunistic investigators to assemble paleontological collections, once they gained permission from Hancock.

Public interest caused problems at La Brea, however. Trespassing and vandalism were major issues. In 1909 William Bebb, secretary of the College of Dentistry at the University of Southern California, wrote to Merriam

▸ FIGURE 14.2. Reporting on the work of the Los Angeles County High School at La Brea. "Class of Girls Reconstruct Giant Sloth," *Los Angeles Examiner*, July 24, 1912. Courtesy of the Bancroft Library.

Class of Girls Reconstruct Giant Sloth

Morotherium Gigias Discovered as Fossil

Girl Student Putting Tooth Into Jaw of Fossil Remains of a Morotherium Gigias (Giant Ground Sloth), a Two-Ton Monster, Which Roamed the Site of Los Angeles Ages Ago.

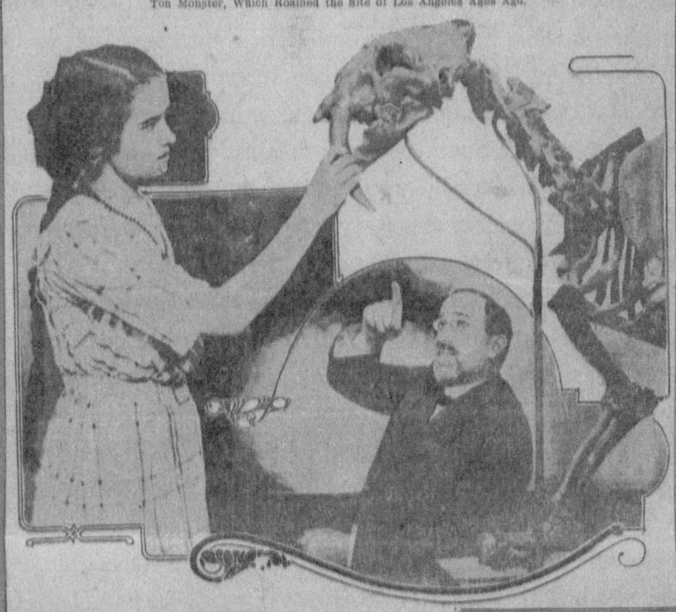

Bones of Prehistoric Two-Ton Monster Found in Asphalt Beds of Rancho La Brea

ONE hundred thousand years ago a morotherium gigias called in museum catalogues, Giant Ground Sloth, laid himself down in the asphalt beds on Rancho La Brea and died. Yesterday a group of young girls, members of Prof. J. Z. Gilbert's physiology class in the Los Angeles High School interested in comparative anatomy and comparative study of teeth, gathered about the mortal remains of this monster that had been dug up under the supervision of Prof. Gilbert and aided the work of reconstruction.

There are in existence only two other whole mounted specimens of this prehistoric beast that so many thousand years ago wandered over this earth. One of these is in the possession of the University of Southern California, and the other in the Academy of Science at Exposition Park. The third and last is now being mounted by Prof. Gilbert and Dr. John Horung assisted by the members of the class.

Wonderfully Preserved

"The remains of this animal," stated Prof. Gilbert, "are the most wonderfully preserved in the world. In no other place in this world are fossil remains found in such a state of perfection, as the asphalt acts as a preservative."

A large number of the bones are missing and the solving of the puzzle that will result in gathering them together and placing them in their proper place, has fallen to the lot of the class under the direction of Prof. Gilbert.

This animal is being reconstructed by the class during the summer term for use during the later months. Last summer the one now in the museum was reconstructed by Prof. Gilbert and Dr. Horung and the students.

The fossil on which the class is now working will be fifteen and a half feet from tip to tip and four feet in diameter when finished. The animal, stated Prof. Gilbert, probably weighed at least two tons. Speaking further of this prehistoric monster, Prof. Gilbert sank into the remotest realms of science and paleontology.

Not First Feat

The animal, according to Prof. Gilbert, was first seen upon the earth in the miocene age, belonged to the class of Edentates and became extinct in the Pleistocene age, which, he supposed, meant that it lived a considerable number of years before America was discovered.

The reconstruction of the animal, according to the professor, will take about four weeks. Already much work has been done by Dr. Horung and himself. The students are only allowed to work when Prof. Gilbert is present, for he declares the fossils which are breakable are to him more precious than their weight in gold.

Some weeks ago the class completed the reconstruction of a sabre tooth tiger, which the scientific books call Similodom Californicus. This animal, according to the professor, preyed upon the giant sloth. Its remains were also found in the asphalt beds of La Brea ranch and there are said to be but five other whole specimens in existence.

that "following the piblication [*sic*] of your article describing the 'Death bed of the ages' curier [*sic*] hunters have been making sad havoc with much valuable material." Bebb offered "to pay the near by farmer a few dollars a month to chase away intruders" to protect the site.[66] Ida Hancock also raised these issues with Merriam, urging in 1912 "your beginning operations at the earliest possible moment as it is almost impossible to prevent vandalism in the asphalt beds. The public is under the impression that marketable treasure is to be found there, and by their surreptitious workings, in their ignorance, may destroy some really valuable bones."[67]

These issues spurred Southern California institutions to consolidate their hold over La Brea, especially around the newly established Los Angeles County Museum of Science and Art (which, like the Perth museum, mixed art gallery, history collections, and natural history museum). The director, Frank S. Daggett, was keen to monopolize work at La Brea. From 1911 he negotiated with Henry and Ida's son, George Allan Hancock, to continue permitting excavations, but to turn control over to his museum. He wrote, "I fully realize that you have plenty of reason for losing enthusiasm over La Brea collection. I feel, however, that the scientific side should not be made to suffer for the general public."[68] Daggett also appealed to family pride, promising the paleontology collections in the Los Angeles Museum, and Rancho La Brea itself, "will always be a monument to the name of your father,"[69] Henry Hancock, who had died in 1883. Local pride was an issue too. Daggett wrote how "since the Museum was opened, thousands of citizens have viewed the exhibit of mounted skeletons, and, without exception, they are astonished at the remarkable find. I am constantly asked 'What steps have been taken to preserve the remainder for southern California?'"[70] As in Nebraska, maintaining material for the locality was an important consideration.

In June 1913 George Allan Hancock gave exclusive rights over La Brea excavations to the Los Angeles museum.[71] This was a significant coup for Daggett. Equally important was patronage by the Los Angeles business community, who funded further excavations. Daggett managed this relationship by writing regular reports, flattering the board, and sending gifts. Richard Woodsworth Pridham, chairman of the Board of Supervisors of Los Angeles County, was sent a specially constructed gavel, with a handle of fossilized wood and the head consisting of "the caudal vertebrae of the Giant Ground Sloth, probably the most remarkable of the many strange forms found compared with modern animals."[72]

Daggett reported of the excavations: "That no employee has suffered from other than colds speaks volumes for the healthiness of the occupation. With the bones slushing around below, and the meadow-larks singing above, the whole bunch of grave diggers are a happy and contented bunch!"[73] This was an overly rosy account of what was challenging work. Daggett wrote that excavating the tar "is a good deal like mining; sometimes we get something, then again we draw a blank."[74] L. E. Wyman, the director of the excavations, recalled problems, describing "the tarry, tedious task of removing these precious relics," where "the gummy asphaltum, comparable to the familiar tangle-foot of sticky fly paper, covered everything," needing to be removed with heated kerosene. Trespassers and theft remained problems, and "the ubiquitous souvenir hunter in all his super-destructiveness and purloining propensities eluded our watchful eyes and did his deadly work,"[75] which led to the team leaving modern animal bones as decoys.

Daggett's reports illustrate values around the site: it allowed the extraction and accumulation of vast amounts of material, described in similar terms to mineral wealth or artistic treasures. At the end of the two-year concession, Daggett wrote that 1,586 skulls and 3,023 boxes of bones had been excavated at a cost of $25,595.34.[76] The bones were overwhelmingly of predators, notably "the saber-toothed tiger which contributed six hundred and thirty perfect skulls, including every possible variation from the tiny kitten to the old 'Tom' with sabers eight inches long," whose "exchange value will add many a rare skull to the comparative collections at the Museum."[77] Large herbivore, bird, and small mammal remains were also found, giving a view of the ancient ecological community. A particularly noteworthy specimen was a complete skeleton of what was termed *Elephas imperator* (imperial mammoth) mounted at the museum. The fossil elephant was a great attraction, with Daggett noting it "will form the board's contribution to the World, to be placed with California's colossal products, together with the Big Trees, the highest water falls, and perhaps our friend Wiggins [*sic*] big pumpkin!"[78] Size was frequently emphasized in US attempts to gain a greater position in the world and was often a key point in the promotion of dinosaurs as well.[79] The Imperial Mammoth and the overall abundance of material from La Brea were further examples of the wonders of scale.

La Brea gained a national and international reputation. Henry Fairfield Osborn wrote to Daggett of "the joy with which I saw the admirable method you have adopted in collecting, preparing, mounting, and exhibiting the

FIGURE 14.3. Workers at La Brea, with oil wells visible in the background. Image RLB89b. Courtesy of The Natural History Museum of Los Angeles County.

FIGURE 14.4. Mounted La Brea animals at the Los Angeles Museum. Image RLB846. Courtesy of The Natural History Museum of Los Angeles County.

precious animals of Rancho La Brea,"[80] and offered to send elephant models in exchange for La Brea material. In 1915 the American Association for the Advancement of Science met in San Francisco and this was seen as such an opportunity to market the site that Daggett negotiated a six-month extension to the concession, so visitors traveling through Los Angeles could see the pits being excavated.[81] La Brea was also visited by political and economic figures, including Presidents Theodore Roosevelt and Taft and General George Washington Goethals and Thomas Edison.[82] Roosevelt wrote to Merriam in 1910, "What an extraordinary and wonderful fauna our continent did hold at that time! Well, I am glad I saw the African fauna at any rate."[83] Clover recalled how Roosevelt "revealed an intimate acquaintance with prehistoric animals," and when shown the bones "was able to bestow their Latin names, which he rattled off with the familiarity of a savant."

However, he was less admiring of "the great ground sloth," and "expressed surprise that it could have survived in such company as the bear, lion, and saber-tooth tiger. He remarked that it couldn't run any faster than a man trots and was powerless to protect itself against the larger carnivore."[84]

While the concession was supposed to last only twenty-four months, Daggett secured several extensions from George Allan Hancock. In 1916 Hancock offered to bequeath the site to the County of Los Angeles to establish a scenic park and museum, although difficulties owing to costs and the extent of required work meant this was not finalized until 1924. To gain support (and funds), Daggett corresponded with paleontological authorities. While some were concerned that building at La Brea might make future excavations difficult, most saw it as an opportunity. William Diller Matthew at the AMNH was particularly enthusiastic, writing how it would be of "great and widespread interest to scientists and to non-scientific people as well," as La Brea "fairly dwarfs to insignificance any other Pleistocene fossil locality, and of quite sensational interest in the method by which the fossils were accumulated. It ought to be familiar not to a few scientists alone but to all educated people."[85] Rancho La Brea and the wonders of the American Pleistocene needed to be secured and publicized. The horrors of the "death trap of the ages," the abundance of material, and the way the site drew connections with past communities of lost animals all constructed a new canonical fossil region away from the traditional centers of paleontological authority.

PART IV

THE END OF THE AGE OF MAMMALS

CHAPTER 15

The Coming of the Age of Man

Loss, Extinction, and Decline

IN THE *AMERICAN MUSEUM JOURNAL* IN 1916, WILLIAM DILLER MATTHEW IMAGINED the life of a *Smilodon* as "the Scourge of the Santa Monica Mountains." He described it perched on peaks above the plains of Pleistocene California, which "all belonged to him by 'weapon-right'—to him and his race, by their strength and activity and the terrible curving sabers that were their favorite weapons." The natural world was defined by conquest and dominance by the strongest predators. The animal was the lord of the region, as "none of the inhabitants of the plain dare dispute their sovereignty." Other creatures were described, including "dun-colored and obscurely striped" horses, "with heavy black manes and zebralike heads" and "the bison, big and black and shaggy-maned," who were "comparative newcomers in this country, immigrants from some distant region who had crossed the mountain passes to the north" (following the paleontological idea that the Pleistocene bison migrated from Asia to North America). Meanwhile "his favorite prey" was "the big clumsy, slow-moving ground sloth that waddled around in such stupid confidence that its heavy hair and thick bone-studded skin made it invulnerable.[1]

Yet the *Smilodon*'s confidence was misplaced. As readers following paleontological news from California might have anticipated, its hunt led it

FIGURE 15.1. An imaginative reconstruction of *Smilodon* at La Brea by Erwin Christman. Matthew, "Scourge of the Santa Monica Mountains," 468. Courtesy of the American Museum of Natural History Library.

to "curious looking pools, each surrounded by a bare black patch on which nothing grew." The more intelligent and superstitious elephants "looked upon the place with horror, and could not be induced to venture into its vicinity" (explaining why their bones were rare in the pits). On leaping onto the grazing ground sloth, the two tumbled into the tar of La Brea, and "the demon of the Black Pools seized him and held him in its dreadful clutch . . . hopelessly doomed to follow his intended victim to an awful and lingering death in the black and sticky depths of the asphalt pool."[2] Even the sovereign of the Santa Monica mountains was powerless against deep, natural forces. The story then zoomed out, describing the new exhibit *Asphalt Group* at the AMNH, built from La Brea fossils. There was a further implication: dominance had again passed by "weapon-right" to a new sovereign: humans.

Another self-imagined "great hunter," former US president Theodore Roosevelt made similar analogies regarding his 1909 trip to East

Africa—sponsored by *Scribner's Magazine* in return for journalistic articles, and by several American museums for animal specimens shot by his party. His reflections were read by hundreds of thousands, if not millions, of people, and tied his experiences to sedimentary views of time, imagining modern Africa as retaining traces of the Age of Mammals. Roosevelt wrote how "for months my companions and I travelled and hunted in the Pleistocene. Man and beasts alike were of types our own world knew only in an incalculably remote past." His African guides were "really men such as those of later Paleolithic times."[3] The animals also reflected deep pasts. Roosevelt's son exclaimed when seeing a rhinoceros: "'Look at him,' said Kermit, 'standing there in the middle of the African plain, deep in prehistoric thought.' Indeed the rhinoceros does seem like a survival from the elder world that has vanished; he was in place in the Pliocene; he would not have been out of place in the Miocene; but nowadays he can only exist at all in regions that have lagged behind, while the rest of the world, for good or for evil, has gone forward."[4] The rhinoceros was regarded as ancient and stupid, either having never developed from a state like the small-brained Dinocerata, or having declined from a greater ancestor. Elephants meanwhile remained magnificent in the mind of the Western scholar-hunter, "unique among the beasts of great bulk in the fact that his growth in size has been accompanied by growth in brain power,"[5] and potentially had a longer-term future. Cope's and Marsh's ideas, that evolution was a morality play marked by the growth of the brain, continued to frame views of nature.

Roosevelt's recollections were of course almost as much a fabrication as Matthew's imagining of *Smilodon* at La Brea. Eastern Africa had been integrated into Indian Ocean trading networks for centuries, and British and German colonialism imposed new plantation and settler systems. Roderick P. Neumann notes that "Roosevelt . . . travelled a path in East Africa that had already become a well-worn tourist circuit,"[6] being the same region where Annie Alexander had traveled with her father on their ill-fated hunting trip several years previously. Roosevelt's account hinted at these changes. While he wrote that "day after day I rode . . . without seeing, from dawn to sundown, a human being save the faithful black followers, hawk-eyed and steel-thewed, who trudged behind me,"[7] he neglected to mention that these "faithful black followers" numbered almost two hundred people. The denial of coevalness was mixed with trepidation. Roosevelt noted how his train struck several animals over the journey, including hyena, giraffe and rhino, "continual mishaps such as could only happen to a railroad in the Pleistocene!"[8] The symbol of progress was literally killing the last representatives of the Age of Mammals.

Roosevelt and Matthew asserted wonder and empathy at mammalian creation. Imagination was necessary to transform the bones of La Brea into the drama of the hunting and dying *Smilodon*, or place humans and animals of 1900s East Africa in the deep past. But these sentiments were compounded by threat and loss. The great beasts had disappeared in most of the world, and the few areas retaining large mammals were threatened. Concerns over changes in the natural world mixed with fin-de-siècle anxieties over modernity, loss, and degeneration. Human society seemed to be reaching a precarious phase, as did nature itself. Reflections on change from the Age of Mammals to the Age of Man were accompanied by feelings of decline.

In these debates humans were separated from the animal world, and their appearance represented an epochal shift. Joseph LeConte described the current era as the "Psychozoic" or "Age of Mental Life," a term widely used by American writers. In LeConte's view, "all previous ages, were reigns of *brute force* and *animal ferocity*." By contrast, the Psychozoic was "characterized by the *reign of mind*," where "dangerous animals decreased in size and number, and the useful animals and plants were introduced, or else preserved by man."[9] The transition implied in earlier discussions returned, that for all their majesty and importance, the decline of great mammals was necessary to ready the world for humans. Yet this triumphal story of human supremacy was predicated on loss. Earlier periods may have been times of "brute force and animal ferocity," but they were also dynamic, majestic, and full of natural wonders disappearing from the modern world.

Understanding these changes was difficult, and troubling. The sheer amount of information about the Age of Mammals assembled over previous decades was difficult to comprehend and structure, especially into easy narratives of progress. And while the Age of Man or "Psychozoic" was often conceptualized through improvement (particularly in the human world), there was persistently a tragic flip side, with the much admired animal world declining. Sometimes, as in Matthew's account, this led to attempts to see the world, either in the present or the past, from the animal's point of view, although loaded with anthropomorphic assumptions. More emotive engagement with nature also sometimes privileged "mythic" ways of understanding the world—whether these be traditions of Indigenous people or reflections on Stone Age engagement with ancient mammals. Insecurity over the ability of science to monopolize interpretations of the great beasts gave space for other ways of knowing, albeit in occluded or modified forms. All revolved around resonant questions: What was preserved from the Age of Mammals, what was lost, and what could potentially still be saved?

Systematizing the Age of Mammals

The significance—but also complexity—of the Age of Mammals can be seen in two near-contemporary systemizing works: Henry Fairfield Osborn's *The Age of Mammals in Europe, Asia and North America* (1910) and William Berryman Scott's *History of Land Mammals in the Western Hemisphere* (1913). These two American paleontologists, with parallel careers and a close relationship, were in prime positions to build general narratives of the fossil mammals. Indeed, Scott and Osborn initially intended to write a single, jointly authored opus of mammalian life. However, the collaboration broke down, not from any theoretical or conceptual disagreements, but due to difficulties coordinating their work. As a result, the two books were presented as complementary rather than competitive.

Scott's and Osborn's books had many common features, most notably the extensive use of artwork. Scott's work contained 32 plates and over 100 drawings by Bruce Horsfall, while Osborn's contained 220 figures, including several reconstructions by Charles R. Knight. The art partly reflected desires to communicate paleontological ideas to nonspecialists. Scott wrote that "to the layman names like *Uintatherium* or *Smilodon* convey no idea whatever, and all that can be done is to attempt to give them a meaning by illustration and description."[10] However, understanding the animals also required artistic sensibility (directed by the scientific master). Scott wrote that it was "the perfectly feasible task, of paleontology to make these dry bones live," and reassured the reader that the restorations were more than "pleasing, graceful or grotesque fancies." Anatomical features like large ears, or "the presence or absence of a proboscis can nearly always be inferred with confidence from the character of the bones of the nose and muzzle."[11] Even for coloration "there is more to go on than sheer guess-work." Analogies with modern animals provided hints. The thick grey skin of elephants and rhinos hinted that large prehistoric mammals like *Titanotherium* and *Uintatherium* may have been similar. Widespread ideas that "the development of the individual is, in some respects at least, an abbreviated and condensed recapitulation of the history of the species"[12] meant that the stripes and spots of young tapirs and pigs could indicate their Eocene or Miocene ancestors were also striped and spotted. Through linking forms of evidence, paleontology could reconstruct the beasts of the past.

The two works faced difficult choices in ordering the diversity of fossil mammals into a coherent structure. Osborn's book was a narrative history, tracing the mammal groups, "their competitions, migrations, and extinctions, and of the times and places of the occurrence of these great events in the world's history" and displaying "this history in all its grandeur."[13] The Age of Mammals was natural history in epic mode, mirroring rise-and-fall narratives of human civilizations. Scott's work meanwhile was organized "zoologically," with opening and closing chapters on paleontological methods and concepts, but most chapters focused on distinct groups. Special attention went to the ungulates and proboscideans, with centerpiece analyses of the horse, camel, and elephant as examples of linear evolution. The middle chapters dealt with completely extinct herbivores from North and South America, before moving on to carnivores and primates. The two final chapters examined edentates and marsupials as the two "lowest" forms of mammal. This structure, partly drawing on the chain of being, but also moving against it, indicates that what "progress" in the mammals meant was becoming less clear.

Scott's and Osborn's works also differed in geographic focus. Both centered North America's evolutionary history, but connected it with other regions. These focuses were partly opportunistic, reflecting the collections in Princeton and New York, but also depended on the authors' valuations of natural history. Osborn drew on the AMNH's vast holdings to construct a grand narrative across Holarctica, the postulated landmass comprising Eurasia and North America, and presented as a single dynamic zone where most significant mammal groups originated. Other parts of the world were brought in, especially Africa, which was "throughout the Age of Mammals a great center of mammalian evolution and contributed its full quota to the world stock of modernized mammals,"[14] including sirenians and proboscideans. South America showed the "migration" of great sloths and glyptodons into North America, and Australia illustrated deeper histories through "primitive" marsupials. However, these were excursions rather than key locations; the main story was always in Holarctica. North America was central to earth's history—an important assertion at a time when US institutions were keen to assert their global significance.

The story of the mammals followed Osborn's principles of adaptive radiation on a grand scale, with an initial flowering of types in the Eocene, and then extinction and refinement across the Tertiary. Like older evolutionary epics, changes occurred on a cooling and harshening earth, with a lush Eocene northern hemisphere becoming colder and more arid as eras progressed. Mammals attuned themselves to the changing environment. Eocene animals were "altogether of very ancient type: they exhibit many primitive characters, such as extremely small brains, simple, triangular teeth, five digits on the hands and feet, prevailing plantigradism . . . the first grand attempts of nature to establish insectivorous, carnivorous and herbivorous groups."[15] Then the Oligocene saw a huge "revolution" in mammalian life, as more familiar creatures invaded new regions, bulky and specialized animals died out in the drying climate, and generalized ones survived. The Miocene saw grassland-dwelling creatures, with "the appearance in great numbers of herbivores with long-crowned or grazing teeth such as the horses (*Protohippus* and *Hipparion*) and the camels (*Procamelus*)."[16] The fauna continued "modernizing" across the Pliocene and the destructive fluctuations of the Pleistocene. While narratives were similar to earlier evolutionary epics, this history was marked by decline and extinction, as well as development and perfection.

Scott's focus on the Western Hemisphere reflected the Princeton collections at Nassau Hall, which contained fossils from the American West excavated during university geological expeditions, and those retrieved by John Bell Hatcher from Patagonia (along with extensive correspondence with Florentino Ameghino). The evolutionary histories of North and South America illustrated biogeography and evolution. North America was sporadically connected with Asia and sometimes Europe, and was the original "home" of dogs, camels, and horses, and saw the cats and elephants reach their full potential in *Smilodon* and the Columbian Mammoth. South America meanwhile was an isolated island continent for much of its history, difficult to conceptualize as "the stock of adjectives, such as 'peculiar,' 'bizarre,' 'grotesque' and the like, already overworked in dealing with northern forms, is quite hopelessly inadequate when everything is strange."[17]

Both authors partly told stories of progress and modernization, but extinction also played a crucial role, especially as extinction was a major

Fig. 118.—Some of the commoner Pampean mammals, reduced to a uniform scale, with a pointer dog (in the frame) to show the relative sizes. 1. †*Dædicurus clavicaudatus.* 2. †*Glyptodon clavipes,* †glyptodonts. 3. †*Macrauchenia patachonica,* one of the †Litopterna. 4. †Pampas Horse (†*Hippidion neogæum*). 5. †*Toxodon burmeisteri,* a †toxodont. 6. †*Megatherium americanum.* 7. †*Mylodon robustus,* †ground-sloths.

FIGURE 15.2. The Pampean mammals. In a link with hunting, a beagle dog is used for scale. Scott, *History of Land Mammals*, 212. Reproduced by kind permission of the Syndics of Cambridge University Library.

concern around 1900, with anxiety over the disappearance of animals like the quagga, American bison, passenger pigeon, and thylacine.[18] Scott argued that the lost South American groups, the toxodonts, litopterns, and astrapotheres, were "of the archaic, non-progressive type and have long been extinct."[19] He also followed the conventional line that Dinocerata, titanotheres and entelodonts had a "deficiency of brain-development [which] was at least one of the factors which led to the early extinction of the group."[20] Beyond this Scott avoided extensive discussion of extinction, even for later periods when he noted how "North America during the Pleistocene was far richer in mammalian life than it was when the continent was first settled by Europeans."[21] Osborn meanwhile saw extinctions moving with mammalian progress, following "the great law of mammalian improvement through the elimination of the least fitted."[22] Extinctions became more finely tuned as geological time progressed, with the Eocene seeing the extinction of whole orders; the Oligocene, Miocene, and Pliocene seeing "less progressive" families die out; and Pleistocene extinctions mainly occurring at the level of species, with the last remnants of many entire lineages disappearing.

Despite these explanations of extinction as natural and often necessary, some instances were difficult to explain. Osborn saw the Pleistocene as a culmination of life, being "the grandest and most varied assemblage of the entire Caenozoic Period on our continent," with "an array of elephants more varied and quite as majestic as those of the Old World," "great herds of large llamas and camels are interspersed with enormous troops of horses," "the varied types of giant sloths," "great armored glyptodonts," and a variety of predatory cats."[23] However, following this abundance, there came a sudden decline with "the ruthless and world-wide extinction of highly adaptive kinds of mammals."[24] A dramatic shift occurred in recent geological history, accompanied by the appearance of humans. Osborn wrote how "the Quaternary is a time of transition, of vast extinction through natural causes, as well as geographic redistribution of life. During this epoch man becomes the destroying angel, who nearly completes the havoc which nature has begun."[25] However, humans were not the root cause, but the final executioner of the great beasts. Extinctions were due to deeper, and potentially more troubling, forces.

Scott and Osborn attempted to synthesize knowledge of the beasts of prehistory around motifs of progress, migration, and extinction. They also illustrated the complexity of the Age of Mammals and uncertainty over paleontological knowledge. This tension is reflected in the reception of the works. While aimed at both scholarly and educated lay audiences, they do not seem to have been widely read. They were favorably reviewed in scientific journals, but reviews in the popular press focused on the illustrations. The *New York Post* wrote of Osborn's work that "the pictures are so numerous, artistic, and instructive that the casual inquirer might gain much from then alone, while a preliminary survey of them will materially aid the comprehension of the text, which is confessedly far from 'light reading.'"[26] The *New York Times* similarly noted that Scott's book "is not for the average, but for the exceptional lay reader," but nevertheless showed "a horde of the bizarre and almost incredible creatures of earlier epochs."[27] Scott noted in his memoirs "that the work has had only a *succès d'estime*," being favorably reviewed, but "the sales have been small."[28] The financially savvy Osborn argued with his publisher for a large advance rather than royalties, on the assumption the book would not sell more than 2,000 copies, being "of a semi-popular, semi-scientific character, quite expensive to publish and which *cannot have a very large sale*."[29] Despite this, Osborn was surprised it sold only 147 copies in its first year, and his editor wrote, "The sales of the book are a great disappointment to us, as I had supposed that a much larger public existed for a really authoritative work of this kind."[30]

The disappointing sales of Scott's and Osborn's books should not imply that interest in deep time had declined. Other books like Henry Neville Hutchinson's *Extinct Monsters* and Wilhelm Bölsche's *Tierwanderungen in der Urwelt* went through multiple editions, sold large numbers of copies, and focused much attention on fossil mammals. Interest in the deep history of animal life remained strong. The relatively small reach of Scott's and Osborn's elite guides indicates the complexity of paleontological work by the 1900s. Systematizing material was difficult, while models of development were entangled with concepts of human history, of rise and fall, progress, and uncertainty for the future. Above this hung extinction and decline. Whether implicit in Scott's narrative, or overtly stated in Osborn's, the Age of Mammals first saw the steady pruning of the tree of mammalian life, and then relentless scything of huge groups over the Quaternary. Losses in the present were situated alongside declines in the past. Did this make modern extinction and animal diminishment a "natural" phenomenon, keeping with general trends in world history, or was it even more tragic, with loss now threatening to overturn all creation? These synthesizing works were marked

FIGURE 15.3. The vitrine of great quaternary carnivores at the Paris Muséum d'Histoire Naturelle. Boule, "Ménagerie," 196. Author's collection.

by doubt and anxiety, rather than being confident, crowning achievements of over a century of debate on the fossil mammals.

Art, Antiquity, and the Return to the Pleistocene

In 1905 a new exhibit was set up at the Gallery of Paleontology at the Jardin des Plantes: The Vitrine of Great Quaternary Carnivores. Behind glass were skeletons of two lions, a cave hyena, a wolf, and three cave bears (one of whom pawed the smashed skull of an aurochs) mounted on hidden armatures as if snarling and clawing at the viewer. The vitrine was funded by Edmond de Rothschild, the backer of the ill-fated French paleontological mission in the Fayum, and an important patron of the Muséum. Marcellin Boule publicized his "menagerie of quaternary animals"[31] in the press, discussing his intentions to show "the first known congeners of wild beasts whose antics still excite us . . . the public—with a little help from the imagination—will get an idea of what the great hunts of thirty or forty thousand years ago may have been like."[32] The vitrine confronted the visitor with "the skeletons of the principal ferocious animals which our prehistoric ancestors struggled against."[33] These had been ancient humanity's rivals. But while they had been dangerous foes in ancient periods, humanity triumphed. The late Pleistocene assured human dominance over the natural world. Humans were hunters and dominators, even in this early state.

Western views of the deep past, and of human relations with extinct animals, depended on comparative analogies, particularly long-running linkages between Stone Age Eurasians and modern peoples judged as "primitive." Works like William Sollas's *Ancient Hunters and Their Modern Representatives* made these links extensively, alternating between chapters on prehistoric and modern populations.[34] These texts argued that the San and Aboriginal Australians represented low stages of development and primitive racial types, more akin to prehistoric peoples than anything in the modern world. A strong racist current, disparaging their societies and cultures and arguing for their inevitable "extinction," was present in many works. However, comparative views sometimes had different agendas. Growing from senses of loss, the idealization of "noble savages," and new currents in ethnography that attempted to understand other cultures more empathically, Indigenous knowledge could be instrumentalized to create new senses of the deep past.

Human dominance over the natural world was conceived of as conceptual as well as technological. Interest in prehistoric art expanded significantly in the years around 1900, with discussion of Paleolithic rock art—especially the painted caves in southwestern Europe. Illustrating hierarchies of knowledge, the first were found in 1868 in Altamira in northern Spain, but were rejected by the French prehistoric establishment as too advanced to be the productions of prehistoric humans, and therefore most likely fakes.[35] It was only in the 1900s that similar sites in France were found, especially at Font-de-Gaume, leading to reevaluations of the Spanish caves. The caves of southern Europe were seen to allow insights into both the Ice Age beasts and human engagement with them. Rather than the snarling predators of the Vitrine of Carnivores, the prehistoric artists of southern Europe mainly depicted medium-size herbivores. A large monograph on Font-de-Gaume by Louis Capitan and Henri Breuil was funded by Prince Albert I of Monaco, who took prehistoric archaeology as an area of patronage. The monograph focused much attention on images of mammoths, the solitary woolly rhino, and a small number of carnivores, but by far the most numerous depictions were of horses, bison, and reindeer, positioning the art within Lartet's Reindeer Age and showing the importance of these animals for ancient humans.

Paleolithic art showed human interaction with ancient creatures. Comparative analogies were frequent. The animals painted in the caves were compared with living creatures, and with the art, tools, and habitation sites of Indigenous people in southern Africa and Australia. The painted caves of France were aligned with "the festivals of the Australians aiming to promote the multiplication of kangaroos."[36] This again relegated the San and Aboriginal Australians to the past, but made their societies crucial for understanding an increasingly esteemed period of prehistory. The animal images were frequently interpreted in terms of spiritual mastery. Salomon Reinach described cave art as "hunting magic," designed to control the animals, with beasts depicted in the hope that their numbers would increase,[37] reinforcing notions that the religion of hunting societies was functional and aimed at subsistence.

Debate on paleolithic art was an international preoccupation. Osborn saw it as representing the creative genius of a particularly "gifted race" of prehistoric humans, the Cro-Magnons (whom he differentiated strongly from other prehistoric peoples). The art demonstrated "a unity of mind and spirit" across long ages among a people who were "the Palæolithic Greeks; artistic observation and representation and a true sense of proportion and of beauty were instinct with them from the beginning."[38] This was not a view of human universalism, or comparison with modern "primitive" populations. The Cro-Magnons in Osborn's view were ancestral to modern white races and showed artistry tempered by interaction with the natural world. Paleolithic art revealed the soul of an elevated race and could be used to understand ancient animals. Charles R. Knight's Pleistocene mural, for example, depicted a herd of mammoths modeled on artworks like the Mammoth of La Madeleine, frozen mammoths from Siberia, and his own observations of elephants. The ancient elephants were simultaneously at their apogee while also marching into a bright oblivion.

It was agreed that early humans had interacted with, hunted, and fought the Pleistocene animals. Yet did this mean that they were also responsible for their extinction? Contemporaries were often unwilling to answer this affirmatively, preferring complex multicausal explanations (recalling Osborn's citing of "man" as "the destroying angel" after longer developments). In *Extinct Animals*, Lankester wrote that while "it is obvious in many cases that another animal, Man, interferes," it was nevertheless the case that "before man appeared on the scene there were changes going on, and different kinds of animals succeeded one another."[39] Extinction was part of nature, rather than something always blamed on humans. The latter stages of the Pleistocene, with huge swings in climate, were stressful periods of reduced food. While prehistoric humans were understood as hunting horses and reindeer, and were often depicted fighting mammoths and cave bears, this was not generally taken as the cause of the animals' extinction. Instead, their decline reflected shifts in climate; humans dominated the environment without responsibility for annihilating the animals.

The prevalence of mammoths, reindeer, and other large beasts in Paleolithic art indicated their importance for early humans, providing food and resources while fulfilling spiritual and cultural roles. The original condition of humanity depended on interaction with these creatures. Many accounts, especially drawing on racialist or degenerational idioms, aimed to retrieve this interaction with ancient animals. At the AMNH Osborn felt murals of Pleistocene creatures could reconnect modern urban humans with their natural faculties.[40] Communing with relics of the mammoth and reindeer in museums could reawaken these senses. While this was an extreme example, the imaginative reveries of ancient animals made the life of "primitive" humans an original condition to return to, rather than something easily disparaged or brushed away.

Modern Extinctions, Loss, and Survival

If "primitive" people were often regarded as incapable of eliminating large animals, this was not the case for agricultural and settled humans. The Age of Man—tied to human civilization and mind—was different, driving the extinction of many creatures. Henry Knipe's *Evolution in the Past* closed with a gloomy prognosis for the future of "brute life," which was currently "at a very low ebb" compared with paleontological history. Elephants were now "reduced to one African and one Indian species. . . . Rhinoceroses, once ranging freely over several continents, are now known only in restricted areas of Africa, India, and the Malayan countries." "Tapirs, once dwellers in Europe, Asia, and both continents of the new world, are to-day found only in isolated districts of Central and South America, and the Malayan countries." While "man . . . has made a wonderful progress,"[41] this was at the expense of the animal world.

Declines of animal life were inevitable, as "civilization" developed and human needs took precedence over nature. As humans cleared land for farming and domesticated animals, wild creatures were driven away. Competing with carnivores in defense of livestock and with wild herbivores for grazing and agricultural space, settled human societies needed to triumph. The historic disappearance of aurochs, bear, wolf, and beaver in Europe continued this process, as did the decline of bison on the North American prairie and the diminishment of rhinos, elephants, and tigers in Asia. Only animals useful to humans could survive. Human triumph over nature was often seen as depending on animal allies, or at least subordinating and "enslaving" domestic creatures. Figuier's expression that mammals were of interest because they supplied humanity's "animal auxiliaries"[42] was a double-edged sword. Utility made mammals interesting in an economic sense, but diminished the wonder that attracted scholars and publics to the beasts of the past and present.

In many cases lost environments were calls to action, rather than just opportunities for emotive reflection. Ideas of decline and transience in the natural world, and the assumed importance of natural systems to early humans, reinforced demands that the remnants of the Age of Mammals should be preserved where they could still be found, or potentially even re-created. The history of conservation and restoration has seen a huge literature, moving beyond earlier hagiographic accounts regarding conservation as connected with preservation—but also racial theory, colonialism, hunting, and eugenics.[43] Paleontology and the new "memory" of ancient landscapes were important constituents of these debates. Fears of loss and "impoverishment" compared to past ages made maintaining what remained of large mammal faunas a pressing concern. That the "racial health" of modern humans also required engagement with large mammals and wild places fed into narratives linking human and animal.

Many late nineteenth-century texts, especially those produced in Europe, contrasted the abundant geological past with modern faunas. Wallace wrote, "We live in a zoologically impoverished world, from which all the hugest, and fiercest, and strangest forms have recently disappeared," admittedly with a double implication for humans that "it is, no doubt, a much better world for us now they have gone."[44] More emphatically, Lankester's essay, "The Effacement of Nature by Man," was a call to preserve animal life. In the context of earth's history, modern humans were a catastrophe:

"If man continues to act in the reckless way which has characterised his behaviour hitherto, he will multiply to such an enormous extent that only a few kinds of animals and plants which serve him for food and fuel will be left on the face of the globe." Humans were like a plague, on track to transform "the gracious earth, once teeming with innumerable, incomparably beautiful varieties of life, into a desert—or, at best, a vast agricultural domain abandoned to the production of food-stuffs for the hungry millions which, like maggots consuming a carcass, or the irrepressible swarms of the locust, incessantly devour and multiply."[45] Wilhelm Bölsche in Germany ended *Tierwanderungen in der Urwelt* with a plea for conservation, as "the last surviving characteristics of the Tertiary period, which were once ours, now only enliven distant steppes and jungles of southern Africa and India, and are murdered for the sake of some ephemeral purpose (to decorate fashionable women's hats or make billiard balls out of elephant's teeth)."[46] These were patrician forms of anger, contrasting the modern rarity of large animals with their abundance in the past, showing how humans destroyed environments and creatures.

National parks aimed to preserve and maintain nature, but other projects sought to bring back the ancient past. J. W. Gregory's use of Aboriginal knowledge and *Diprotodon* fossils to dream of water-engineering projects was an extreme manifestation. More usually, attempts to restore Pleistocene animals focused on domestic breeding. Understanding that the ancestors of horses, cattle, sheep, and dogs were powerful creatures adapted to varied conditions ensured that restoration was sometimes regarded as a matter of utility. Just as modern humans needed to remain in touch with primitive values to stave off degeneration, returning domestic creatures to their Pleistocene state would prevent their decline. In a period where antiquity and the primordial were valued, breeding back ancient species reified "national" or "regional" animals. The most dramatic example were the Heck brothers in Germany, who attempted to breed back the aurochs as symbols of ancient Germanness.[47] Restoring animals to these "original" states connected modern breeding with ancient pasts.

Yet there was another side to debates on extinction. The years around 1900 saw increasing engagement with another question: Could some of the beasts of prehistory (or their close relatives) still live in places little known to European and American science? The idea that great prehistoric animals might survive somewhere in the world was rarely engaged with for much of the nineteenth century (with a couple of exceptions, such as discussions that sea serpent stories derived from encounters with surviving marine reptiles, an idea vigorously discussed in the 1850s and 1860s).[48] Indeed, the implausibility of large animals like mammoths and giant sloths surviving when so much of the world had been explored and when so many Indigenous stories were catalogued was a central argument used by Cuvier to cement the idea of extinction. However, in the 1900s possibilities of survival were increasingly raised.

The new credibility of prehistoric survivals drew from some well-publicized animals. In the 1900s the okapi, while well known to West and Central African people, was hunted, caught, and made known to Europe and North America. Efforts also went into acquiring specimens of the pygmy hippopotamus, which had been known to Western scholars since the 1840s, but rarely observed or caught.[49] Unknown large mammals therefore could exist in regions of the world only partly "explored," even if there were fewer of these territories left (and indeed the idea of "unexplored" lands was very much a European confection by the 1900s). Interpretations of these animals also echoed characterizations of early mammals. The okapi seemed to chimerically mix giraffe- and horselike features, mapping onto characterizations of Tertiary animals like *Sivatherium* as linking different modern creatures. And the pygmy hippo was often interpreted as an undifferentiated marsh dweller, much like the Eocene beasts studied by Cuvier a century before.

Increased interest in new animals also drew from interaction with Indigenous people. As discussed throughout this book, Western naturalists were often dependent on local people for finding fossil localities, learning how to travel in particular territories, and for labor in digging, packing, and transportation. Around 1900 there was also an increased valuation of Indigenous knowledge as potentially offering entry points to the ancient animal world. Stories of animals and lost landscapes linked with the deep past. Carl Hagenbeck publicized an expedition to the Congo aiming to find a living dinosaur inspired by stories of the Mokele-mbembe—although it is difficult to assess how much this was a publicity stunt for a more standard animal-collecting trip. In his later career, Charles William Andrews referred to East African legends of the Nandi Bear, a mysterious clawed animal which only emerged at night, as possibly related to the clawed chalicotheres.[50]

An especially high-profile search for prehistoric survivors occurred in Patagonia. In 1895 a landowner and explorer of German origin, Hermann Eberhard, located a hide and bones in a cave at Última Esperanza in southwestern Patagonia on Chilean territory. Some of the remains seemed human and others animal. While the human remains were, according to the Swedish zoologist Einar Ljönnberg, "instantly burned to drive away the spooks (!), I suppose,"[51] by the local residents, the hide was kept and preserved. News of the site spread around scholarly networks: a team of Swedish surveyors under Otto Nordenskjöld brought fragments of hair and skin from the cave back to Uppsala, leading to reports in English and German (aimed at international audiences, indicating the assumed significance of the specimen).[52]

In November 1897, Francisco Moreno visited the area while supervising surveyors demarcating the Argentinian–Chilean border. Moreno claimed to have seen, by chance, the hide hanging from a tree, saying it "attracted my attention most strangely, as I could not determine to what class of Mammalia it could belong, more especially because of the resemblance of the small incrusted bones it contained to those of the Pampean *Mylodon*." This account seems overly serendipitous, and it is more likely that Moreno was informed of the hide by his surveyors. Moreno's assertion that "the inhabitants of the locality looked upon it as an interesting curiosity, some of them believing that it was the hide of a cow incrusted with pebbles, and others asserting that it was the skin of a large Seal belonging to a hitherto unknown species," constructed a story of the scientific master bringing true knowledge to a remote place. Part of the skin was taken to La Plata, and analyzed as likely from a giant sloth (from the presence of bony ossicles in the hide), and compared with frozen mammoths and rhinos in Siberia, preserved moa in New Zealand, and an ancient "mummified human body painted red, with the head still covered in part with its short hair wonderfully preserved" that Moreno described in Patagonia in 1877.[53]

Both Moreno and Florentino Ameghino used the hide from Última Esperanza to build influence, linking the specimen with the antiquity of the southern continents. However (possibly following their rivalry and different places within international networks), they came to different conclusions. Ameghino connected the hide with a story told by Ramón Lista, an Argentinian military officer and surveyor in Patagonia who claimed to have shot a strange animal, thinking it to be a pangolin. Ameghino linked this with Indigenous Patagonian stories of the *iemisch* or *hyimché*, a "strange creature, with long claws and a terrifying appearance, impossible to kill because it has a body impenetrable alike to firearms and missiles." Ameghino argued this could be "the last representative of a group which was believed to be quite extinct, a gravigrade edentate related to *Mylodon*."[54] He named it *Neomylodon listai* in honor of the now dead Lista and urged expeditions to seek it out. Tournouër similarly claimed to have heard stories of the iemisch from the Tehuelche, and recalled having once shot at an unidentifiable puma-size animal. In 1900 he wondered, "Does the future hold in store for me, in my next exploration, the chance to capture this mysterious animal, despite the difficulties of the enterprise? I very much desire it in the interests of science."[55] Patagonia was further marked as a land of mysteries, where antiquity and the present merged.

In 1899 Rudolf Hauthal, a German-born naturalist engaged in geological and geographic surveys in Patagonia, decided to conduct excavations at the cave at Última Esperanza, having heard that another Swedish team had been working there. He hired four laborers and spent a week excavating, locating more giant sloth fossils, the remains of a big cat, smaller herbivores, a dog, and an apparent human shoulder blade, which he brought to the La Plata Museum. That year's *Revista* of the museum featured Hauthal, Santiago Roth, and Robert Lehmann-Nitsche all analyzing the finds under the heading of "the Mysterious Mammal of Patagonia." While following Moreno in arguing the bones belonged to long-dead animals, their analysis of the site led to more extraordinary conclusions. Many bones bore cut marks, there was dried hay in the cave, and the stems of plants in preserved sloth dung seemed cut. The sloth also had wounds on its head. Lehmann-Nitsche deduced "it was humans who killed the great edentate, took its hide, tore it to pieces, and ate it raw."[56] This constructed a tentative narrative for the site: humans arrived in Patagonia in a warm interglacial period and kept and fed the great sloth in their cave, before slaughtering and eating it and using its hide as a blanket. The story was inscribed in the terminology, with the animal named *Grypotherium domesticum*.[57]

The La Plata scholars also engaged with Indigenous stories around animals, and particularly the iemisch. Roth thought *Grypotherium* was an inoffensive plant eater and so could not be the iemsich. The stories seemed to describe "an animal of the character of the cats,"[58] so it might relate to felid remains found at Última Esperanza. Roth went on to say that the Tehuelche chief Kankel, his sometime guide, had told him that his grandfather had known a mysterious creature which killed numerous horses around Lake Buenos Aires. Roth continued that "as all Indians are very superstitious and have many tales of this kind, I did not pay attention to him," but when he visited the lake, Kankel made fearful excuses not to go farther. This convinced Roth there was actual danger, most likely this great cat, which "the present-day Indians" misunderstood, as they "know it only from traditional tales, and combine the characters of different animals into one, confusing them with each other."[59] This was enough for Roth to tentatively name the fossil cat bones *Iemisch listae*, a name both enshrining Indigenous stories, and the animal's possible survival. However, Hatcher opposed it, writing that names of "barbarous origin" like *Iemisch* should be "discouraged," and "in *Iemisch listai* we have an instance in Zoological Science, which if not unique, it surely ought to be, of a species in which the original type may be fairly said to consist of traditions collected among an entirely uncivilized people."[60]

Moreno sent a notice on the skin to Britain, building on his connections with Lydekker and Arthur Smith Woodward, and seeking to inspire a British-Argentinian expedition across Patagonia, Australasia, and Antarctica to connect the histories of the southern lands. The hide was reported by the British press, and an article in the *Daily Express* led its founder, Cyril Arthur Pearson, to finance the British hunter and journalist Hesketh Hesketh-Prichard to travel to Patagonia to search for the animal—although Hesketh-Prichard's eventual book on his travels noted "during the whole time I spent in Patagonia I came upon no single scrap of evidence of any kind which would support the idea of the survival of the Mylodon."[61]

The debates around *Neomylodon*, whose hide was found in newly surveyed regions and aligned with Indigenous belief systems, brings us back to the first case traced in this book—the preserved body of the mammoth in northern Siberia. These two occurrences, at almost opposite ends of the earth and a century apart, were exceptional, being rare instances where parts of animal bodies other than bones were interpreted as the remains of lost creatures. However, they show important consistencies in how the deep animal past was made known across the nineteenth century. Animal material was inscribed into landscapes and territories, linking the past and present and inspiring meditations on the modern animal world, which were consistent across the century and constant underlying motifs within paleontological discourse.

In some respects, paleontological knowledge re-enchanted and restored the ancient earth. It also incurred tension. Arguments for the persistence of lost animals reinforced tropes around the Age of Mammals. It was a great and magnificent era in earth's history and key to understanding the present, but could only be known through interconnected evidence—modern animals, Indigenous stories, fossils, geological strata, philosophical reflections on creation and development, and engagement with the landscape. It had also not entirely receded. Many parts of the world and many creatures retained the imprint of the Age of Mammals, though by the 1900s these were rarer, more shadowy, and distant. Assertions that great sloths and chalicotheres could perhaps be living in distant parts of the world drew from understandings that time, development, and history moved unevenly. Ancient beasts only fully receded when human civilization overran their environments. While the movement of the eras ground on, not all relics of the Age of Mammals had disappeared. Perhaps they survived, perhaps they still thrived, and the wonderful inhabitants of the antique earth still lived on. Or perhaps they did not, and humans were increasingly alone in their Age of Man.

CONCLUSION

> Ladies and gentlemen, we arrive here at the end of a very long voyage into the past. . . . I hope that all these terrible creatures do not trouble your sleep. I would be sorry if tonight, you believe yourself transported to the land of the ceratosaurs, mastodons and *Machairodus*.
>
> I would prefer to think that you will leave with a good idea of the truly exciting interest presented by paleontology.
>
> It not only illuminates the past, but it reveals the future by showing that all changes obey a law of progress. . . .
>
> And so it is, with these old skeletons and petrified bones, through this nature which has been dead for thousands of centuries, that we see the seeds, which develop and germinate, under the light of science, into the noble flower of the ideal!

WITH THESE WORDS, MARCELLIN BOULE CLOSED HIS 1902 SUNDAY LECTURE AT THE PARIS Muséum d'Histoire Naturelle on "Giant Creatures of Other Times." Boule led his audience around the Gallery of Paleontology, following the teleological arrangement "to present from among the thousands of ancient creatures revived by science, the ones which are most striking to our imagination." These culminated in the great mammals, who on the extinction of the reptiles "would become in their turn majestic creatures; they were the new Lords of Creation."[1] The fossil mammals were wondrous, part of a dramatic narrative of life. However, this was not the main lesson Boule wanted to impart. He argued that life was linked across the ages, and that deep-time sciences made a series of lost worlds relevant to the present. The modern world itself bore traces of long-term developmental forces. The life of the past showed the world was old and suffused with history, which mirrored myth, religion, and the human past, and explained the character of modern animals and environments.

One contention of this book is that to understand nineteenth-century engagement with animals and environments, we need to bring in paleontology. The Age of Mammals was made known at exactly the same time that Western scholarship reinterpreted and sought to master the world and its animal inhabitants, and by many of the same institutions and people. The deep past was not just about monsters and strangeness (even if this could be part of its appeal), but the origin of the modern world. Fossils explained familiar creatures like horses and elephants, giving these privileged animals illustrious histories as evocative as those of human empires. Many modern animals were also understood through paleontology, whether these be the "ancient" tapir, the widespread hyena, the "declined" sloths, or the "primitive" marsupials. Paleontology marked out particular places and landscapes as tied to the deep past, which could include dramatic badlands, open pampas, and mysterious caverns, but also mines and building sites. The field demonstrated the antiquity of nature and potentially showed its future, which was often conceptualized as a future of human dominance. The entire world held traces of the past, if the rigor and imagination of the paleontological scholar—and the labor of excavators—were brought to bear.

Multiple skills and practices were required for fossil work. These were often based on discipline and hierarchy, as scholars built institutions and gave meaning to fossils. Nevertheless, numerous actors were critical for paleontological work: hunters, agricultural workers, miners, builders, preparators, artists, and others. These people were often subordinate but not powerless—whether they were prospectors and excavators arguing for wages, landowners demanding credit for excavation on their territories, preparators and craftsmen producing studiable fossils, or inhabitants of particular places taking fossils for their own commercial, religious, or political purposes. Fossil work required many people with many skills, not all of whom were even interested in the fossils as relics of the deep past. The field depended on work and labor, as well as erudition.

Constructing the Age of Mammals also depended on relations between places. Understanding the last era of earth's history before humans required research across the world, wherever mammal fossils were found. This was a world-building project, defining modern territories through their fossil past and bringing the modern world and its inhabitants—whether human or animal—under the sway of visions of past environments and development. These concepts mobilized numerous ways of knowing and traditions, and much of their power was in their ability to incorporate older ideas into novel histories.

The world-building project of paleontology was especially important because it was deeply integrated within the expansion of global power structures and expanding industrial and commercial economies. A persistent trend has been that work on fossils followed exploitation and control of territory. Paleontology was defined by economies and empires. The deep past connected the mastery of lands and their resources with new ideas about the history of life, a conceptual dominance over environments, with fossils being valued because they could be positioned within the overall narrative of life.

Yet geographies of paleontological science moved in unpredictable ways, showing the complexity of imperial and economic connections across the nineteenth century. Some projects discussed in this book are potentially easy to slot into narratives of colonialism and science: fossil research followed hot on the heels of settler colonialism and the dispossession of Indigenous peoples in the Americas and Australasia; India served as a territory for scientific work by British scholars, but with considerable mobilization of Indian knowledge and expertise; and paleontological science in Egypt operated as a complex codominance between Western interest groups and assertive (but sometimes hidden) Egyptian workers. However, other examples have been less straightforward, showing different ways that geographies of scientific knowledge could operate. Scholars in Australia, Argentina, and the western United States using their control of particular sites to build positions reinforces current historiographic trends questioning clear centers and peripheries. Additionally, paleontology and fossil work often grew from more informal scholarly and commercial influence—from the French and German work in Greece to foreign scholars in South America. Finally, work in Europe often depended on similar conceptual and economic control as occurred in colonial contexts.

Paleontology rested on patterns of power—although power that moved variably, and was often contested. The Age of Mammals itself, and its implications for the modern world, could potentially build a comforting image of improvement across the ages. Mammals were privileged as diverse, intelligent, emotional, and social. Seeing them as the summit of animal life, and as presaging "human" qualities, was central to their importance. Valuing mammals cemented the ideas of progress often associated with the nineteenth century and reworked older notions of scales of life. While many scholars attempted to deconstruct the chain of being or argue that life was more complicated than linear models implied, the chain continued to undergird understandings. Defining the mammals allowed fossils and paleontology to reinforce older notions.

But the history of mammals was never entirely a comforting story of linear progress. It often destabilized simple teleological or hierarchical ideas. The narratives of the Age of Mammals showed a general upward direction, but moved through a series of worlds with their own qualities, disrupted by large events like the volcanoes and continental movements of the Miocene, as well as deluges or ice ages. As the superiority of mammals was often thought to rest on their diversity, constructing clear scales was often difficult, even fruitless. The way paleontologists constructed the Age of Mammals as a period of wonder and grandeur increasingly led to fear of the declined state of the modern world. While this was often understood in vague sense of Providence, as the world was made "fit" for humans by the decline of huge herbivores and fearsome predators, the links between humans and animals exacerbated senses of loss and melancholy. The modern world was as much an impoverished place as one where humans formed the "crown of creation."

This book ends around the time of the First World War, with the early twentieth century being something of a break in paleontology. Reflecting in 1927 on his trip to the Bridger Formation with Osborn fifty years before, William Berryman Scott stated, "That day is gone, never to return. Fossils are there in abundance still, as they always will be, but the adventure and the romance have departed, flying from the wire fence and the Ford car."[2] Yet the loss of the "romance of the frontier" was something of an illusion: paleontology was always a field which took advantage of new accessibility and technology, and the motorcar and expansion of settlement were continuations of long-running trends. More significant was the impact of the war itself, which saw a sundering of the international and colonial networks around fossil work, and the more fragmented and blocked period of early twentieth-century paleontology deserves a separate study. Interest in the Age of Mammals continued into these years, even as dinosaurs and human ancestors began to loom larger in public imaginations and scholarly research.

The legacies of the nineteenth-century Age of Mammals still persist. Despite constant refrains by paleontologists and evolutionary biologists to understand natural diversity in nonhierarchical manners, common language still refers to mammals as "higher" creatures, and a norm in the animal world. References to animals like the rhinoceros and hippopotamus as "prehistoric," and megafauna (both living and extinct) being valued as national or local icons, all persist from these nineteenth-century debates.

More widely, modern ideas of the Anthropocene recapitulate notions of the epochal shift of the Age of Man or Psychozoic, indicating that, as argued by Christophe Bonneuil and Jean-Baptiste Fressoz, that this is hardly a novel concept.[3] Calls to conserve or restore landscapes to their "natural" state can be placed in a genealogy with the feelings of crisis presented by John Walter Gregory and Theodore Roosevelt. Just as nineteenth-century naturalists saw traces of the Age of Mammals in distant parts of the world, traces of older conceptualizations can be seen in current debates.

The structures of inequality and power that were so important for the development of nineteenth-century paleontology have also persisted. Much of this book has traced how large collections developed through taking advantage of extraterritorial and colonial links, or through working with building, mining, and economic expansion. Many of the institutions examined in this book—the British Museum of Natural History, the Muséum d'Histoire Naturelle in Paris, and the American Museum of Natural History in New York—still present themselves as central locations for the discussion of the natural world. However, this study has also shown gaps in this process, and how scholars, institutions, and people throughout the world played important roles in building deep histories. Paleontology grew out of colonial and semicolonial links, but its aggrandizing world-building project also provided space for others to build their positions. This drew from the field's key tensions: rhetorically based on accumulation in huge collections, but with the task of ever-increasing expansion often destabilizing these same projects. The construction of the deep past built a powerful new history for the natural world, based on visions of drama in the past and the present, but beset by uncertainty and mystery. The lost mammals and their deep history showed the ambitions, but also the limits, of human mastery over the world and its inhabitants. The world of the beasts may have receded, but it cast a long shadow over the present.

NOTES

Introduction: Constructing an Age of Mammals

1. J. Fabian, *Time and the Other*, 31–35. See also Qureshi, "Dying Americans"; and A. Buckland, "Inhabitants," for its specific use in nineteenth-century contexts.

2. Rudwick, *Earth's Deep History*, 1–3.

3. Chakrabarti, *Inscriptions*; and Chakrabarti and Bashford, "Towards a Modern History."

4. See for example classic works by Desmond, *Archetypes and Ancestors*; and Young, *Darwin's Metaphor*.

5. Yusoff, *Billion Black Anthropocenes*, 12.

6. Rieppel, *Assembling the Dinosaur*.

7. Daunton, *Organisation of Knowledge*; and Lightman and Zon, *Victorian Culture*, discuss the potentialities, and difficulties, within this process.

8. Sepkoski, *Rereading*. Gould, "G.G. Simpson," presents a participant's view of these developments.

9. The usually conservative politics around early twentieth-century paleobiology is discussed in Nieuwland, *American Dinosaur*; and Rieppel, "Othenio Abel."

10. Cuvier's and William Buckland's careers have been examined in Rudwick, *Worlds before Adam*; Outram, *Georges Cuvier*; and Rupke, *Great Chain*.

11. The links with stratigraphy are discussed in Rudwick, *Meaning of Fossils*, 139–42, 149–50, 264.

12. A major theme in Yusoff, *Billion Black Anthropocenes*; and studied in Rieppel "Prospecting for Dinosaurs."

13. The limited visibility of workers in early accounts is highlighted in Barnett, *After the Flood*.

14. Wylie, *Preparing Dinosaurs*; Brinkman, "Modernizing"; and Brown, "Development."

15. Cohen, *Méthode de Zadig*, 25. All translations in the book are my own, unless otherwise stated.

16. O'Connor, *Earth on Show*.

17. The field of "paleoart" has been especially engaged with in Rudwick, *Scenes from Deep Time*; and Witton, Naish, and Conway, "State of the Paleoart."

18. Daston and Galison, "Image of Objectivity."

19. A point made in Nieuwland, *American Dinosaur Abroad*, 37.

20. This point is made especially in Pickstone, "Museological Science?"

21. A common thread in museum history; see especially Alberti, *Nature and Culture*; and Rader and Cain, *Life on Display*.

22. Secord, "Knowledge in Transit."

23. Fan, "Circulating Material Objects," 210.

24. A term taken from Krige, "Hybrid Knowledge," 338–39, who argues for the need to

write transnational histories of science in ways that still take account of power differentials and concentrations.

25. Brinkman, *Second Jurassic Dinosaur Rush*; Nieuwland, *American Dinosaur Abroad*; Rieppel, *Assembling the Dinosaur*; and Fallon, *Reimagining Dinosaurs*.

26. Heumann et al., *Dinosaurierfragmente*.

27. Black, *Written in Stone*; and Rains Wallace, *Beasts of Eden*.

28. Semonin, *American Monster*; Cohen, *Destin du mammouth*; McKay, *Discovering the Mammoth*; Pimentel, *Rhinoceros and Megatherium*; and Rauch, "Sins of Sloths."

29. Fallon, *Reimagining Dinosaurs*, 63–98.

30. Rieppel, *Assembling the Dinosaur*, 140–77.

31. Fallon, *Reimagining Dinosaurs*, 20–26.

32. See esp. Sutter, "World with Us"; Tyrrell, *Crisis of the Wasteful Nation*; Spiro, *Defending the Master Race*; Powell, *Vanishing America*; Crosby, *Ecological Imperialism*; and Grove, *Green Imperialism*.

33. Schama, *Landscape and Memory*, 7.

34. Chakrabarti, "Towards a Modern History"; and Zizzamia, "Making the West" and "Restoring the Paleo-West."

35. For especially important works, see Ritvo, *Animal Estate* and *Noble Cows*; Cowie, *Exhibiting Animals*; Fudge, *Perceiving Animals*; and Fudge, *Renaissance Beasts*.

36. As well as the above, these issues have been especially well examined in Saha, *Colonizing Animals*; and Burton and Mawani, *Animalia*.

37. Ritvo, *Animal Estate*, 4.

38. Burton and Mawani, *Animalia*.

39. Flower and Lydekker, *Introduction*, 1.

40. Figuier, *Mammifères*, 1.

41. Barton, "Haast and the Moa."

42. O'Connor, *Earth on Show*, 159.

43. There are virtually no specialist studies on the history of nineteenth-century invertebrate or fish paleontology, which is a major gap in the literature. The only exceptions would be Allmon, "Invertebrate Paleontology"; and some sections of Bowler, *Life's Splendid Drama*, on invertebrate taxonomy, 97–140, and vertebrate origins, 141–202.

44. A point made in Flack, "Continental Creatures."

45. Bowler, *Invention of Progress*.

46. Lovejoy, *Great Chain of Being*.

47. Ritvo, *Platypus and Mermaid*, 30.

48. Scott, *History of Land Mammals*, 2.

49. Kete, *Beast in the Boudoir*; and McShane and Tarr, *Horse in the City*.

50. Jørgensen, *Recovering Lost Species*.

Chapter 1: The Cave and the Drift

1. Cohen, *Destin du mammouth*, 74–85.

2. Ariew, "Leibniz."

3. Barnett, *After the Flood*; and presaged in Rossi, *Dark Abyss*.

4. Sommer, "Romantic Cave?"; and Ziolkowski, *German Romanticism*.

5. Winterer, *American Enlightenments*, 41–72.

6. Mayor, *First Fossil Hunters* and *Fossil Legends*.

7. For early expansionism in the Americas and connections with scholarship, see esp. Delbourgo and Dew, *Science and Empire*; and Winterer, *American Enlightenments*. For Russia, see Burbank and Hagen, *Russian Empire*; and Jones, *Empire of Extinction*.

8. Cohen, *Destin du mammouth*; McKay, *Discovering the Mammoth*; and Semonin, *American Monster*.

9. Tolmachoff, "Carcasses," vii.

10. Ides, *Three Years Travels*, 25–26.

11. Adams, "Some Account." The story has also been discussed in McKay, *Discovering the Mammoth*.

12. Adams, "Some Account," 127.

13. Adams, "Some Account," 122.

14. Adams, "Some Account," 126.

15. Adams, "Some Account," 121.

16. Adams, "Some Account," 128–29.

17. Tilenau, *On the Mammoth*, 15.

18. Tilenau, *On the Mammoth*, 5–6.

19. Blumenbach's image and description are reproduced in Reich and Gehler, "Giant's Bones," 44–45.

20. Cooley, "Giant Remains."

21. Catesby, *Natural History*, 1:vii.

22. Morris, "Geomythology."

23. Hedeen and Faragher, *Big Bone Lick*; and McKay, *Discovering the Mammoth*.

24. Peale has generated a large literature, including Rigal, *American Manufactory*; Harvey, "Founding Landscape"; and Sellers, *Peale's Museum*.

25. Peale, *Account of the Skeleton*, 18.

26. Peale, *Account of the Skeleton*, 22.

27. Semonin, *American Monster*.

28. Koch, *Description of the Missourium*.

29. O'Connor, *Earth on Show*, 31–46, discusses the reception of the mastodon in London.

30. Peale, "Advertisement," *Historical Disquisition*.

31. Richard Owen, "Report on the Missourium," 689.

32. Peale, *Account of the Skeleton*, 44–45. The quote was drawn from Hunter, "Observations," 45.

33. See esp. Rudwick, *Great Devonian Controversy*, *Bursting the Limits*, and *Worlds before Adam*.

34. W. Buckland, *Geology and Mineralogy*; and discussed in Rupke, *Great Chain*; and Topham, *Reading the Book of Nature*.

35. Lyell, *Principles of Geology*.

36. W. Buckland, *Reliquiæ diluvianæ*, 173–74.

37. W. Buckland, *Reliquiæ diluvianæ*, 61–67.

38. F. Buckland, *Curiosities*, 91–92.

39. Simons, *Obaysch*; and Flack, "'Illustrious Stranger.'"

40. R. Owen, *History of British Fossil Mammals*, xxiii.

41. R. Owen, *History of British Fossil Mammals*, xxiv.

42. Penny, "German Polycentrism."

43. Almost all the entries in Beck and Joger, *Paleontological Collections*, had their origins in dynastic collections.

44. Phillips, *Acolytes of Nature.*

45. Noted in Ziolkowski, *German Romanticism*, 27, 36.

46. Goldfuß, *Umgebungen von Muggendorf*, v.

47. Goldfuß, *Umgebungen von Muggendorf*, 39.

48. Goldfuß, *Umgebungen von Muggendorf*, 63.

49. Anthony, "Making Historicity."

50. Goldfuß, *Umgebungen von Muggendorf*, 11–12.

51. Goldfuß, *Umgebungen von Muggendorf*, 64.

52. Goldfuß, *Umgebungen von Muggendorf*, 277.

53. Wiwjorra, *Germanenmythos*; Crane, *Collecting*; and Williamson, *Longing for Myth.*

54. Hagen, *Der Nibelungen Lied*, 99 (lines 3761–62).

55. Quoted in Goldfuß, "Osteologische Beiträge," 502.

56. Hibbert, "Additional Contributions," 314.

57. Olsen, *Upside Down World*; Moyal, *Bright and Savage Land*; Plumb, "Spectacle of the Kangaroo"; and Hubber and Cowley, "Distinct Creation."

58. Lang, "Account," 366.

59. Lang, "Account," 368.

60. Mitchell, *Three Expeditions.*

61. Mitchell, "Account," 321.

62. Mitchell, *Three Expeditions*, 2:7.

63. These points are strongly made in Olsen and Russell, *Australia's First Naturalists*; and Clarke, *Aboriginal Plant Collectors.*

64. Douglas, "Time Beneath," 258–85.

65. Lang, "Account," 366. The reprint in *Asiatic Journal* 5 (1831): 27, noted that "Bail is a native negation."

66. Similar case studies are discussed in Edmonds, "Bunyip"; and Clark, "Indigenous Spirit."

67. Pentland, "On the Fossil Bones," 303.

68. Clift, "Report," 394; and Jameson, "On the Fossil Bones," 393.

69. Pentland, "On the Fossil Bones," 307.

70. R. Owen, "Mastodontoid Pachyderm," 1.

71. Clift, "Report," 396.

72. R. Owen, "Mastodontoid Pachyderm," 11.

73. R. Owen, "Extinct Mammals of Australia," 240.

Chapter 2: Defining the Mammals

1. Cited in Pimentel, *Rhinoceros and Megatherium*; Semonin, *American Monster*; and Cohen, *Destin du mammouth.*

2. Figuier, *Terre avant le déluge*, 238.

3. Miller, *Testimony*, 93.

4. Rudwick, *Georges Cuvier* and "Georges Cuvier's Paper Museum"; Appel, *Cuvier-Geoffroy Debate*; Outram, *Georges Cuvier*; and Dawson, *Show Me the Bone.*

5. The status and singularity of this institution means it will be referred to as "the Muséum" throughout this book, to match this contemporary usage.

6. For the early history of the Muséum, see esp. Spary, *Utopia's Garden.*

7. Latour, *Science in Action*, 215–57.

8. Rudwick, "Georges Cuvier's Paper Museum."
9. Cuvier, "Mémoire sur les espèces d'éléphans," 442.
10. Cuvier, "Mémoire sur les espèces d'éléphans," 444.
11. Sellier, *Curiosités*, 24–25.
12. Cuvier, "Extract," 37.
13. Cuvier, "Sur les espèces d'animaux: 5. Mémoire," 273.
14. Cuvier, "Extract," 37.
15. Initially published in Cuvier and Brongniart, "Essai sur la géographie minéralogique," and then a much longer volume, as Cuvier and Brongniart, *Essai sur la géographie minéralogique.*
16. Cuvier and Brongniart, "Essai," 312–19.
17. Cuvier, "Introduction," in *Recherches*, 3:4.
18. Dawson, *Show Me the Bone*; and Cohen, *Méthode de Zadig.*
19. Cuvier, "Introduction," in *Recherches*, 3:3–4.
20. Cuvier, "Sur les espèces d'animaux: 1. Mémoire," 275–76.
21. Some key works in the relatively limited history of taxonomy are Pavlinov, *Taxonomic Nomenclature*; Knight, *Ordering the World*; and Ritvo, *Platypus and Mermaid.*
22. Cuvier, "7. Mémoire," in *Recherches*, 3:74–75. Several of these species were later renamed and reclassified, most notably *Anoplotherium medium*, which first became *Anoplotherium gracile* and then *Xiphodon gracile.*
23. Cuvier, "6. Mémoire," in *Recherches*, 3:70.
24. Cuvier, "Description ostéologique," 122.
25. Cuvier, "Description ostéologique," 124–26.
26. Cuvier, "Sur quelques dents."
27. Cuvier, "Description ostéologique," 132.
28. Early accounts primarily discussed South American tapirs. Reports of the Malayan tapir reached Europe in the early 1770s, but a live specimen was bought only in 1818.
29. Buffon, *Histoire naturelle*, 11:444–50.
30. Buffon, *Histoire naturelle, Supplément* 6:1.
31. Buffon, *Histoire naturelle, Supplément* 6:17–23.
32. Cuvier, "7. Mémoire," in *Recherches*, 3:63.
33. Cuvier, "7. Mémoire," in *Recherches*, 3:66–69.
34. See Sepkoski, *Catastrophic Thinking*; Barrow, *Nature's Ghosts*; Qureshi, "Dying Americans"; and Semonin, *American Monster.*
35. Ritvo, *Platypus and Mermaid.*
36. In English contexts these have been especially investigated in Fudge, *Perceiving Animals* and *Renaissance Beasts.*
37. Sloan, "John Locke."
38. Linnaeus, *Systema naturæ.*
39. Schiebinger, "Why Mammals."
40. Müller-Wille, "Linnaeus and Four Corners."
41. Müller-Wille and Charmantier, "Natural History"; and Müller-Wille, "Walnuts."
42. Pratt, *Imperial Eyes*, 31.
43. Cuvier, *Règne animal* (1817), 1:70.
44. R. Owen, *Odontography*, 1:i–ii.
45. Jacyna, "We Are Veritable Animals."
46. Martin, *Natural History of Quadrupeds*, 7.

47. A. Fabian, *Skull Collectors*; Poskett, *Materials*; and Stepan, *Idea of Race*.
48. Geoffroy Saint-Hilaire, *Cours*.
49. Desmarest, *Mammalogie*.
50. The only detailed study of Blainville is Appel, "Henri de Blainville." Blainville's initial taxonomy is presented in Blainville, "Prodrome."
51. Appel, "Henri de Blainville."
52. Ritvo, *Platypus and Mermaid*; and Moyal, *Platypus*.
53. Hubber and Cowley, "Distinct Creation"; and Moyal, *Bright and Savage Land*.
54. Rudwick, *Georges Cuvier*, 68–73.
55. R. Owen, *On the Classification*, 29.
56. Lawrence, "Exotic Origins."
57. McCook, "It May Be Truth"; and Olivier-Mason, "'These Blurred Copies.'"
58. Rothfels, *Elephant Trails*.
59. Pliny, *Natural History*, Book 8, 1:3.
60. Putnam, "Captive Audiences."
61. The complex position of whales in nineteenth-century thought is presented in Burnett, *Trying Leviathan*; and Davis, Gallman, and Gleiter, *In Pursuit of Leviathan*.
62. Rieppel, "Albert Koch's Hydrarchos Craze."
63. Figuier, *Mammifères*, 8.
64. Figuier, *Mammifères*, 293.
65. R. Owen, *History of British Fossil Mammals*, 77.
66. Page, *Past and Present*, 243–44.
67. Biographical studies of Lartet include Michard, Mille, and Tassy, *Secret l'Archeobelodon*; and Hamy, *Edouard Lartet*.
68. Hamy, *Edouard Lartet*, 3.
69. Lartet, *Notice*, 3.
70. Lartet, *Notice*, 8.
71. Rapport sur la collection d'ossemens fossils de M. Lartet, 8 Avril 1845, AJ/15/557/1, Archives Nationales de France, Paris (hereafter AN).
72. Lartet, *Notice*, 4; and the history of the site and the mastodon is described in depth in Mille, Michard and Tassy, *Secret l'Archeobelodon*.
73. Gervais, *Zoologie et paléontologie*, 1:1.
74. Miller, *Testimony*, 100.

Chapter 3: Great and Terrible Beasts

1. Mieg, *Paseo por el Gabinete*, 441.
2. Mieg, *Paseo por el Gabinete*, 442–43.
3. Included in Kaup and Klipstein, *Beschreibung*.
4. Kaup and Klipstein, *Beschreibung*, iii–iv.
5. Rieppel, "Plaster Cast Publishing."
6. A point strongly made in Nieuwland, *American Dinosaur Abroad*, 3–5.
7. Pimentel, *Rhinoceros and Megatherium*; Podgorny, "Bureaucracy, Instructions, and Paperwork"; and Rauch, "Sins of Sloths."
8. Podgorny, "Bureaucracy, Instructions, and Paperwork."
9. Bru de Rámon, *Colección de láminas*.
10. Piñero, "Juan Bautista Bru."

11. Thomson, *Legacy of the Mastodon*, 348.

12. Poliquin, *Breathless Zoo*, 11–42.

13. Bru de Rámon, "Descripción del esqueleto," 1.

14. Hoffstetter, "Rôles respectifs," 538.

15. Boyd, "Megalonyx," 431.

16. Cuvier, "Notice sur le squelette."

17. Cuvier, "Notice sur le squelette," 310.

18. Vos, "Natural History"; and Cowie, *Conquering Nature.*

19. Garriga, "Prólogo," in *Descripción del esqueleto*, n.p.

20. Garriga, *Descripción del esqueleto*, i–ii.

21. Garriga, *Descripción del esqueleto*, xii.

22. Hoffstetter, "Rôles respectifs," 542.

23. Piñero, "Juan Bautista Bru," 160.

24. Schmitt, "From Eggs to Fossils."

25. *Naturphilosophie* is discussed in Richards, *Romantic Conception*; and Gambarotto, "Lorenz Oken."

26. Pander and d'Alton, *Riesen-Faulthier*, 3.

27. Pander and d'Alton, *Riesen-Faulthier*, 6.

28. Pander and d'Alton, *Riesen-Faulthier*, 5.

29. Pander and d'Alton, *Riesen-Faulthier*, 10. Buffon's description of the sloths is found in his *Histoire naturelle*, 13:34–48.

30. Boyd, "Megalonyx"; and Thomson, *Legacy of the Mastodon.*

31. Jefferson, "Memoir," 251.

32. Wistar, "Description."

33. Cuvier, "Sur le Megalonyx."

34. Hodgson, *Memoir*, 10.

35. Hamilton Couper, "On the Geology," 38.

36. For Couper and the canal, see Smith, *Slavery*, 60; and Jennison, *Cultivating Race*, 252.

37. R. Owen, "Observations."

38. Mitchell, "Observations," 61.

39. Podgorny, "Fossil Dealers."

40. Levine and Novoa, *¡Darwinistas!*, 85–95.

41. For biographic studies of Lund, see Lopes, "'Scenes from Deep Time'"; and Holten and Sterll, "Danish Naturalist" and "Peter Wilhelm Lund."

42. Lund, "Nouvelles observations," 207.

43. Brinkman, "Charles Darwin's Beagle Voyage" and "Looking Back"; and Lister, *Darwin's Fossils.*

44. Parish, *Buenos Ayres*, 172.

45. Clift, "Some Account," 439.

46. "The Great Skeleton of the Megatherium," *Penny Magazine of the Society for the Diffusion of Useful Knowledge*, August 4, 1832, 180–81.

47. Discussed and quoted in O'Connor, *Earth on Show*, 191–93.

48. R. Owen, *Description*, 3–4.

49. R. Owen, *Description*, 15.

50. R. Owen, *Description*, 14.

51. R. Owen, *Description*, 36.

52. R. Owen, *Description*, 162.

53. R. Owen, *Description*, 157–58.

54. R. Owen, *Memoir*, 12.

55. Ward, *Notice*, 3.

56. Ward, *Catalogue*, 14.

57. Pimentel, *Rhinoceros and Megatherium*, 283.

58. Hodgson, *Memoir*, 24.

59. "Great Skeleton," 181.

60. Scheffel, *Gaudeamus!*, 9.

61. Scheffel, *Gaudeamus!*, 10.

62. Sorignet, *Cosmogonie*, 352, notes it was already there in 1854.

63. Michel-Eugène Chevreul and Alexandre Brongniart to the Minister of Public Instruction, May 25, 1837, AJ/15/841, AN.

64. Podgorny, "Camino de los fósiles."

65. Gervais, "Sur une nouvelle collection."

66. "Un nouveau mégatherium," *La Liberté*, November 14, 1873.

67. "Le Mégathérium," 305–7.

68. Kaup, "*Deinotherium giganteum*." In the original text, Kaup called the creature "*Deinotherium*" but reverted to the spelling *Dinotherium* in later studies—so the latter is the term predominantly used in this book.

69. A claim initially suggested in Cuvier, "Sur quelques dents," and then developed in subsequent editions of Cuvier, *Recherches*.

70. Kaup, "*Deinotherium giganteum*," 403.

71. Kaup and Scholl, *Catalogue*, 3–4.

72. Kaup and Scholl, *Catalogue*, 9.

73. Kaup, "Préface," in *Descriptions*, 1:n.p.

74. William Buckland, *Geology and Mineralogy*, 1:137–38.

75. Klipstein, "Geognostischer Theil," 17.

76. Kaup, "Zoologischer Theil," 2–3.

77. Kaup, "Zoologischer Theil," 3.

78. Blainville, "Note," 421–23.

79. Geoffroy Saint-Hilaire, "Communication," 429–30.

80. Kaup, "Sur la place," 527–29.

81. Wagner, "Sitzung," 468.

82. Schmidt, *Petrefactenbuch*, 171.

83. The skeleton had apparently been exhibited in London prior to this, according to Simpson and Tobien, "Rediscovery."

84. Andrews, "Note," 526.

85. Musil, "*Dinotherium* Skeleton."

86. The closest account of later *Dinotherium* research is in Osborn, *Proboscidea*, 1:95–103.

Chapter 4: Uncovering Siwalik and Pikermi

1. Davy, *Collected Works*, 7:43–44.

2. Díaz-Andreu, *World History*; and Schnapp, *Conquête*.

3. The cross-disciplinary connections between knowledge of the earth and the human past are a core feature in Chakrabarti, *Inscriptions*.

4. The Siwalik excavations are examined in Nair, "'Eyes and No Eyes'"; Geer, Dermitzakis, and Vos, "Fossil Folklore"; and Chakrabarti and Sen, "'World Rests.'"

5. Pikermi has been much less extensively studied than Siwalik; see Buffetaut, *Short History*; 114–19; and Tassy, *Gaudry*, 27–47.

6. Nair, "'Eyes and No Eyes'"; Geer, Dermitzakis, and Vos, "Fossil Folklore"; and Mayor, *First Fossil Hunters*.

7. Arnold, "Plant Capitalism."

8. Murchison, "Biographical Sketch," xxiii–liv.

9. Arnold, *Technology*; Edney, *Mapping an Empire*; and Raj, *Relocating*.

10. Ratcliff, "East India Company's Museum."

11. Raj, *Relocating*.

12. Geer, Dermitzakis, and Vos, "Fossil Folklore."

13. Davidson, *Diary*, 1:157.

14. Raj, *Relocating*, 60–94.

15. Chakrabarti, *Inscriptions*.

16. Falconer, *Paleontological Memoirs*, 1:4.

17. A trajectory traced in Trautmann, *Aryans and British India*.

18. Chakrabarti and Sen, "'World Rests.'"

19. Brown, "Memoir," 198.

20. Cautley, "Letter" (1834), 528.

21. Nair, "'Eyes and No Eyes.'"

22. Falconer, "Letter," 57.

23. Cautley, "Letter" (1835), 586–87.

24. Shapin, "Invisible Technician," 558–59.

25. Cautley, "On the Structure," 293.

26. Cautley, "Letter" (1834), 593.

27. Lyell, "Organic Remains," 249.

28. Falconer and Cautley, "Notice on the Remains," 308.

29. Chakrabarti and Sen, "'World Rests.'"

30. Cautley, "On the Structure," 293; and Chakrabarti, *Inscriptions*, 59.

31. Falconer and Cautley, "Notice on the Remains," 305.

32. Falconer and Cautley, "Note on the Fossil Camel," 130.

33. Falconer and Cautley, "Note on the Fossil Camel," 130–31.

34. Falconer and Cautley, "*Sivatherium giganteum*," 39.

35. Falconer and Cautley, "*Sivatherium giganteum*," 47.

36. Falconer and Cautley, "*Sivatherium giganteum*," 38.

37. Falconer and Cautley, "*Sivatherium giganteum*," 50.

38. Falconer, *Paleontological Memoirs*, 1:li.

39. O'Connor, *Earth on Show*.

40. Cautley, "Note on a Fossil Ruminant," 168–69.

41. Geoffroy Saint-Hilaire, "Sur le nouveau genre," 54.

42. Geoffroy Saint-Hilaire, "Sur le nouveau genre," 59.

43. Blainville, "Sur le chameau fossile," 76.

44. Murchison, "Biographical Sketch," xxxiv–xxxv.

45. Falconer and Cautley, *Fauna Antiqua Sivalensis*, 2.

46. Examined in Beaton and Ricks, *Modern Greece*.

47. Northern European philhellenism has been studied in Marchand, *Down from Olympus*; and Hagerman, *Britain's Imperial Muse*.

48. Herzfeld, *Ours Once More*; and Hamilakis, *Nation and Its Ruins*.

49. Lindermayer, "Fossilen Knochenreste," 111–12.

50. George Finlay to Charles Lyell, February 25, 1837, in Finlay, "Memoir," 137.

51. Lindermayer, "Fossilen Knochenreste," 112.

52. Lindermayer, "Fossilen Knochenreste," 112.

53. Solounias, "Mammalian Fossils," 236.

54. Wagner, "Zur Berichtigung," 650–51.

55. Wagner, "Urweltliche Säugthier-Ueberreste," 336.

56. Wagner and Roth, "Fossilen Knochenüberreste," 377.

57. Wagner and Roth, " Fossilen Knochenüberreste," 376.

58. Heracles Mitzopoulos, "Ossements fossiles découverts dans la montagne du Pentélique de Grèce: Correspondance et dessin (1852–1855)," AJ/15/840, AN.

59. Todd, *Velvet Empire.*

60. Gaudry, *Recherches scientifiques*, 78–79.

61. Gaudry, "Sur le Monte Pentélique," 611–13.

62. Duvernoy, "Sur les ossements," 252.

63. Gaudry, "Mission géologique," 506.

64. Gaudry, "Résultats," 292.

65. Gaudry, "Mission géologique," 506–7.

66. Gaudry, *Animaux fossiles et géologie*, 1:17.

67. Gaudry, *Animaux fossiles aux environs*, 6.

68. Lindermayer, "Fossilen Knochenreste," 117–18.

69. Gaudry, "Résultats," 291–93.

70. Gaudry, *Animaux fossiles et géologie*, 1:16.

71. Gaudry, "Lettre," 502.

72. Fan, "Circulating Material Objects."

73. Gaudry, *Animaux fossiles et géologie*, 1:331.

74. Gaudry, *Animaux fossiles et géologie*, 1:260.

75. Gaudry, "Note sur les carnassiers fossiles," 527.

76. Gaudry, *Animaux fossiles et géologie*, 1:327.

77. Gaudry, *Animaux fossiles et géologie*, 1:337.

78. Gaudry, "Sur la position géologique," 501–2.

79. Gaudry, "Mission géologique," 528.

80. Gaudry, "Mission géologique," 529.

81. Gaudry, *Animaux fossiles et géologie*, 1:6.

82. Gaudry, *Animaux fossiles aux environs*, 11–12.

Chapter 5: A Tale of Two Elephants

1. Gaudry, "Paléontologie," 698.

2. Gaudry, "L'éléphant de Durfort," 328.

3. Maurice Daubin, "La nouvelle galerie de paléontologie au Jardin des Plantes," *Le journal de la jeunesse*, 1885, 282.

4. Stanislas Meunier, "Galerie Paléontologique," *La nouvelle revue*, 1885, 302.

5. Gaudry, "Paléontologie," 699.

6. Gaudry, "L'éléphant de Durfort," 328–31.

7. Cazalis de Fondouce, "Sur la recontre."

8. Stahl, "Nouveau procédé."

9. Bennett, *Pasts beyond Memory.*

10. The messiness of this growth is argued especially in Alberti, "Objects."

11. For histories of the Muséum, see Limoges, "Development"; Burkhardt, "Leopard in the Garden"; Blanckaert, *Muséum*; and Schnitter, "Développement."

12. *Le Muséum d'Histoire Naturelle*, guidebook, 1884, 3, AJ/15/515, AN.

13. *Le Muséum d'Histoire Naturelle*, 6.

14. The few studies of later periods bear this out: Limoges, "Development"; and Schnitter, "Développement."

15. Report from Michel-Eugène Chevreul, director of Muséum d'Histoire Naturelle, February 29, 1849, F17/13566, AN.

16. Alcide d'Orbigny to Ministry of Public Instruction, May 5, 1853, F17/13566, AN.

17. Rapport de la Commission Relative à la Chaire de Paléontologie, January 30, 1872, F17/13566, AN.

18. Gaudry's later career has been discussed in Gaudant, "Albert Gaudry"; Tassy, *Evolution*; and Buffetaut, *Short History.*

19. Frédéric Montargis, "Le combat du Jardin des Plantes," *Le Rappel*, January 22, 1885.

20. Albert Gaudry to Alphonse Milne-Edwards, April 3, 1894, Ms 2473, Muséum Nationale d'Histoire Naturelle, Bibliothèque Centrale, Paris (hereafter MNHN-BC).

21. "Nouvelle galerie," 464.

22. "La nouvelle galerie de paléontologie," *Le petit moniteur universel*, March 20, 1885.

23. Argus, "Chronique," *La semaine des familles*, April 4, 1885, 15.

24. Argus, "Chronique," 15.

25. Jean-Camille Fulbert-Dumonteil, "Un nouveau musée," *La Lanterne*, April 25, 1886. Fulbert-Dumonteil is referring to Jacques Callot, a seventeenth-century illustrator best known for his series "The Great Miseries of War."

26. Gaudry, *Ancêtres*, 288.

27. Gaudry, *Ancêtres*, 296.

28. The early history of the British Museum is discussed in Delbourgo, *Collecting the World.*

29. R. Owen, *Palaeontology*, 2.

30. British Museum, *Guide*, 24–34.

31. Jerrold, *How to See the British Museum*, 82.

32. Masson, *British Museum*, 261.

33. Nead, *Victorian Babylon*, esp. 22–56; and Zimmerman, *Excavating Victorians.*

34. "Remains of Extinct Mammalia at Charing Cross," *Illustrated London News*, January 13, 1883.

35. Clifford, *West Ham.*

36. Manley Hopkins, "Essex Elephants," *Once a Week*, July 7, 1860, 53–54.

37. George, *Sir Antonio Brady.*

38. Brady, "Introduction," viii.

39. Brady, "Introduction," ix.

40. Brady, "Introduction," viii.

41. H. Woodward, "Excursion to Ilford," 274.

42. Dawkins, "On the Range," 277.

43. Walker, "Day's Elephant Hunting," 30–31.
44. Walker, "Day's Elephant Hunting," 36.
45. Walker, "Day's Elephant Hunting," 36–37.
46. H. Woodward, "How the Skull of the Mammoth," 93.
47. H. Woodward, "On the Curvature," 540–43.
48. Brady, "Introduction," xiv.
49. Antonio Brady to Alfred Waterhouse, February 4, 1874, included within the copy of William Davies, *Catalogue of Pleistocene Vertebrata in the Collection of Sir Antonio Brady*, 1874, Paleontology Manuscripts 423285–1001, Natural History Museum Library and Archives, London (hereafter NHM).
50. Richard Owen, "Departments of Natural History," *Annual Returns for 1872*, 22, DF940/1/1, NHM.
51. Discussed effectively in Rupke, "Road to Albertopolis"; and Yanni, *Nature's Museums*.
52. R. Owen, *On the Extent and Aims*, 6.
53. R. Owen, *On the Extent and Aims*, 126.
54. R. Owen, *On the Extent and Aims*, 113.
55. R. Owen, *On the Extent and Aims*, 68.
56. The architectural composition is discussed in Yanni, *Nature's Museums*.
57. H. Woodward, *Guide*, 15.

Chapter 6: Beasts from the West

1. Edward Drinker Cope, "The Monster of Mammoth Buttes," *Pennsylvania Monthly*, August 1873, 529.
2. Cope, "Monster," 526.
3. Cope, "Monster," 521–22.
4. Cope, "Monster," 529.
5. Cope, "Monster," 534.
6. See for example Lanham, *Bone Hunters*; Rains Wallace, *Bonehunters' Revenge*; and Jaffe, *Gilded Dinosaur*.
7. Vetter, *Field Life*.
8. Cerney, *Badlands National Park*, 8.
9. Walker, *Lakota Belief*, 108. These stories are linked with fossils in Mayor, *Fossil Legends*, 221–38.
10. D. Owen, *Report*, 1:197.
11. Prout, "Description," 248.
12. "Gigantic Paleotherium," 288–89.
13. D. Owen, *Report*, 1:197.
14. D. Owen, *Report*, 1:198.
15. Cassidy, *Ferdinand V. Hayden*.
16. Leidy, *Extinct Mammalian Fauna*, 8.
17. Leidy, *Extinct Mammalian Fauna*, 355.
18. Leidy, *Extinct Mammalian Fauna*, 72–73.
19. Leidy, *Extinct Mammalian Fauna*, 31.
20. "Palaeontologie—Nouveau genre," 254.
21. Leidy, *Extinct Mammalian Fauna*, 39.
22. "Donations to the Museum," 109.

23. Leidy, "Dr. Leidy's Memoir," 552.

24. As well as the books on the feud previously mentioned, the closest biography of Marsh remains Schuchert, *Biographical Memoir.*

25. Cope's career is discussed in Frazer, "Life and Letters."

26. Davis, "Sutler."

27. William A. Carter to Othniel Charles Marsh, August 1, 1870, Correspondence box 6, folder 225: Carter, William A., 1870–1872, MS343: Othniel Charles Marsh Papers, Yale Peabody Museum Archives, Yale University, New Haven, CT (hereafter MS343: Marsh Papers, YPMA).

28. Leidy, "On Some New Species," 167–68.

29. Leidy, "On Some New Species," 169.

30. Charles Betts, "The Yale College Expedition of 1870," *Harper's New Monthly Magazine,* October 1871, 663–71.

31. The ancient West being an immense sea was a major trope of the era, discussed in Zizzamia, "Making the West," 239–308.

32. Betts, "Yale College Expedition," 671.

33. "Stated Meeting," 515.

34. The timeline of the dispute is outlined in Wheeler, "Uintatheres."

35. The history of the concept of "priority" is another neglected area in the history of science, but has been treated in Pavlinov, *Taxonomic Nomenclature,* esp. 150–78.

36. Daston and Galison, "Image of Objectivity."

37. Marsh, "Fossil Mammals," 151–53.

38. Marsh, "On the Dates," 306.

39. James van Allen Carter to Marsh, June 2, 1872, box 6, folder 222, MS343: Marsh Papers, YPMA.

40. Mayor, *Fossil Legends.*

41. Discussed in depth in Bradley, *Dinosaurs and Indians*; and Schuller, "Fossil and Photograph."

42. The best account of Marsh's complex relations with Native Americans is in Bradley, *Dinosaurs and Indians,* 38–58; Cope is covered in 33–38.

43. Sam Smith to Marsh, April 16, 1874, box 6, folder 1279, MS343: Marsh Papers, YPMA. The original spelling used by Marsh's collectors has been retained in these quotes.

44. Smith to Marsh, April 5, 1883, box 6, folder 1282, MS343: Marsh Papers, YPMA.

45. Marsh to Smith, telegram, April 10, 1883, box 6, folder 1282, MS343: Marsh Papers, YPMA.

46. Murphey and Evanoff, *Stratigraphy,* 17.

47. Ervin Devendorf to Marsh, July 24, 1874, box 8, folder 357, MS343: Marsh Papers, YPMA.

48. Smith to Marsh, May 3, 1875, box 6, folder 1280, MS343: Marsh Papers, YPMA.

49. Devendorf to Marsh, November 12, 1873, box 8, folder 356, MS343: Marsh Papers, YPMA.

50. Devendorf to Marsh, September 5, 1874, box 8, folder 357, MS343: Marsh Papers, YPMA.

51. Smith to Marsh, September 2, 1882, box 6, folder 1282, MS343: Marsh Papers, YPMA.

52. Smith to Marsh, December 3, 1882, box 6, folder 1282, MS343: Marsh Papers, YPMA.

53. Rieppel, *Assembling the Dinosaur,* 119; and Vetter, *Field Life,* 109.

54. Devendorf to Marsh, September 5, 1874, box 8, folder 357, MS343: Marsh Papers, YPMA.

55. Aubert, *German Roots*, 20–26; Gundlach, *Process and Providence*; and Regal, *Osborn*, 36–38.

56. Van Allen Carter to Marsh, June 27, 1877, box 6, folder 224, MS343: Marsh Papers, YPMA.

57. John Chew to Marsh, June 22, 1877, box 6, folder 247, MS343: Marsh Papers, YPMA.

58. William Berryman Scott, Memoirs, Typescript, 128, box 7, folder 1, C0265: William Berryman Scott Papers, Manuscripts Division, Department of Special Collections, Princeton University Library, Princeton, NJ (hereafter C0265: Scott Papers, PUL).

59. Arnold Guyot to Marsh, October 16, 1877, box 13, folder 559, MS343: Marsh Papers, YPMA.

60. Osborn, "Memoir upon *Loxolophodon* and *Uintatherium*," 9.

61. Osborn, "Memoir upon *Loxolophodon* and *Uintatherium*," 43–44.

62. Zittel, "Museums," 194.

63. Zittel, "Museums," 194.

64. Marsh, *Dinocerata*, 1.

65. Marsh, *Dinocerata*, xvii.

66. Marsh, *Dinocerata*, 167–78.

67. Marsh, *Dinocerata*, 190.

68. Barbour, "Notes," 389.

69. Barbour, "Notes," 394.

70. Barbour, "Notes," 390.

71. Dwight, "Presentation," 77–79.

72. Walcott and Langley, "Correspondence," 22.

73. Barbour, "Notes," 388.

74. William Henry Flower to Marsh, April 11, 1885, box 11, folder 474, MS343: Marsh Papers, YPMA.

75. Gaudry, "Restauration," 1292.

76. Gaudry to Marsh, April 12, 1889, box 12, folder 496, MS343: Marsh Papers, YPMA.

77. Henry Woodward, quoted in Schuchert, *Biographical Memoir*, 11.

Chapter 7: Narratives of the Tertiary

1. The literature here is vast, but particularly relevant works include Berger and Donovan, *Writing National Histories*; Confino, *Germany*; Toews, *Becoming Historical*; Koditschek, *Liberalism*; and A. Buckland and Qureshi, *Time Travelers*.

2. The development of this visual language is discussed in Rudwick, "Emergence."

3. Lyell, *Principles of Geology*, 3:239.

4. Rudwick, *Great Devonian Controversy*.

5. O'Connor, *Earth on Show*, 55–57. Much drawing on James Hutton's reference to "the dark abyss of time" opened by geological reasoning. See esp. Rudwick, *Bursting the Limits*, 169; and Rossi, *Dark Abyss*.

6. Burrow, *Evolution and Society*; and many of the essays in Sera-Shriar, *Historicizing Humans*. These are also issues I engaged with in my earlier work; see Manias, *Race, Science and Nation*.

7. See esp. Bevir, *Historicism*; and Beiser, *German Historicist Tradition*.

8. Flower, "Extinct Animals," 104.

9. O'Connor, "From the Epic"; and Rudwick, *Earth's Deep History*.

10. For studies of the evolutionary epic, see O'Connor, "From the Epic"; Hesketh, "Evolutionary Epic"; and Lightman, *Victorian Popularizers*, 219–94.

11. Keene, *Science in Wonderland*, 21–53.

12. Zimmerman, *Wunder der Urwelt*, preface. The work was authored by Carl Gottfried Wilhelm Vollmer under the pseudonym W. F. A. Zimmerman; the pseudonym has been used in this book for clarity.

13. Rudwick, *Earth's Deep History*, 4.

14. Fraas, *Vor der Sündfluth!*, vi.

15. Figuier, *Terre avant le déluge*, xv.

16. Figuier, *Terre avant le déluge*, xvi.

17. Chambers, *Vestiges*, 123–24.

18. Flammarion, *Monde*, 20.

19. Somerset, "Textual Evolution."

20. The multifaceted discourses around progress and development in the nineteenth century are discussed in Burrow, *Evolution and Society*; and Bowler, *Invention of Progress*.

21. Page, *Past and Present*, 54.

22. Schaffer, "Nebular Hypothesis."

23. Figuier, *Terre avant le déluge*, 27.

24. Zimmerman, *Wunder der Urwelt*, 44.

25. See Rudwick, *Earth's Deep History*, 65–66 for Buffon's calculations and 232–33 for Kelvin's.

26. Zittel, *Aus der Urzeit*, 69.

27. Stepan, *Picturing Tropical Nature*.

28. Gross, *Scientific Sublime*.

29. Marsh, *Introduction*, 3.

30. Page, *Past and Present*, 82.

31. Adelman, "*Eozoön*."

32. Yuval-Naeh, "Cultivating the Carboniferous."

33. William Berryman Scott, "Kainozoic Era," box 10, folder 3: Geology Lectures to the Junior Class (1882–1883), C0265: Scott Papers, PUL.

34. Fraas, *Vor der Sündfluth!*, 330.

35. Flammarion, *Monde*, 671.

36. Fraas, *Vor der Sündfluth!*, 339.

37. Flammarion, *Monde*, 667.

38. Dawkins, *Early Man*, 31–32.

39. Marsh, *Introduction*, 25.

40. Flammarion, *Monde*, 718.

41. Page, *Past and Present*, 158.

42. Figuier, *Terre avant le déluge*, 273.

43. Flammarion, *Monde*, 691.

44. Flammarion, *Monde*, 712–13.

45. Fraas, *Vor der Sündfluth!*, 399–400.

46. Page, *Past and Present*, 161.

47. Flammarion, *Monde*, 715.

48. Page, *Past and Present*, 159.
49. Flammarion, *Monde*, 730.
50. Krüger, *Discovering*.
51. Howorth, *Mammoth*, xix–xx.
52. Figuier, *Terre avant le déluge*, 339–40.
53. Geikie, *Great Ice Age*, 106.
54. Zittel, *Aus der Urzeit*, 501.
55. Lyell, *Geological Evidences*, 180.
56. William Boyd Dawkins, "The British Lion," *Popular Science Monthly*, November 1882, 74.
57. Riper, *Men among the Mammoths*; Grayson, *Establishment*; and Gamble, *Making Deep History*.
58. Lubbock, *Pre-historic Times*; Figuier, *Homme primitif*; and Dawkins, *Early Man*.
59. Koditschek, "Narrative Time"; Reybrouck, *From Primitives to Primates*; and A. Buckland, "Inhabitants."
60. Dawkins, *Cave Hunting*, 393.
61. Dawkins, *Cave Hunting*, 395.
62. Dawkins, *Cave Hunting*, 399.
63. Lartet, "Nouvelles recherches," 231–32.
64. Figuier, *Homme primitif*, 131.
65. Jones, *Reliquiæ Aquitanicæ*, 206–7.
66. Baer, "Neue Auffindung," 291.

Chapter 8: Development, Origins, and Distribution

1. Gould, "G. G. Simpson," 153–54. Indeed, statements like Gould's were likely rhetorical attempts to bolster his "paleobiological" research program, discussed in Sepkoski, *Rereading the Fossil Record*.
2. Bowler, *Splendid Drama*.
3. Brinkman, "Charles Darwin's Beagle Voyage" and "Looking Back."
4. Darwin, *Origin*, 287.
5. Thomas Henry Huxley, "The Rise and Progress of Palaeontology," *Popular Science Monthly*, December 1881, 166.
6. T. Huxley, "Rise and Progress," 173.
7. Marsh, *Dinocerata*, 58.
8. The tensions between these models has been discussed in Bowler, *Invention of Progress* and *Progress Unchained*.
9. Kovalevsky, "On the Osteology," 19–20.
10. Harvey, *Almost a Man*.
11. Gaudry, *Ancêtres*, 32.
12. Gaudry, *Ancêtres*, 17.
13. Gaudry, *Essai*, 30.
14. Gaudry, *Ancêtres*, 234.
15. Pfeifer, "Genesis"; and Bowler, "Edward Drinker Cope" and "American Paleontology."
16. Cope, *Origin*, 16.
17. Cope, *Primary Factors*, 74–75.
18. Cope, *Origin*, 23.

19. Cope, *Origin*, 280.
20. Schuller, *Biopolitics of Feeling*.
21. Such oscillations are discussed in Moser, *Ancestral Images*.
22. Haeckel, *Natürliche Schöpfungsgeschichte*, 538.
23. Knipe, *Nebula to Man*.
24. T. Huxley, "On the Application," 653.
25. Hutchinson, *Creatures*, 87.
26. Cohen, "Mr. Bain and Dr. Atherstone."
27. R. Owen, "Report on the Reptilian Fossils," 233.
28. Seeley, *Dragons*.
29. Seeley, "Croonian Lecture," 341–42.
30. Seeley, "Note," 2.
31. Ochev and Surkov, "History of Excavation."
32. Broderip, "Observations on the Jaw."
33. Buffetaut, *Short History*, 88–91.
34. R. Owen, *Monograph of Fossil Mammalia*; and Osborn, "On the Structure and Classification."
35. R. Owen, *Monograph of Fossil Mammalia*, 111–12.
36. Marsh, "Discovery I," 83.
37. Marsh, "Discovery III," 250.
38. Osborn, "Review of the Cretaceous Mammalia," 134.
39. Marsh, "Note on Mesozoic Mammalia," 237.
40. Osborn, "Reply," 775.
41. Henry Fairfield Osborn to Robert Broom, April 11, 1904, box 17, folder 46: Robert Broom 1, VPA1, American Museum of Natural History, Department of Vertebrate Paleontology Archives, New York (hereafter AMNH-DVP).
42. The principal studies have been Browne, *Secular Ark* and "Science of Empire"; Bowler, *Life's Splendid Drama*, 371–418; Camerini, "Evolution, Biogeography, and Maps"; Vetter, "Wallace's Other Line"; and Greer, "Avian Imperial Archive."
43. Bölsche, *Tierwanderungen*; and Lydekker, *Geographical History*.
44. Browne, "Science of Empire," 453.
45. Lydekker, *Geographical History*, 6.
46. Wallace, *Geographical Distribution*, 1:5.
47. Sclater, "On the General Geographical Distribution"; and Greer, "Geopolitics."
48. Sclater and Sclater, *Geography of Mammals*, 2.
49. Wallace, *Geographical Distribution*, 1:57.
50. Ramaswamy, *Lost Land of Lemuria*.
51. Osborn, "Geological and Faunal Relations." The implications of this idea for Osborn's racial conceptions have been discussed in Regal, *Henry Fairfield Osborn*.

Chapter 9: Building and Contesting Collections

1. Sixth Annual Report from Department of Vertebrate Paleontology (1896), VPA 2: 1.1: Annual Reports, Vol. 3: 1891–1903, AMNH-DVP.
2. William Berryman Scott, Memoirs, Manuscript, Notebook 4, 243–44, box 5, C0265: Scott Papers, PUL.
3. Rainger, *Agenda*.

4. For Osborn's career and views, see Bender, *American Abyss*; Sommer, *History Within*; Regal, *Osborn*; and Cain, "Direct Medium."

5. Rieppel, *Assembling the Dinosaur.*

6. Matthew, *Fossil Vertebrates*, 8–11.

7. These were further expanded over the 1910s: dinosaur fossils were placed in an extended Great Hall of the Dinosaurs, and proboscidean and South American Pleistocene fossils were moved to the Hall of the Age of Man.

8. Matthew, *Fossil Vertebrates*, 18–19.

9. The AMNH is prominent in Rieppel, *Assembling the Dinosaur*; Brinkman, *Second Dinosaur Rush*; and Rader and Cain, *Life on Display.*

10. For an international history of the new museum movement, see Sheets-Pyenson, *Cathedrals of Science.*

11. Emphasized in Rieppel, *Assembling the Dinosaur*; Nieuwland, *American Dinosaur Abroad*; and Brinkman, *Second Dinosaur Rush.*

12. Alberti, "Objects"; Yanni, *Nature's Museums*; and Forgan, "Building the Museum."

13. Goode, "Relationships and Responsibilities," 201.

14. Flower, *Essays*, 13.

15. Schwartz, *Spectacular Realities.*

16. Rieppel, *Assembling the Dinosaur*, esp. "Accounting for Dinosaurs," 110–39.

17. AMNH, *Annual Report, 1900*, 12.

18. Annual Report for 1901, DF940/1/5: Annual Returns 1901–1905, NHM.

19. Bennett, *Pasts beyond Memory.*

20. Bender, *American Abyss*; and Rader and Cain, *Life on Display.*

21. Nyhart, *Modern Nature.*

22. Annual Report of the Director for the Year Ending March 31, 1899, Carnegie Museum of Natural History, Archives, Pittsburgh (hereafter CMNH).

23. A term promoted by Alexander, *Museum Masters.*

24. Shindler, *Discovering Dorothea*; and Berta and Turner, *Rebels, Scholars, Explorers*, 36–54.

25. Milner, *Knight*; Sommer, "Seriality"; and Cain, "Direct Medium."

26. A concept developed by Shteir, "Botany," and noted as a major feature of paleontological work by Fallon, *Reimagining Dinosaurs.*

27. Hermann, "Laboratory Methods"; and Bather, "Preparation and Preservation."

28. Henry Fairfield Osborn to Henry Woodward, January 28, 1901, DF105/41: Donations and Exchange of Duplicates with Institutions: In North and South America, 1889–1901, NHM.

29. Dingus, *Hatcher.*

30. Sternberg, *Fossil Hunter*, xiii.

31. W. D. Matthew, Tenth Annual Report for Department of Vertebrate Paleontology (1900), VPA 2: 1.1: Annual Reports, vol. 3: 1891–1903, AMNH-DVP.

32. Plan for Rearrangement of the Museum by A. Smith Woodward, 1901, DF107/3, NHM.

33. See for example McClelland, *State, Society, and University*; Pietsch, *Empire of Scholars*; and Thelin, *American Higher Education.*

34. Mayr, "Zittel," 15.

35. Tamborini, "Wurzeln" and "Reception of Darwin."

36. Dehm, "Beiträge," 14.

37. Karl Alfred von Zittel to Wilhelm von Gümbel, February 1, 1866, quoted in Mayr, "Zittel," 17.

38. Zittel, *Geschichte.*

39. Zittel, *Handbuch* and *Grundzüge.*

40. Tamborini, "Wurzeln."

41. Zittel, *Grundzüge*, 739.

42. Mayr, "Zittel," 45–47.

43. Schlosser, *Führer*, 29.

44. Manias, "Terra Incognita."

45. Schlosser, *Führer*, 4–5.

46. Scott, *Memories.* He produced earlier manuscript versions for his family which are now held in the Princeton University Library in C0265: Scott Papers, PUL.

47. Scott, Memoirs, Manuscript, Notebook 11, 726-7, box 6, C0265: Scott Papers, PUL.

48. Scott, "Morphological Discipline," 182.

49. Stuart Patton to Mother, August 17, 1886, box 27, folder 1, AC012: Scientific Expeditions, PUL (hereafter AC012, PUL).

50. Cornelius Rea Agnew, diary, box 2, folder 3, AC012, PUL.

51. John Bell Hatcher to Scott, April 24, 1894, box 27, folder 3, AC012, PUL.

52. Hatcher to Scott, August 19, 1894, box 27, folder 3, AC012, PUL.

53. Flower, "Modern Museums," in *Essays*, 30–53.

54. Albert Gaudry to Marcillin Boule, September 11, 1894, MS BOU1, Fonds Boule, MNHN-BC.

55. Gaudry to Boule, June 14, 1896, MS BOU1, Fonds Boule, MNHN-BC.

56. Seurat, "Nouvelles galeries," 37.

57. Alphonse Milne-Edwards, "Discours," in *Inauguration des nouvelles galeries d'anatomie comparée, de paléontologie et d'anthropologie, le 21 juillet 1898*, pamphlet (Paris: Masson et Cie, 1889), vi, AJ/15/843, AN.

58. Albert Gaudry quoted in "Nouvelles galeries," 322.

59. Gaudry, "Nouveau musée," 817.

60. "Nouvelle galerie," 322–23.

61. Gindhart, "Cormon's Painting Cycle."

62. Louis Liard, "Discours," in *Jubilé*, 19–20.

63. Gaudry, "Nouveau musée," 800.

64. *Exposition Universelle: Palais de l'optique: Guide*, 1900, Ms BOU 30: Travaux, 1891–1913, Fonds Boule, MNHN-BC.

65. *Jubilé*, 1.

66. Edmond Perrier, "Discours," in *Jubilé*, 6.

67. Marcillin Boule, "Discours," in *Jubilé*, 10.

68. Albert Gaudry, "Réponse," in *Jubilé*, 24.

69. Charles Depéret to Muséum d'Histoire Naturelle, January 1903, F17/13566: Chaires du Muséum, AN.

70. Boule's mode of operating is discussed in Bont, "Prehistoric Man"; and Sommer, "Mirror, Mirror."

71. "Titres et travaux scientifiques de M. Marcellin Boule (1902)," Ms BOU 6, MNHN-BC.

72. Keith, "Boule," 42–43.

73. Sommer, "Mirror, Mirror," and Hammond, "Expulsion."

74. Boule, Séance du 24 Octobre 1907, F/17/13560 Assemblée des Professeurs, 1899–1934, AN.

75. Nieuwland, *American Dinosaur Abroad*, 136–39.

76. Gaudry to Milne-Edwards, January 19, 1900, Ms 2473, MNHN-BC.

77. Director of Muséum to minister of public instruction, July 25, 1905, AJ/15/515, AN.

Chapter 10: The Story of the Horse

1. Flower, *Horse*, 4.
2. Flower, *Horse*, 7.
3. Matthew, *Evolution*, 11.
4. Henry Fairfield Osborn, "The Evolution of the Horse in America. Fossil: Wonders of the West," *Century Magazine*, November 1904, 3.
5. Gould, "Case."
6. Scott, *Theory of Evolution*, 98–99.
7. T. Huxley, "Palaeontology and Evolution," 360–61.
8. L. Huxley, *Life and Letters*, 1:495.
9. Data from Dorré, *Victorian Fiction*; and McShane and Tarr, *Horse in the City*.
10. Lydekker, *Horse*, v.
11. Flower, *Horse*, x.
12. Huth, *Works*.
13. Flower, *Horse*, 95–96.
14. Tyrrell, "Bred for Race"; and Derry, *Bred for Perfection*.
15. Matthew, "Exhibit," 120.
16. Matthew, "Exhibit," 118.
17. Osborn, *Horse, Past and Present*, 13.
18. Osborn, *Horse, Past and Present*, 12.
19. Matthew, "Exhibit," 120.
20. Bowler, *Eclipse of Darwinism*; and Bowler, "Revisiting."
21. Levit and Olsson, "'Evolution'"; and MacFadden et al., "Fossil Horses."
22. This interpretation was later to be turned around as an example of why orthogenesis was incorrect in Gould, "Origin and Function."
23. Loomis, "Momentum," 840. This was not a universal view—other scholars saw the saber-toothed cats as powerful dominant predators, as we will see in later chapters.
24. Flower, *Horse*, 18–26.
25. R. Owen, "Description," 203.
26. R. Owen, "Description," 205.
27. Lull, "Evolution," 163–64.
28. Lull, "Evolution," 170.
29. Lydekker, *Horse*, 239.
30. Matthew, "Exhibit," 12.
31. Bölsche, *Tierwanderungen*, 73.
32. Gilman, "Similar Cases," 165.
33. William Berryman Scott, Lecture, December 11, 1920, box 3, AC139: Department of Geological and Geophysical Sciences Records 1880–1994, PUL.
34. Lydekker, *Horse*, 4.
35. Lydekker, *Horse*, 239.
36. Matthew, *Evolution*, 12.
37. Lydekker, *Horse*, 277.
38. Matthew, *Evolution*, 3.
39. McShane and Tarr, *Horse in the City*.
40. Lull, "Evolution," 165.
41. Matthew, *Evolution*, 27.

42. Lull, "Evolution," 168.
43. Osborn, "Evolution," 6.
44. Osborn, "Evolution," 7.
45. Lydekker, *Horse*, 281.
46. Gidley and Osborn, "Revision," 865.
47. Osborn, "Law of Adaptive Radiation," 355.
48. Wells, *Tales*, 109–10.
49. Swart, "Zombie Zoology."
50. J. C. Ewart, "Introduction," in Zalensky, *Prjevalsky's Horse*, vii.
51. Lull, "Evolution," 182.
52. Matthew, *Evolution*, 8.
53. Bölsche, *Tierwanderungen*, 76.
54. Osborn, *Origin and History*, 1.
55. Flower, *Horse*, 151.
56. Flower, *Horse*, 203–4.
57. Flower, *Horse*, 151.

Chapter 11: Ordering the Pampas and Patagonia

1. Schmitt, *Darwin and Memory*; and Stepan, *Picturing Tropical Nature*.
2. Lopes and Podgorny, "Shaping of Latin American Museums"; Levine and Novo, *¡Darwinistas!*; and Sheets-Pyenson, *Cathedrals of Science*.
3. Wallace, *Geographical Distribution*, 2:5.
4. Lydekker, *Mostly Mammals*, 69–70.
5. Lydekker, *Mostly Mammals*, 72.
6. Brown, "Global History."
7. Maybury-Lewis, "New World Dilemma"; and Kerr, "From Savagery to Sovereignty."
8. Levine and Novoa, *¡Darwinistas!*, 97–112.
9. Burmeister, *Anales*.
10. Levine and Novoa, *¡Darwinistas!*, 113–23; Andermann, *Optic of the State*; and Sheets-Pyenson, *Cathedrals of Science*.
11. Moreno, "Museo de La Plata," 50.
12. "Documentos," vii.
13. Flower, "Los museos"; and Moreno, "Museo de La Plata," 29.
14. Andermann, *Optic of the State*, 45–57.
15. Urgell, *Arte*.
16. Moreno, "Museo de La Plata," 39.
17. Moreno, "Museo de La Plata," 52–53.
18. Moreno, "Museo de La Plata," 41.
19. "Documentos," xiii.
20. Lydekker, *Mostly Mammals*, 77.
21. Ward, "Museos Argentinos," 149.
22. Ward, "Museos Argentinos," 151.
23. The Ameghinos have recently been the subject of two recent studies by Podgorny, *Florentino Ameghino* and *Argentinos vienen de los peces*.
24. This construction has been well traced in Podgorny, "Daily Press" and *Florentino Ameghino*, 293–312.

25. Kerr, "From Savagery to Sovereignty."
26. Roth, "Beobachtungen," 446.
27. Roth, "Beobachtungen," 461.
28. Podgorny, "Bones and Devices," 267.
29. Moreno, *Viaje*, vi–vii.
30. Dixie, *Across Patagonia*, 1.
31. Dixie, *Across Patagonia*, 2–3.
32. Podgorny, "Human Origins."
33. Many of these letters are included within the series edited by Alfredo Torcelli, *Obras completas*.
34. Florentino Ameghino to Carlos Ameghino, June 1, 1895, in Torcelli, *Obras completas*, 21:168.
35. For example, James Secord, "Global Geology"; and Podgorny, "Human Origins."
36. William Diller Matthew, "Florentino Ameghino," *Popular Science Monthly*, March 1912, 307.
37. Karl von Zittel to Florentino Ameghino, January 21, 1892, in Torcelli, *Obras completas*, 21:394.
38. Florentino Ameghino to Zittel, June 4, 1892, in Torcelli, *Obras completas*, 21:385.
39. Florentino Ameghino to Zittel, December 7, 1892, in Torcelli, *Obras completas*, 21:462.
40. Florentino Ameghino to Zittel, July 7, 1899, in Torcelli, *Obras completas*, 21:761–63.
41. Boule and Thevenin, *Mammifères fossiles*.
42. Tournouër's expeditions have recently been reconstructed in Buffetaut, "Frenchman in Patagonia."
43. Gaudry, "Fossiles de Patagonie," 5.
44. Tournouër, "Recherches paléontologiques," 541.
45. Gaudry, "Fossiles de Patagonie," 5.
46. Florentino Ameghino to Carlos Ameghino, November 29, 1902, in Torcelli, *Obras completas*, 22:15.
47. André Tournouër to Florentino Ameghino, May 25, 1903, in Torcelli, *Obras completas*, 22:467.
48. Albert Gaudry to Florentino Ameghino, June 15, 1899, in Torcelli, *Obras completas*, 21:767.
49. Gaudry, "Sur la marche."
50. Tournouër, "Notes," 12.
51. Hatcher, *Reports*, 3.
52. Hatcher, *Reports*, 14.
53. Hatcher, *Reports*, 86.
54. Hatcher, *Reports*, 84–85.
55. Hatcher, *Reports*, 52–53.
56. Hatcher, *Reports*, 37.
57. William Berryman Scott, Memoirs, Manuscript, Notebook 11, 765–66, box 6, C0265: Scott Papers, PUL.
58. Scott to Florentino Ameghino, October 16, 1901, in Torcelli, *Obras completas*, 22:375.
59. Florentino Ameghino to Scott, October 20, 1901, in Torcelli, *Obras completas*, 22:377–78.

Chapter 12: Lands of the *Diprotodon*

1. "The Great Wombat," *Evening News* (Sydney), June 2, 1909.
2. Haeckel, "Zur Phylogenie," v.

3. Robin, *How a Continent.*

4. Cathcart, *Water Dreamers.*

5. Byrne, "Deep Nation."

6. Bashford, "Anthropocene," 345.

7. Brantlinger, *Dark Vanishings.*

8. Minard, *All Things Harmless.*

9. Byrne, "Deep Nation"; and Robin, *How a Continent.*

10. Minutes of the meeting of the Australian Museum trustees, January 5, 1871, Vol. 2: 01/04/1863–01/07/1874, AMS 1: Trustees Minutes, Australian Museum Archive, Sydney.

11. Australian Museum, *Report for 1884*, 29.

12. Australian Museum, *Report for 1884*, 1.

13. Minard, "'Making the Marsupial Lion'"; and Dawson, *Show Me the Bone*, 333–35.

14. "The Diprotodon," *Melbourne Leader*, May 10, 1862.

15. McEntee, "Why Mulligan."

16. Pledge, "Fossils," 68.

17. Douglas, "Pictures," 299.

18. Stirling, "Recent Discovery I," 186.

19. Stirling, "Recent Discovery II," 211.

20. Douglas, "Pictures," 292–93.

21. Stirling, "Recent Discovery II," 207.

22. Stirling, "Recent Discovery I," 186.

23. Stirling, "Recent Discovery II," 210–11.

24. "Geological—The Fossiliferous Deposits at Lake Mulligan," *Adelaide Observer*, December 9, 1893.

25. "The Adelaide Museum. Fossils from a Wilderness and a Wilderness of Fossils," *South Australian Register*, March 22, 1894.

26. Professor Stirling's Report on His Visit to Museums in America and Europe, January 6, 1902, GRG19/399, State Records of South Australia, Adelaide (hereafter SRSA).

27. Stirling, "Recent Discovery I," 185.

28. McEntee, "Why Mulligan."

29. E. Ray Lankester to Edward Charles Stirling, December 14, 1898, GRG19/394, SRSA.

30. Lankester to Stirling, December 14, 1898.

31. Lankester to Lord Tennyson, February 27 [no year], 104539, GRG19/394, SRSA.

32. Stirling to Lankester, February 14, 1899, GRG19/394, SRSA.

33. Henry Woodward to Stirling, March 28, 1899, GRG19/394, SRSA.

34. Stirling's Report.

35. Stirling, "Reconstruction," 544.

36. "An Australian Fossil," *Sydney Mail*, November 27, 1907.

37. Memorandum from Hately Waddell Marshall, General Secretary to Edgar Ravenswood Waite, museum director, September 22, 1921, AA398, South Australian Museum, Archives, Adelaide (hereafter SAM).

38. Barton, "Haast and the Moa," esp. 251–53.

39. Stirling's Report.

40. *Statistical Register*, 5.

41. Western Australian Museum and Art Gallery, *Report for 1897–8*, 97.

42. Bernard Woodward, "Fossil Marsupials," 9.

43. "Interesting Fossil Remains. Discovered in the Mammoth Cave," *West Australian*, February 2, 1910.
44. John Sharp to Bernard Woodward, May 6, 1908, A230–75–1, Western Australian Museum Archive, Perth (hereafter WAM).
45. William Ponton to Bernard Woodward, May 13, 1914, A230–75–1, WAM.
46. Glauert, "Fossil Remains," 55–56.
47. "The South-West Caves. How They Are Being Improved. A Year's Progress," *Morning Herald* (Perth), August 9, 1905.
48. Glauert, "Mammoth Cave," 248.
49. Ludwig Glauert, "Extinct Marsupials," *Our Rural Magazine*, October 1929, 415.
50. "Interesting Fossil Remains. Discovered in the Mammoth Cave," *West Australian*, February 2, 1910.
51. "Illustrated Interviews. W. A.'s Diprotodon. Bones 100,000 Years Old," *Sunday Times* (Perth), May 24, 1914.
52. "Fossil Marsupials," *Western Mail*, June 19, 1914.
53. "A New Australian Fossil," *Western Mail*, February 6, 1909.
54. J. G. Hay, "The Marsupial Fossils," *West Australian*, February 23, 1914.
55. Hay, "Marsupial Fossils."
56. Australian Museum, *Report for 1912*, 6.
57. Etheridge, *Australian Museum*, 5.
58. Fletcher, "*Palorchestes*," 363.
59. Mackness, "Reconstructing *Palorchestes*," 21–36.
60. Scott, *History of Land Mammals*, 626.
61. Dollo, "Ancêtres."
62. "Great Wombat."
63. Lucas and Le Souef, *Animals of Australia*, 13–14.
64. Glauert, "Extinct Marsupials," 412.
65. Gregory, *Dead Heart*, 3–4.
66. Gregory, *Dead Heart*, 151.
67. Gregory, *Dead Heart*, 343–44.
68. Daisy M. Bates, "Evolution of Australia. The Recent Discovery of Fossil Remains," *West Australian*, February 5, 1910.

Chapter 13: Africa as the Source of Life

1. Lankester, *Extinct Animals*, 120–24.
2. Lankester, *Extinct Animals*, 132.
3. Saha, "Colonizing Elephants"; and Heintzman, "E Is for Elephant."
4. Lydekker, *Geographical History*, 255.
5. Bölsche, *Tierwanderungen*, 58–65.
6. Reid, "Egyptian Geographical Society."
7. Lyons, *Cadastral Survey*.
8. "The Eclipse Expeditions to Assuan," *The Times* (London), August 28, 1905.
9. Jeffreys, *Views of Ancient Egypt*; and MacDonald and Rice, *Consuming Ancient Egypt*.
10. Reid, *Whose Pharaohs?*
11. Beadnell, *Topography and Geology*.
12. Andrews, "Preliminary Note II," 444.

13. Andrews, "Preliminary Note I," 403.
14. Beadnell, *Preliminary Note*, 3.
15. Beadnell, *Topography and Geology*, 13.
16. Charles William Andrews to Arthur Smith Woodward, March 4, 1903, DF PAL/100/34, NHM.
17. Frank Barlow to Smith Woodward, December 5, 1903, DF PAL/100/34, NHM.
18. Hume, *Catalogue*, 24.
19. Lankester, "A New Extinct Monster," *The Sphere*, September 12, 1903, 238.
20. Andrews, *Descriptive Catalogue*, xv–xviii.
21. Andrews, *Guide*, 6.
22. Andrews, *Guide*, 4.
23. Andrews to Smith Woodward, March 4, 1903, DF PAL/100/34, NHM.
24. Discussed in Schaffer et al., *Brokered World*.
25. Stromer von Reichenbach, "Richard Markgraf," 287.
26. Stromer von Reichenbach, "Richard Markgraf," 287.
27. Eberhard Fraas to the administration of the Royal Naturalienkabinett, January 8, 1906, folder: Prof. Dr. Eberhard Fraas III (Reisen), Staatliches Museum für Naturkunde Stuttgart, Archives, Stuttgart.
28. Fraas, "Wüstenreise," 263–65.
29. Fraas, "Wüstenreise," 266–67.
30. Fraas, "Wüstenreise," 267.
31. Fraas, "Wüstenreise," 267.
32. Nothdurft and Smith, *Lost Dinosaurs*.
33. Stromer von Reichenbach, "Eine geologische Forschungsreise," 110.
34. Stromer von Reichenbach, "Bericht," 342.
35. Stromer von Reichenbach, "Eine geologische Forschungsreise," 109.
36. Riggs, *Photographing Tutankhamun*, 141–72; Stevenson, *Scattered Finds*; and Quirke, *Hidden Hands*.
37. Schlosser, "Beiträge," 151.
38. Schlosser, "Beiträge," 103–4.
39. Stevenson, *Scattered Finds*, 29–35; and Drower, "Gaston Maspero."
40. Reid, "French Egyptology."
41. Claude Gaillard to Albert Gaudry, February 27, 1904, ARCH PAL 84(2), Muséum Nationale d'Histoire Naturelle, Archives du Laboratoire de Paléontologie, Paris (hereafter MNHN-ALP).
42. René Fourtau to Muséum d'Histoire Naturelle, March 4, 1903, ARCH PAL 84(2), MNHN-ALP.
43. Marcillin Boule to His Excellency the Minister of Public Works in Cairo, October 11, 1903, ARCH PAL 84(2), MNHN-ALP.
44. Maspero to Boule, December 2, 1903, ARCH PAL 84(2), MNHN-ALP.
45. Fourtau to Boule, January 4, 1904, ARCH PAL 84(2), MNHN-ALP.
46. Fourtau to Boule, March 3, 1904, ARCH PAL 84(2), MNHN-ALP.
47. Boule to Fourtau, March 12, 1904, ARCH PAL 84(2), MNHN-ALP.
48. Fourtau to Boule, March 24, 1904, ARCH PAL 84(2), MNHN-ALP.
49. Fourtau to Boule, March 24, 1904, ARCH PAL 84(2), MNHN-ALP.
50. Fourtau to Boule, June 11, 1904, ARCH PAL 84(2), MNHN-ALP.

51. Boule to Fourtau, June 10, 1904, ARCH PAL 84(2), MNHN-ALP.

52. Fourtau to Boule, June 24, 1904, ARCH PAL 84(2), MNHN-ALP.

53. Fourtau to Boule, June 24, 1904, ARCH PAL 84(2), MNHN-ALP.

54. Boule, "Visite de M. Andrews," memo, May 25, 1904, ARCH PAL 84(2), MNHN-ALP.

55. Henry George Lyons to Boule, February 6, 1907, ARCH PAL 84(2), MNHN-ALP.

56. Delmer and Fernández, "Género *Arsinoitherium*."

57. Henry Fairfield Osborn to Fraas, May 8, 1906, VPA1: Correspondence, box 33, folder 72: Fraas, Eberhard, AMNH-DVP.

58. Osborn to Andrews, October 8, 1906, VPA1: Correspondence, box 3, folder 35: Andrews, C.W., AMNH-DVP.

59. Morgan and Lucas, "Notes," provides a detailed account of the Fayum expeditions, and material from Granger's personal archive (now apparently untraceable).

60. Walter Granger, "Report on the Expedition to the Fayum, Egypt, 1907," VPA 2: 1.1: Annual Reports, Vol. 4: 1904–1912, AMNH-DVP.

61. Granger, "Report."

62. Granger, "Report."

63. Granger, "Report."

64. Osborn, "Fayûm Expedition," 514–15.

65. Osborn to Andrews, April 5, 1907, Correspondence, box 3, folder 35: Andrews, C.W., AMNH-DVP.

66. Granger, "Report."

67. Osborn, "Fayûm Expedition," 515–16.

68. Henry Fairfield Osborn, "Hunting the Ancestral Elephant in the Fayûm Desert," *Century Magazine*, October 1907, 816–17.

69. Osborn, "Hunting the Ancestral Elephant," 819–20.

70. Charles R. Knight, Report for the Year 1907 on Restorations, VPA 2: 1.1: Annual Reports, Vol. 4: 1904–1912, AMNH-DVP.

71. Osborn to Knight, October 31, 1907, VPA1: Correspondence, box 51, folder 57: Knight, Charles R., AMNH-DVP.

72. Osborn, "Hunting the Ancestral Elephant," 835.

73. Osborn to Richard Markgraf, November 26, 1907, VPA1: Correspondence, box 58, folder 9: Markgraf, Richard, AMNH-DVP.

74. Markgraf to Osborn, March 24, 1908, VPA1: Correspondence, box 58, folder 9: Markgraf, Richard, AMNH-DVP.

75. Osborn, "Hunting the Ancestral Elephant," 819.

76. Swart, "Writing Animals."

77. Bölsche, *Tierwanderungen*, 78–79.

Chapter 14: New Communities

1. Zizzamia, "Making the West."

2. Vetter, *Field Life.*

3. Brinkman, "Establishing Vertebrate Paleontology."

4. Samuel Williston to Erwin Hinckley Barbour, November 11, 1906, box 78: Barbour Correspondence, 1905–1906, 32-01-01: Erwin H. Barbour Papers, UNL, University of Nebraska-Lincoln, Archives & Special Collections, Lincoln (hereafter 32-01-01: Barbour Papers, UNL).

5. Brinkman, *Second Jurassic Dinosaur Rush.*

6. Kohler, *All Creatures.*

7. Barbour to George Baur, April 4, 1892, box 72: Barbour Correspondence, 1892–1897, 32-01-01: Barbour Papers, UNL.

8. Barbour to Chancellor Samuel Avery, April 3, 1913, Barbour Correspondence 1913–1914, 32-01-01: Barbour Papers, UNL.

9. Barbour to R. E. Barney, March 19, 1894, box 72: Barbour Correspondence, 32-01-01: Barbour Papers, UNL.

10. Barbour to Barney, May 1, 1894, box 72: Barbour Correspondence, 32-01-01: Barbour Papers, UNL.

11. Barbour to John Dryden, March 20, 1894, box 72: Barbour Correspondence, 32-01-01: Barbour Papers, UNL.

12. Morrill, *Morrils,* 34.

13. Morrill, *Morrils,* 1.

14. Unnamed to Charles Schuchert, July 31, 1911, Barbour Correspondence, 1911–1912, 32-01-01: Barbour Papers, UNL.

15. Barbour to Avery, November 18, 1914, Barbour Correspondence, 1913–1915, 32-01-01: Barbour Papers, UNL.

16. The Cooks and the Agate ranch have been discussed in Vetter, "Cowboys, Scientists, and Fossils"; Hunt, *Agate Hills*; Evans-Hatch, *Centuries*; and Cockrell, *Bones of Agate.*

17. Cook's own biography contains information on his career and self-construction: J. Cook, *Fifty Years.*

18. H. Cook, *Tales,* 185.

19. H. Cook, *Tales,* 185–86.

20. Harold J. Cook to Barbour, May 26, 1905, box 77: Barbour Correspondence, 1904–1905, 32-01-01: Barbour Papers, UNL.

21. Barbour to Harold Cook, n.d., box 77: Barbour Correspondence, 1904–1905, 32-01-01: Barbour Papers, UNL.

22. Harold Cook to Barbour, May 30, 1905, box 77: Barbour Correspondence, 1904–1905, 32-01-01: Barbour Papers, UNL.

23. Barbour to Avery, April 28, 1913, Barbour Correspondence, 1913–1915, 32-01-01: Barbour Papers, UNL.

24. James Cook to John Adams, September 1, 1906, VPA1: Correspondence, box 24, folder 37: Cook, James H., AMNH-DVP.

25. William J. Holland to Henry Fairfield Osborn, July 17, 1907, VPA1: Correspondence, box 44, folder 13: Holland, W. J., AMNH-DVP.

26. Osborn to Harold J. Cook, July 22, 1908, VPA1: Correspondence, box 23, folder 34: Cook, Harold J., 1, AMNH-DVP.

27. William Utterback to Douglas Stewart, April 24, 1908, box: Utterback, William H.: Correspondence, CMNH.

28. James Cook to Albert Thomson, November 4, 1907, VPA1: Correspondence, box 24, folder 37: Cook, James H., AMNH-DVP.

29. This episode is discussed in Evans-Hatch, *Centuries,* 180–81.

30. Holland, *Twelfth Annual Report,* 29.

31. Discussed in more depth in Manias, "Reconstructing."

32. Holland and Peterson, "Osteology."

33. A major theme in Nieuwland, *American Dinosaur Abroad*.

34. Holland to Barbour, October 7, 1908, box 81: Barbour Correspondence, 1908–1911, 32-01-01: Barbour Papers, UNL.

35. Harold J. Cook to William Diller Matthew, May 12, 1914, VPA1: Correspondence, box 23, folder 34: Cook, Harold J., 1, AMNH-DVP.

36. Holland and Peterson, "Osteology," 196.

37. Harold J. Cook to Osborn, April 7, 1914, Barbour Correspondence, 1913–1915, 32-01-01: Barbour Papers, UNL.

38. Vetter, "Cowboys, Scientists, and Fossils."

39. Harold J. Cook to Osborn, April 7, 1914, Barbour Correspondence, 1913–1915, 32-01-01: Barbour Papers, UNL.

40. Harold J. Cook to Barbour, April [no day], 1914, Barbour Correspondence, 1913–1915, 32-01-01: Barbour Papers, UNL.

41. Osborn to Harold J. Cook, May 11, 1914, VPA1: Correspondence, box 23, folder 34: Cook, Harold J., 1. AMNH-DVP.

42. Osborn to Harold J. Cook, May 11, 1914.

43. Osborn to Harold J. Cook, May 11, 1914.

44. Harold J. Cook to Osborn, May 29, 1914, VPA1: Correspondence, box 23, folder 34: Cook, Harold J., 1, AMNH-DVP.

45. Harold J. Cook to Matthew, May 12, 1914, VPA1: Correspondence, box 23, folder 34: Cook, Harold J., 1. AMNH-DVP.

46. Vetter, "Cowboys, Scientists, and Fossils."

47. Osborn to James Cook, July 8, 1917, VPA1: Correspondence, box 24, folder 37: Cook, James H., AMNH-DVP.

48. See Beesley, *Crow's Range*; and Speece and Sutter, *Defending Giants*.

49. LeConte, "Race Problem," 349–50.

50. LeConte, "Race Problem," 360–61.

51. Stein, *On Her Own Terms*.

52. Annie Alexander to Guy C. Earl, October 4, [1930–1931?], box 1, Series 1: Correspondence, 1901–1949, Miscellany, A-F., Annie Alexander Montague Papers, University of California Museum of Paleontology, Archival Collections, University of California, Berkeley (hereafter Alexander Papers, UCMP).

53. Alexander to Martha Beckwith, November 10, 1901, quoted in Stein, *On Her Own Terms*, 29.

54. This museum has formed the classic case study of "boundary work" in science; see Star and Griesemer, "Institutional Ecology."

55. John C. Merriam to Alexander, January 13, 1912, box 3, Series 2: Administrative and Financial Records of the Department and Museum of Paleontology, 1906–1948, Alexander Papers, UCMP.

56. Deverell and Hise, *Land of Sunshine*, 81–83.

57. There is surprisingly little on the history of La Brea, but some works include McNassor, *Los Angeles's La Brea*; Clover, *Pioneer Heritage*; and Laurence, "Pleistocene Park."

58. Orcutt, "Discovery," 338.

59. Clover, *Pioneer Heritage*, 66.

60. Orcutt, "Discovery," 339.

61. Merriam to Benjamin Ide Wheeler, April 19, 1913, box 2, Series 1: Correspondence, 1901–1949, K-Z, Alexander Papers, UCMP.

62. John Merriam, Report of Work in Paleontology, Jan 1, 1912 to April 1, 1913, box 3, Series 2: Administrative and Financial Records of the Department and Museum of Paleontology, 1906–1948, Alexander Papers, UCMP.

63. John Merriam, "Death Trap of the Ages: Tragedies of Aeons Ago: . . . A Pathetic Page from the Prehistoric Past," *Sunset Magazine*, October 1908, 465–75.

64. Editors' Note, in Merriam, "Death Trap," 465–66.

65. "Class of Girls Reconstruct Giant Sloth: Moretherium Gigias Discovered as Fossil," *Los Angeles Examiner*, July 24, 1912.

66. William Bebb to Merriam, January 21, 1909, folder 1.34, William Bebb, Carton 1, MSS C-B 970: John C. Merriam Papers, Bancroft Library (hereafter MSS C-B-970: Merriam Papers, BL).

67. Ida Hancock Ross to Merriam, July 30, 1912, folder: Hancock, Ida Ross, Carton 4, MSS C-B-970: Merriam Papers, BL.

68. Frank S. Daggett to George Allan Hancock, April 1, 1911, folder: Correspondence—G. Allan Hancock 1913–14 [*sic*], box: Rancho La Brea, 1913–1924, Los Angeles County Museum of Natural History, Archives & Library, Los Angeles (hereafter box: RLB, LACMNH).

69. Daggett to George Allan Hancock, November 22, 1912, folder: Correspondence—G. Allan Hancock 1913–14 [*sic*], box: RLB, LACMNH.

70. Daggett to George Allan Hancock, April 30, 1913, folder: Correspondence, 1916 [*sic*], box: RLB, LACMNH.

71. George Allan Hancock to Daggett, June 23, 1913, folder: Correspondence—G. Allan Hancock 1913–14, box: RLB, LACMNH.

72. Daggett to R. W. Pridham, July 2, 1914, folder: Correspondence 1913–1919, box: RLB, LACMNH.

73. Daggett to Board of Governors, February 28, 1914, folder: Progress Reports 1913–15, box: RLB, LACMNH.

74. Daggett to Sidney A. Butler, July 6, 1914, folder: Correspondence—Board of Supervisors 1914, box: RLB, LACMNH.

75. Wyman, "La Brea in Retrospect," *Museum Graphic*, January 1927, 84–86.

76. Daggett, Final Report Covering Operations at Rancho La Brea, from July 1, 1913, to July 31 1915, November 9, 1915, folder: Progress Reports, 1913–15, box: RLB, LACMNH.

77. Untitled report signed by Frank S. Daggett, November 27, 1915, folder: Correspondence—La Brea—Newspapers, 1916–18 [*sic*], box: RLB, LACMNH.

78. Daggett to Board of Governors, November 30, 1914, folder: Progress Reports, 1913–15, box: RLB, LACMNH. Frank Wiggins was secretary of the Los Angeles Chamber of Commerce, and presumably the "big pumpkin" referred to was grown by him.

79. Discussed in Rieppel, *Assembling the Dinosaur*, 6.

80. Osborn to Daggett, May 4, 1915, folder: Correspondence 1912–1917, box: RLB, LACMNH.

81. Daggett to George Allan Hancock, March 29, 1915, folder: Correspondence—G. Allan Hancock 1913–14 [*sic*], box: RLB, LACMNH.

82. Untitled report, November 27, 1915.

83. Theodore Roosevelt to Merriam, September 24, 1910, Carton 4, MSS C-B-970: Merriam Papers, BL.

84. Clover, *Pioneer Heritage*, 74.

85. Matthew to Daggett, May 17, 1916, folder: Correspondence, 1916, box: RLB, LACMNH.

Chapter 15: The Coming of the Age of Man

1. Matthew, "Scourge," 469–70.
2. Matthew, "Scourge," 470–71.
3. Roosevelt, *Book-Lover's Holidays*, 204–5.
4. Roosevelt, *African Game Trails*, 206–8.
5. Roosevelt, *African Game Trails*, 284.
6. Neumann, "Through the Pleistocene," 63.
7. Roosevelt, *Book-Lover's Holidays*, 197.
8. Roosevelt, *African Game Trails*, 18.
9. LeConte, *Elements*, 586. Italics in the original.
10. Scott, *History of Land Mammals*, 51.
11. Scott, *History of Land Mammals*, 41–44.
12. Scott, *History of Land Mammals*, 45–46.
13. Osborn, *Age of Mammals*, vii.
14. Osborn, *Age of Mammals*, 68.
15. Osborn, *Age of Mammals*, 96.
16. Osborn, *Age of Mammals*, 297.
17. Scott, *History of Land Mammals*, 261.
18. Swart, "Zombie Zoology"; Isenberg, *Destruction*; Fuller, *Passenger Pigeon*; and Freeman, *Paper Tiger*.
19. Scott, *History of Land Mammals*, 279.
20. Scott, *History of Land Mammals*, 445.
21. Scott, *History of Land Mammals*, 210.
22. Osborn, *Age of Mammals*, 172.
23. Osborn, *Age of Mammals*, 434.
24. Osborn, *Age of Mammals*, 172.
25. Osborn, *Age of Mammals*, 374.
26. "Literary News and Reviews—Mammals," *New York Post*, March 30, 1911.
27. "Some Land Mammals in America," *New York Times*, November 30, 1913.
28. Scott, *Some Memories*, 300.
29. Henry Fairfield Osborn to Brander Matthews, February 23, 1910, box 88: Age of Mammals, O835: Henry Fairfield Osborn Papers, Central Archives, American Museum of Natural History, New York (hereafter O835: Osborn Papers, AMNH-CA). Emphasis in the original.
30. George Brett to Osborn, September 6, 1912, box 88: Age of Mammals. O835: Osborn Papers, AMNH-CA. A further letter in the same series, Brett to Osborn, March 13, 1917, noted that the work had sold 1,473 copies by 1916, but this was still some way below Osborn's initial low prediction and Macmillan's hoped for sales.
31. Boule, "Ménagerie."
32. "Une ménagerie fossile," *Le Matin*, February 13, 1905.
33. Boule, "Ménagerie," 195.
34. Sollas, *Ancient Hunters*; discussed in Sommer, "Ancient Hunters."
35. Abadía, "Thinking about the Concept of Archive."
36. Capitan and Breuil, *Caverne*, 13.
37. Reinach, "L'art et la magie"; and Palacio-Pérez, "Salomon Reinach."
38. Osborn, *Men of the Old Stone Age*, 316.
39. Lankester, *Extinct Animals*, 28–29.

40. Cain, "'Direct Medium.'"
41. Knipe, *Evolution*, 214–15.
42. Figuier, *Mammifères*, 1.
43. Spiro, *Defending the Master Race*; Tyrrell, *Crisis*; and Allen, "'Culling the Herd.'"
44. Wallace, *Geographical Distribution*, 1:114.
45. Lankester, *Science from an Easy Chair*, 366.
46. Bölsche, *Tierwanderungen*, 94.
47. Lorimer and Driessen, "From 'Nazi Cows.'"
48. Rieppel, "Albert Koch's Hydrarchos Craze"; and Lyons, *Serpents, Spirits, and Skulls*.
49. Swart, "O Is for Okapi"; and Robinson, "Looking for Enigmas."
50. Manias, "Reconstructing."
51. Ljönnberg, "On Some Remains," 149.
52. Nordenskjöld, "Neue Untersuchungen," 335–36; and Ljönnberg, "On a Remarkable Piece of Skin."
53. Moreno and Woodward, "On a Portion of Mammalian Skin," 144–46.
54. Ameghino, "Existing Ground Sloth," 324–25.
55. Tournouër, "Sur le *Neomylodon*," 344.
56. Lehmann-Nitsche, "Coexistencia," 471.
57. Roth, "Descripción," 431.
58. Roth, "Descripción," 444.
59. Roth, "Descripción," 445.
60. Hatcher, "Mysterious Mammal," 815.
61. Hesketh-Prichard, *Through the Heart of Patagonia*, xiii–xiv.

Conclusion

1. Marcellin Boule, "Les créatures géants d'autrefois: Conférence du dimanche, 1902," Ms BOU 14: Conférences Diverses, 1894–1922, MNHN-BC.
2. Scott, "Development of American Palaeontology," 428–29.
3. Bonneuil and Fressoz, *Shock of the Anthropocene*.

BIBLIOGRAPHY

Archives

AM	Australian Museum Archive, Sydney
AMNH-CA	American Museum of Natural History, Central Archives, New York
AMNH-DVP	American Museum of Natural History, Department of Vertebrate Paleontology Archives, New York
AN	Archives Nationales de France, Paris
BL	The Bancroft Library, University of California, Berkeley
CMNH	Carnegie Museum of Natural History, Archives, Pittsburgh.
LACMNH	Los Angeles County Museum of Natural History, Archives and Library, Los Angeles
MM	Manchester Museum, Archives, Manchester, UK
MNHN-ALP	Muséum Nationale d'Histoire Naturelle, Archives du Laboratoire de Paléontologie, Paris
MNHN-BC	Muséum Nationale d'Histoire Naturelle, Bibliothèque Centrale, Paris,
NHM	British Museum (Natural History) (now Natural History Museum), Library and Archives, London
PUL	Princeton University Library, Manuscripts Division, Department of Special Collections, Princeton, NJ
SAM	South Australian Museum, Archives, Adelaide
SMNS	Staatliches Museum für Naturkunde Stuttgart, Archives, Stuttgart
SRSA	State Records of South Australia, Adelaide
UCMP	University of California Museum of Paleontology, Archival Collections, University of California, Berkeley
UNL	University of Nebraska-Lincoln, Archives and Special Collections, Lincoln
WAM	Western Australian Museum Archive, Perth
YPMA	Yale Peabody Museum of Natural History, Yale University Archives, New Haven, CT

Newspapers and Magazines

Adelaide Observer
Century Magazine
Daily Telegraph (Sydney)
Evening News (Sydney)
Graphic
Harper's New Monthly Magazine
Illustrated London News

Le Journal de la Jeunesse
La Lanterne
La Liberté
Los Angeles Examiner
Le Matin
McClure's Magazine
Melbourne Leader
Morning Herald (Perth)
Museum Graphic (Los Angeles)
New York Post
New York Times
La Nouvelle Revue
Once A Week
Our Rural Magazine
Pennsylvania Monthly
The Penny Magazine of the Society for the Diffusion of Useful Knowledge
Le Petit Moniteur Universel
La Petite Presse
Popular Science Monthly
Punch, or The London Charivari
Le Rappel
La Semaine des Familles
Sphere
Sunday Times (Perth)
Sunset Magazine
Sydney Mail
Times (London)
West Australian
Western Mail

Primary and Secondary Sources

Abadía, Oscar Moro. "Thinking about the Concept of Archive: Reflections on the Historiography of Altamira." *Complutum* 24, no. 2 (2013): 145–52.

Abel, Othenio. *Grundzüge der Palaeobiologie der Wirbeltiere*. Stuttgart: Schweizerbart, 1912.

Adams, Michael. "Some Account of a Journey to the Frozen-Sea, and of the Discovery of the Remains of a Mammoth." *Philadelphia Medical and Physical Journal* 3, no. 1 (1808): 120–37.

Adelman, Juliana. "*Eozoön*: Debunking the Dawn Animal." *Endeavour* 31, no. 3 (2007): 94–98.

Alberti, Samuel. *Nature and Culture: Objects, Disciplines and the Manchester Museum*. Manchester, UK: Manchester University Press, 2009.

Alberti, Samuel. "Objects and the Museum." *Isis* 96, no. 4 (2005): 559–71.

Alexander, Edward. *Museum Masters: Their Museums and Their Influence*. Nashville, TN: American Association for State and Local History, 1983.

Allen, Garland. "'Culling the Herd': Eugenics and the Conservation Movement in the United States, 1900–1940." *Journal of the History of Biology* 46, no. 1 (2013): 31–72.

Allmon, Warren. "Invertebrate Paleontology and Evolutionary Thinking in the US and Britain, 1860–1940." *Journal for the History of Biology* 53, no. 3 (2020): 423–50.

Ameghino, Florentino. "An Existing Ground-Sloth in Patagonia." *Natural Science* 13 (1898): 324–26.

American Museum of Natural History. *Annual Report of the President, Treasurer's Report, List of Accessions, Act of Incorporation, Constitution, By-Laws and List of Members for the Year 1900*. New York: American Museum of Natural History, 1901.

American Museum of Natural History. *Forty-Sixth Annual Report of the American Museum of Natural History for the Year 1914*. New York: American Museum of Natural History, 1915.

Andermann, Jens. *The Optic of the State: Visuality and Power in Argentina and Brazil*. Pittsburgh: University of Pittsburgh Press, 2007.

Andrews, Charles William. *A Descriptive Catalogue of the Tertiary Vertebrata of the Fayûm, Egypt*. London: Printed by order of the Trustees of the British Museum, 1906.

Andrews, Charles William. *A Guide to the Elephants (Recent and Fossil) Exhibited in the Department of Geology and Palæontology in the British Museum (Natural History)*. London: Printed by order of the Trustees of the British Museum, 1908.

Andrews, Charles William. "Note on the Skull of *Dinotherium giganteum* in the British Museum." *Proceedings of the Zoological Society of London* 91 (1921): 525–34.

Andrews, Charles William. "Notes on an Expedition to the Fayum, Egypt, with Descriptions of Some New Mammals." *Geological Magazine*, decade 4, 10 (1903): 337–43.

Andrews, Charles William. "Preliminary Note on Some Recently Discovered Extinct Vertebrates from Egypt I." *Geological Magazine*, decade 4, 8 (1901): 400–409.

Andrews, Charles William. "Preliminary Note on Some Recently Discovered Extinct Vertebrates from Egypt II." *Geological Magazine*, decade 4, 8 (1901): 436–44.

Anthony, Patrick. "Making Historicity: Paleontology and the Proximity of the Past in Germany, 1775–1825." *Journal of the History of Ideas* 82, no. 2 (2021): 231–56.

Appel, Toby. *The Cuvier-Geoffroy Debate: French Biology in the Decades before Darwin*. Oxford: Oxford University Press, 1987.

Appel, Toby. "Henri de Blainville and the Animal Series: A Nineteenth-Century Chain of Being." *Journal of the History of Biology* 13, no. 2 (1980): 291–319.

Ariew, Roger. "Leibniz on the Unicorn and Various Other Curiosities." *Early Science and Medicine* 3, no. 4 (1998): 267–88.

Arnold, David. "Plant Capitalism and Company Science: The Indian Career of Nathaniel Wallich." *Modern Asian Studies* 42, no. 5 (2008): 899–928.

Arnold, David. *Science, Technology and Medicine in Colonial India*. Cambridge: Cambridge University Press, 2000.

Aubert, Annette. *The German Roots of Nineteenth-Century American Theology*. Oxford: Oxford University Press, 2013.

Australian Museum. *Report of the Trustees for 1884*. Sydney: Australian Museum, 1885.

Australian Museum. *Report of the Trustees for the Year Ended 30th June 1912*. Sydney: Australian Museum, 1912.

Baer, Karl-Ernst von. "Neue Auffindung eines vollständigen Mammuths, mit der Haut und den Weichttheilen, in Eisboden Sibiriens." *Bulletin de l'Académie Impériale des Sciences de St.-Pétersbourg*, ser. 3, 10 (1866): 230–96.

Barbour, Erwin Hinckley. "Notes on the Paleontological Laboratory of the United States Geological Survey under Professor Marsh." *American Naturalist* 24 (1890): 388–400.

Barnett, Lydia. *After the Flood: Imagining the Global Environment in Early Modern Europe*. Baltimore: John Hopkins University Press, 2019.

Barnett, Lydia. "Showing and Hiding: The Flickering Visibility of Earth Workers in the Archives of Earth Science." *History of Science* 58, no. 3 (2020): 245–74.

Barrow, Mark. *Nature's Ghosts: Confronting Extinction from the Age of Jefferson to the Age of Ecology*. Chicago: University of Chicago Press, 2009.

Barton, Ruth. "Haast and the Moa: Reversing the Tyranny of Distance." *Pacific Science* 54, no. 3 (2000): 251–63.

Bashford, Alison. "The Anthropocene Is Modern History: Reflections on Climate and Australian Deep Time." *Australian Historical Studies* 44, no. 3 (2013): 341–49.

Bather, F. A. "The Preparation and Preservation of Fossils." *Museum Journal* 8 (1908): 76–90.

Beadnell, Hugh. *A Preliminary Note on* Arsinoitherium zitteli, *Beadn. from the Upper Eocene Strata of Egypt*. Cairo: National Printing Department, 1902.

Beadnell, Hugh. *The Topography and Geology of the Fayum Province of Egypt*. Cairo: National Printing Department, 1905.

Beaton, Roderick, and David Ricks, eds. *The Making of Modern Greece: Nationalism, Romanticism, and the Uses of the Past (1797–1896)*. Farnham, UK: Ashgate, 2009.

Beck, Lothar, and Ulrich Joger, eds. *Paleontological Collections of Germany, Austria and Switzerland: The History of Life of Fossil Organisms at Museums and Universities*. Cham, Switzerland: Springer, 2018.

Beesley, David. *Crow's Range: An Environmental History of the Sierra Nevada*. Reno: University of Nevada Press, 2008.

Beiser, Frederick. *The German Historicist Tradition*. Oxford: Oxford University Press, 2015.

Bender, Daniel. *American Abyss: Savagery and Civilization in the Age of Industry*. Ithaca, NY: Cornell University Press, 2009.

Bennett, Tony. *Pasts beyond Memory: Evolution Museums Colonialism*. London: Routledge, 2004.

Berger, Stefan, and Mark Donovan, eds. *Writing National Histories: Western Europe since 1800*. London: Routledge, 2002.

Berta, Annalisa, and Susan Turner. *Rebels, Scholars, Explorers: Women in Vertebrate Paleontology*. Baltimore: John Hopkins University Press, 2020.

Bevir, Mark. *Historicism and the Human Sciences in Victorian Britain*. Cambridge: Cambridge University Press, 2020.

Black, Riley. *Written in Stone: The Hidden Secrets of Fossils*. London: Icon Books, 2012.

Blainville, Henri de. "Notice sur la tête de *Dinotherium giganteum*, actuellement à Paris." *Comptes rendus de l'Académie des Sciences* 4 (1837): 421–27.

Blainville, Henri de. "Prodrome d'une nouvelle distribution systématique du règne animal." *Bulletin des Sciences* 8 (1812): 113–24.

Blainville, Henri de. "Sur le chameau fossile, et sur le *Sivatherium* des Sous-Himalayas méridionaux." *Comptes rendus de l'Académie des Sciences* 4 (1837): 71–76.

Blanckaert, Claude, Claudine Cohen, Pietro Corsi, and Jean-Louis-Fischer, eds. *Le Muséum au premier siècle de son histoire*. Paris: Muséum National d'Histoire Naturelle, 1997.

Bölsche, Wilhelm. *Tierwanderungen in der Urwelt*. Stuttgart: Kosmos, 1914.

Bommeli, Rudolf. *Die Geschichte der Erde*. Stuttgart: J. H. W. Dietz, 1890.

Bonneuil, Christophe, and Jean-Baptiste Fressoz. *The Shock of the Anthropocene: The Earth, History and Us*. London: Verso, 2016.

Bont, Raf de. "The Creation of Prehistoric Man: Aimé Rutot and the Eolith Controversy, 1900–1920." *Isis* 94, no. 4 (2003): 604–30.

Boule, Marcellin. "Une ménagerie d'animaux quaternaires." *La Nature* 33, no. 1 (1905): 195–98.

Boule, Marcellin, and A. Thevenin. *Mammifères fossiles de Tarija*. Paris: Imprimerie nationale, 1920.

Bowler, Peter. "American Paleontology and the Reception of Darwinism." *Studies in the History and Philosophy of Science, Part C* 66 (2017): 3–7.

Bowler, Peter. *The Eclipse of Darwinism: Anti-Darwinian Evolution Theories in the Decades around 1900*. Baltimore: Johns Hopkins University Press, 1983.

Bowler, Peter. "Edward Drinker Cope and the Changing Structure of Evolutionary Theory." *Isis* 68, no. 2 (1977): 249–65.

Bowler, Peter. *The Invention of Progress: The Victorians and the Past*. Oxford: Blackwell, 1989.

Bowler, Peter. *Life's Splendid Drama: Evolutionary Biology and the Reconstruction of Life's Ancestry, 1860–1940*. Chicago: University of Chicago Press, 1996.

Bowler, Peter. *Progress Unchained*. Cambridge: Cambridge University Press, 2021.

Bowler, Peter. "Revisiting the Eclipse of Darwinism." *Journal of the History of Biology* 38, no. 1 (2005): 19–32.

Boyd, Julian. "The Megalonyx, the Megatherium, and Thomas Jefferson's Lapse of Memory." *Proceedings of the American Philosophical Society* 102, no. 5 (1958): 420–35.

Bradley, Lawrence. *Dinosaurs and Indians: Paleontology Resource Dispossession from Sioux Lands*. Parker, CO: Outskirts Press, 2014.

Brady, Antonio. "Introduction." In William Davies, *Catalogue of the Pleistocene Vertebrata, from the Neighbourhood of Ilford, Essex, in the Collection of Sir Antonio Brady*, vii–xvii. London: Printed for private circulation, 1874.

Brantlinger, Patrick. *Dark Vanishings: Discourse on the Extinction of Primitive Races, 1800–1930*. Ithaca, NY: Cornell University Press, 2003.

Brinkman, Paul. "Charles Darwin's Beagle Voyage, Fossil Vertebrate Succession, and 'The Gradual Birth & Death of Species,'" *Journal of the History of Biology* 43, no. 2 (2010): 363–99.

Brinkman, Paul. "Establishing Vertebrate Paleontology at Chicago's Field Columbian Museum, 1893–1898." *Archives of Natural History* 27, no. 1 (2000): 81–114.

Brinkman, Paul. "Looking Back with 'Great Satisfaction' on Charles Darwin's Vertebrate Paleontology." In *The Cambridge Encyclopedia of Darwin and Evolutionary Thought*, edited by Michael Ruse, 56–63. Cambridge: Cambridge University Press, 2013.

Brinkman, Paul. "Modernizing American Fossil Preparation at the Turn of the Twentieth Century." In *Methods in Fossil Preparation: Proceedings of the First Annual Fossil Preparation and Collections Symposium*, edited by Matthew A. Brown, John F. Kane, and William G. Parker, 21–34. Petrified Forest, 2009. http://preparation.paleo.amnh.org/assets/FPCSvolume-Final.pdf.

Brinkman, Paul. *The Second Jurassic Dinosaur Rush Museums and Paleontology in America at the Turn of the Twentieth Century*. Chicago: University of Chicago Press, 2010.

British Museum. *A Guide to the Exhibition Rooms of the Departments of Natural History and Antiquities*. London: Printed by order of the Trustees of the British Museum, 1861.

Broderip, William. "Observations on the Jaw of a Fossil Mammiferous Animal, Found in the Stonesfield Slate." *Zoological Journal* 3 (1828): 408–12.

Brown, Joyce. "A Memoir of Colonel Sir Proby Cautley, F. R. S. 1802–1871, Engineer and Paleontologist." *Notes and Records of the Royal Society of London* 34 (1980): 185–225.

Brown, Matthew. "The Development of 'Modern' Paleontological Laboratory Methods: A Century of Progress." *Earth and Environmental Science: Transactions of the Royal Society of Edinburgh* 103, no. 3–4 (2012): 205–16.

Brown, Matthew. "The Global History of Latin America." *Journal of Global History* 10, no. 3 (2015): 365–86.

Browne, Janet. "A Science of Empire: British Biogeography before Darwin." *Revue d'histoire des sciences* 45, no. 4 (1992): 453–75.

Browne, Janet. *The Secular Ark: Studies in the History of Biogeography*. New Haven, CT: Yale University Press, 1983.

Bru de Ramón, Juan Bautista. *Colección de láminas que representan los animales y monstruos del Real Gabinete de Historia Natural de Madrid*. 2 vols. Madrid: Andres de Sotos, 1784.

Bru de Ramón, Juan Bautista. "Descripción del esqueleto en particular, según las observaciones hechas al tiempo de armale y colocarle en este Real Gabinete." In *Descripción del esqueleto de un quadrúpedo muy corpulento y raro que se conserva en el Real Gabinete de Historia Natural de Madrid*, edited by José Garriga, 1–16. Madrid, 1796.

Buckland, Adelene. "'Inhabitants of the Same World': The Colonial History of Geological Time." *Philological Quarterly* 97, no. 2 (2018): 219–40.

Buckland, Adelene, and Sadiah Qureshi, eds. *Time Travelers: Victorian Encounters with Time and History*. Chicago: University of Chicago Press, 2020.

Buckland, Francis. *Curiosities of Natural History: Second Series*. London: Richard Bentley, 1860.

Buckland, William. *Geology and Mineralogy Considered with Reference to Natural Theology*. 2 vols. London: William Pickering, 1836.

Buckland, William. *Reliquiæ diluvianæ; or, Observations on the Organic Remains Contained in Caves, Fissures, and Diluvial Gravel, and on Other Geological Phenomena, Attesting the Action of an Universal Deluge*. London: J. Murray, 1823.

Buffetaut, Eric. "A Frenchman in Patagonia: The Paleontological Expeditions of André Tournouër (1898–1903)." *Colligo* 3, no. 3 (2020): 1–14.

Buffetaut, Eric. *A Short History of Vertebrate Paleontology*. London: Croom Helm, 1987.

Buffon, Georges-Louis Leclerc. *Histoire naturelle, générale et particulière, avec la description du Cabinet du Roi*. 15 vols. Paris: Imprimerie Royale, 1749–1767.

Buffon, Georges-Louis Leclerc. *Histoire naturelle, générale et particulière, servant de suite à l'histoire des animaux quadrupèdes. Supplément* 6. Paris: Imprimerie Royale, 1782.

Burbank, Jane, and Mark von Hagen, eds. *Russian Empire: Space, People, Power, 1700–1930*. Bloomington: Indiana University Press, 2007.

Burkhardt, Richard, Jr. "The Leopard in the Garden: Life in Close Quarters at the Muséum d'Histoire Naturelle." *Isis* 98, no. 4 (2007): 675–94.

Burmeister, Herman. *Anales del Museo Público de Buenos Aires*. 3 vols. Buenos Aires: La Tribuna, 1864–1881.

Burnett, D. Graham. *Trying Leviathan: The Nineteenth-Century New York Court Case That Put the Whale on Trial and Challenged the Order of Nature*. Princeton, NJ: Princeton University Press, 2007.

Burrow, John. *Evolution and Society: A Study in Victorian Social Theory*. Cambridge: Cambridge University Press, 1968.

Burton, Antoinette, and Renisa Mawani, eds. *Animalia: An Anti-Imperial Bestiary for Our Times*. Durham, NC: Duke University Press, 2020.

Byrne, Denis. "Deep Nation: Australia's Acquisition of an Indigenous Past." *Aboriginal History* 20 (1996): 82–107.

Cain, Victoria. "The Art of Authority: Exhibits, Exhibit-Makers, and the Contest for Scientific Status in the American Museum of Natural History, 1920–1940." *Science in Context* 24, no. 2 (2011): 215–38.

Cain, Victoria. "'The Direct Medium of the Vision': Visual Education, Virtual Witnessing and the Prehistoric Past at the American Museum of Natural History, 1890–1923." *Journal of Visual Culture* 9, no. 3 (2010): 284–303.

Camerini, Jane. "Evolution, Biogeography, and Maps: An Early History of Wallace's Line." *Isis* 84, no. 4 (1993): 700–727.

Capitan, Louis, and Henri Breuil. *La caverne de Font-de-Gaume aux Eyzies (Dordogne)*. Monaco: A. Chêne, 1910.

Carrasquero, Silvia. "Naturalistas Suizos en el Museo de La Plata. Siglos XIX-XX." *Revista del Museo de La Plata* 1 (2016): 55–60.

Cassidy, James G. *Ferdinand V. Hayden: Entrepreneur of Science*. Lincoln: University of Nebraska Press, 2000.

Catesby, Mark. *The Natural History of Carolina, Florida and the Bahama Islands*. 2 vols. London: Printed at the expense of the author, 1731–1743.

Cathcart, Michael. *The Water Dreamers: How Water and Silence Shaped Australia*. Melbourne: Text Publishing, 2009.

Cautley, Proby. "Letter to the Asiatic Society about the Discovery of Fossil Deposits." *Journal of the Asiatic Society of Bengal* 3 (1834): 527–29, 592–94.

Cautley, Proby. "Letter to the Asiatic Society about the Discovery of Fossil Deposits." *Journal of the Asiatic Society of Bengal* 4 (1835): 279–81, 585–87.

Cautley, Proby. "Note on a Fossil Ruminant Genus Allied to Giraffidae in the Siwalik Hills." *Annals and Magazine of Natural History* 1 (1839): 167–69.

Cautley, Proby. "On the Structure of the Sevàlik Hills and the Organic Remains Found in Them." *Madras Journal of Literature and Science* 12 (1840): 292–304.

Cerney, Jan. *Badlands National Park*. Charleston, SC: Arcadia Publishing, 2004.

Chakrabarti, Pratik. "Gondwana and the Politics of Deep Past." *Past & Present* 242, no. 1 (2019): 119–53.

Chakrabarti, Pratik. *Inscriptions of Nature: Geology and the Naturalization of Antiquity*. Baltimore: Johns Hopkins University Press, 2020.

Chakrabarti, Pratik, and Alison Bashford. "Towards a Modern History of Gondwanaland." *Journal of the British Academy* 9, S6 (2021): 5–26.

Chakrabarti, Pratik, and Joydeep Sen. "'The World Rests on the Back of a Tortoise': Science and Mythology in Indian History." *Modern Asian Studies* 50, no. 3 (2016): 808–40.

Chambers, Robert. *Vestiges of the Natural History of Creation*. London: John Churchill, 1844.

Clarke, Philip. *Aboriginal Plant Collectors: Botanists and Australian Aboriginal People in the Nineteenth Century*. Kenthurst, Australia: Rosenberg, 2008.

Clarke, Philip. "Indigenous Spirit and Ghost Folklore of 'Settled' Australia." *Folklore* 118, no. 2 (2007): 141–61.

Clifford, Jim. *West Ham and the River Lea: A Social and Environmental History of London's Industrialized Marshland, 1839–1914*. Vancouver: UBC Press, 2017.

Clift, William. "Report by Mr Clift, of the College of Surgeons, London, in Regard to the Fossil Bones Found in the Caves and Bone Breccia of New Holland." *Edinburgh New Philosophical Journal* 10 (1831): 394–96.

Clift, William. "Some Account of the Remains of the Megatherium Sent to England from

Buenos Ayres by Woodbine Parish, Jun. Esq. F.G.S. F.R.S." *Transactions of the Geological Society of London* 2–3 (1835): 437–50.

Clover, Sam. *A Pioneer Heritage*. Los Angeles: Saturday Night Publishing, 1932.

Cockrell, Ron. *Bones of Agate: An Administrative History of Agate Fossil Beds National Monument, Nebraska*. US National Park Service Publications and Papers. Omaha: National Park Service, 1986.

Cohen, Alan. "Mr. Bain and Dr. Atherstone: South Africa's Pioneer Fossil Hunters." *Earth Sciences History* 19, no. 2 (2000): 175–91.

Cohen, Claudine. *Le destin du mammouth*. Paris: Seuil, 1994.

Cohen, Claudine. *La méthode de Zadig: La trace, le fossile, la preuve*. Paris: Seuil, 2011.

Confino, Alon. *Germany as a Culture of Remembrance: Promises and Limits of Writing History*. Chapel Hill: University of North Carolina Press, 2006.

Cook, Harold J. *Tales of the 04 Ranch: Recollections of Harold J. Cook, 1887–1909*. Lincoln: University of Nebraska Press, 1968.

Cook, James H. *Fifty Years on the Old Frontier: Cowboy, Hunter, Guide, Scout, and Ranchman*. New Haven: Yale University Press, 1923.

Cooley, Mackenzie. "The Giant Remains: Mesoamerican Natural History, Medicine, and Cycles of Empire." *Isis* 112, no. 1 (2021): 45–67.

Cope, Edward Drinker. *The Origin of the Fittest: Essays on Evolution*. New York: D. Appleton, 1887.

Cope, Edward Drinker. *The Primary Factors of Organic Evolution*. Chicago: Open Court, 1896.

Cowie, Helen. *Conquering Nature in Spain and Its Empire, 1750–1850*. Manchester, UK: Manchester University Press, 2011.

Cowie, Helen. *Exhibiting Animals in Nineteenth-Century Britain: Empathy, Education, Entertainment*. Basingstoke, UK: Palgrave Macmillan, 2014.

Crane, Susan. *Collecting and Historical Consciousness in Early Nineteenth-Century Germany*. Ithaca, NY: Cornell University Press, 2000.

Crosby, Alfred. *Ecological Imperialism: The Biological Expansion of Europe, 900–1900*. Cambridge: Cambridge University Press, 1986.

Cuvier, Georges. "Description ostéologique du tapir." *Annales du Muséum d'Histoire Naturelle* (1804): 122–31.

Cuvier, Georges. "Extract from a Memoir on an Animal of Which the Bones Are Found in the Plaster Stone [*pierre à plâtre*] around Paris, and Which Appears No Longer to Exist Alive Today. Read at the public session of the National Institute on 15 Vendémaire, Year VII [6 October 1798]." In *Georges Cuvier, Fossil Bones, and Geological Catastrophes: New Translations and Interpretations of the Primary Texts*, edited and translated by Martin Rudwick, 35–41. Chicago: University of Chicago Press, 1997.

Cuvier, Georges. "Mémoire sur les espèces d'éléphans tant vivantes que fossiles." *Magasin encyclopédique* 2 (1796): 440–45.

Cuvier, Georges. "Notice sur le squelette d'une très grande espèce de quadrupède inconnue jusqu'à présent, trouvé au Paraguay et déposé au cabinet d'histoire naturelle de Madrid." *Magasin encyclopédique* 2 (1796): 303–10.

Cuvier, Georges. *Recherches sur les ossemens fossiles de quadrupèdes: Où l'on rétablit les caractères de plusieurs espèces d'animaux que les révolutions du globe paroissent avoir détruites*. 4 vols. Paris: Deterville, 1812.

Cuvier, Georges. *Le règne animal distribué d'après son organisation*. 4 vols. Paris: Deterville, 1817.

Cuvier, Georges. *Le règne animal distribué d'après son organisation, pour servir de base à l'histoire naturelle des animaux et d'introduction à l'anatomie comparée*. 11 vols. Paris: Fortin, Masson, 1836–1849.

Cuvier, Georges. "Sur le Megalonyx." *Annales du Muséum d'Histoire Naturelle* 5 (1804): 358–75.

Cuvier, Georges. "Sur les espèces d'animaux dont proviennent les os fossiles répandus dans la pierre à plâtre des environs de Paris: 1. Mémoire, restitution de la tête." *Annales du Muséum d'Histoire Naturelle* 3 (1804): 275–303.

Cuvier, Georges. "Sur les espèces d'animaux dont proviennent les os fossiles répandus dans la pierre à plâtre des environs de Paris. 5. Mémoire: Description de deux squelettes presque entiers d'*Anoplotherium Commune*." *Annales du Muséum d'Histoire Naturelle* 9 (1807): 272–82.

Cuvier, Georges. "Sur quelques dents et os trouvés en France, qui paroissent avoir appartenu à des animaux du genre de tapir." *Annales du Muséum d'Histoire Naturelle* 3 (1804): 132–43.

Cuvier, Georges, and Alexandre Brongniart. "Essai sur la géographie minéralogique des environs de Paris." *Annales du Muséum d'Histoire Naturelle* 11 (1808): 293–326.

Cuvier, Georges, and Alexandre Brongniart. *Essai sur la géographie minéralogique des environs de Paris, nouvelle édition*. Paris: Baudouin, 1811.

Darwin, Charles. *On the Origin of Species by Means of Natural Selection, or the Preservation of Favoured Races in the Struggle for Life*. London: John Murray, 1859.

Daston, Lorraine, and Peter Galison. "The Image of Objectivity." *Representations* 40 (1992): 81–128.

Daunton, Martin, ed. *The Organisation of Knowledge in Victorian Britain*. Oxford: Oxford University Press, 2005.

Davidson, Charles. *Diary of Travels and Adventures in Upper India*. 2 vols. London: Henry Colburn, 1843.

Davies, William. *Catalogue of the Pleistocene Vertebrata, from the Neighbourhood of Ilford, Essex, in the Collection of Sir Antonio Brady*. London: Printed for private circulation, 1874.

Davis, Lance, Robert Gallman, and Karin Gleiter. *In Pursuit of Leviathan: Technology, Institutions, Productivity, and Profits in American Whaling, 1816–1906*. Chicago: University of Chicago Press, 1997.

Davis, W. N., Jr. "The Sutler at Fort Bridger." *Western Historical Quarterly* 2, no. 1 (1971): 37–54.

Davy, John, ed. *The Collected Works of Sir Humphry Davy*. Vol. 7. London: Smith, Elder, 1840.

Dawkins, William Boyd. *Cave Hunting: Researches on the Evidence of Caves Respecting the Early Inhabitants of Europe*. London: Macmillan, 1874.

Dawkins, William Boyd. *Early Man in Britain and His Place in the Tertiary Period*. London: Macmillan, 1880.

Dawkins, William Boyd. "On The Range of the Mammoth." *Popular Science Review* 7 (1868): 275–86.

Dawson, Gowan. *Show Me the Bone: Reconstructing Prehistoric Monsters in Nineteenth-Century Britain and America*. Chicago: University of Chicago Press, 2016.

Dehm, Richard. "Zur Geschichte der Bayerische Staatssammlung und Universitäts-Institut für Paläontologie und historische Geologie in München." *Freunde der Bayerischen Staatssammlung für Paläontologie und historische Geologie München, Jahresbericht und Mitteilungen* 6 (1978): 13–46.

Delbourgo, James. *Collecting the World: Hans Sloane and the Origins of the British Museum*. Cambridge, MA: Harvard University Press, 2017.

Delbourgo, James, and Nicholas Dew. *Science and Empire in the Atlantic World*. New York: Routledge, 2008.

Delmer, Cyrille, and Jorge Mondéjar Fernández. "El género *Arsinoitherium*: Catálogo de la colección inédita del Muséum National d'Histoire Naturelle de París y el problema del número de especies." *Paleontological nova* 8 (2008): 291–304.

Derry, Margaret. *Bred for Perfection: Shorthorn Cattle, Collies, and Arabian Horses since 1800*. Baltimore: John Hopkins University Press, 2003.

Desmarest, Anselme Gaëtan. *Mammalogie, ou, Description des espèces de mammifères*. 3 vols. Paris: Encyclopédie méthodique, 1820–1822.

Desmond, Adrian. *Archetypes and Ancestors: Palaeontology in Victorian London, 1850–1875*. Chicago: University of Chicago Press, 1984.

Desmond, Adrian. *The Hot-Blooded Dinosaurs*. London: Blond & Briggs, 1975.

Deverell, William, and Greg Hise. *Land of Sunshine: An Environmental History of Metropolitan Los Angeles*. Pittsburgh: University of Pittsburgh Press, 2006.

Díaz-Andreu, Margarita. *A World History of Nineteenth-Century Archaeology: Nationalism, Colonialism, and the Past*. Oxford: Oxford University Press, 2007.

Dingus, Lowell. *King of the Dinosaur Hunters: The Life of John Bell Hatcher and the Discoveries That Shaped Paleontology*. New York: Pegasus Books, 2018.

Dixie, Florence. *Across Patagonia*. London: Richard Bentley and Son, 1880.

"Documentos." *Revista del Museo de La Plata* 1 (1891): vii–xv.

Dollo, Louis. "Les ancêtres des marsupiaux étaient-ils arboricoles?" *Travaux de la station zoologique de Wimereux* 7 (1899): 188–203.

"Donations to the Museum in September and October, 1848." *Proceedings of the Academy of Natural Sciences of Philadelphia* 4 (1850): 109–10.

Dorré, Gina. *Victorian Fiction and the Cult of the Horse*. Aldershot, UK: Ashgate, 2006.

Douglas, Kirsty. "'Pictures of Time Beneath': Science, Landscape, Heritage and the Uses of the Deep Past in Australia, 1830–2003." PhD dissertation, Australian National University, 2004.

Drower, Margaret. "Gaston Maspero and the Birth of the Egypt Exploration Fund (1881–3)." *Journal of Egyptian Archaeology* 68 (1982): 299–317.

Duvernoy, Georges. "Sur les ossements des mammifères fossiles découvert à Pikermi." *Comptes rendus de l'Académie des Sciences* 38 (1854): 251–57.

Dwight, Timothy. "Presentation of Professor Marsh's Collections to Yale University." *Science* 7 (1898): 77–79.

Edmonds, Perry. "The Bunyip as Uncanny Rupture: Fabulous Animals, Innocuous Quadrupeds and the Australian Anthropocene." *Australian Humanities Review* 63 (2018): 80–98.

Edney, Matthew. *Mapping an Empire: The Geographical Construction of British India, 1765–1843*. Chicago: University of Chicago Press, 1997.

Evans-Hatch, Gail. *Centuries along the Upper Niobrara: Historic Resource Study: Agate Fossil Beds National Monument*. Lincoln, NE: Midwest Region National Park Service, 2008.

Fabian, Ann. *The Skull Collectors: Race, Science, and America's Unburied Dead*. Chicago: University of Chicago Press, 2020.

Fabian, Johannes. *Time and the Other: How Anthropology Makes Its Object*. New York: Columbia University Press, 2014.

Falconer, Hugh. "Letter to the Asiatic Society about Further Fossil Discoveries." *Journal of the Asiatic Society of Bengal* 4 (1835): 57.

Falconer, Hugh. "On Some Fossil Remains of Anoplotherium and Giraffe from the Sewalik Hills in the North of India." *Proceedings of the Geological Society* 4 (1846): 235–49.

Falconer, Hugh. *Paleontological Memoirs and Notes of H. Falconer, with a Biographical Sketch of the Author.* 2 vols. London: Robert Hardwicke, 1868.

Falconer, Hugh, and Proby Cautley. *Fauna Antiqua Sivalensis, Being the Fossil Zoology of the Sewalik Hills in the North of India.* London: Smith, Elder, 1846.

Falconer, Hugh, and Proby Cautley. "Note on the *Felis Cristata*, a New Fossil Tiger, from the Sivâlik Hills." *Asiatic Researches* 19 (1836): 135–42.

Falconer, Hugh, and Proby Cautley. "Note on the Fossil Camel of the Sivâlik Hills." *Asiatic Researches* 19 (1836): 115–34.

Falconer, Hugh, and Proby Cautley. "Note on the Fossil Hippopotamus of the Sivâlik Hills." *Asiatic Researches* 19 (1836): 39–53.

Falconer, Hugh, and Proby Cautley. "Notice on the Remains of a Fossil Monkey from the Tertiary Strata of the Sewalik Hills in the North of Hindoostan." *Madras Journal of Literature and Science* 12 (1840): 304–9.

Falconer, Hugh, and Proby Cautley. "*Sivatherium giganteum*, a New Fossil Ruminant Genus, from the Valley of Markanda, in the Sivàlik Branch of the Sub-Himalayan Mountains." *Journal of the Asiatic Society of Bengal* 5 (1836): 38–50.

Fallon, Richard. *Reimagining Dinosaurs in Late Victorian and Edwardian Literature.* Cambridge: Cambridge University Press, 2021.

Fan, Fa-ti. "Circulating Material Objects: The International Controversy over Antiquities and Fossils in Twentieth-Century China." In *The Circulation of Knowledge between Britain, India, and China: The Early-Modern World to the Twentieth Century*, edited by Bernard Lightman, Gordon McOuat, and Larry Stewart, 209–36. Leiden: Brill, 2013.

Farber, Paul Lawrence. *Finding Order in Nature: The Naturalist Tradition from Linnaeus to E. O. Wilson.* Baltimore: John Hopkins University, 2000.

Figuier, Louis. *L'homme primitif.* Paris: Hachette, 1870.

Figuier, Louis. *Les mammifères.* Paris: Hachette, 1869.

Figuier, Louis. *La terre avant le déluge.* Paris: Hachette, 1863.

Finlay, George. "Memoir Read at a Meeting of the Natural History Society of Athens on the 1/13 December 1836 on Presenting a Collection of Fossil Bones Found in Attica (1836)." Reprinted in Nektarios Karadimas. "O George Finlay Ως Προϊστορικός Αρχαιολόγος Και η Πρώτη, Παλαιοντολογική Έρευνα Στην Ελλάδα (Πικέρμι, 1836)." In *Proceedings of the 14th Scientific Meeting of SE Attica*, edited by Athanasios Stefanis, 127–38. Kalyvia Thorikou: Society for the Study of Southeast Attica, 2013.

Flack, Andrew. "Continental Creatures: Animals and History in Contemporary Europe." *Contemporary European History* 27, no. 3 (2018): 517–29.

Flack, Andrew. "'The Illustrious Stranger': Hippomania and the Nature of the Exotic." *Anthrozoös* 26, no. 1 (2013): 43–59.

Flammarion, Camille. *Monde avant la création de l'homme: Origines de la terre, origines de la vie, origines de l'humanité.* Paris: C. Marpon & E. Flammarion, 1886.

Fletcher, Harold. "*Palorchestes*—Australia's Extinct Giant Kangaroo." *Australian Museum Magazine* 8 (1945): 361–65.

Flower, William Henry. *Essays on Museums and Other Subjects Connected with Natural History.* London: Macmillan, 1898.

Flower, William Henry. "The Extinct Animals of North America." *Notices of the Proceedings at the Meetings of the Members of the Royal Institution of Great Britain* 8 (1879): 103–25.
Flower, William Henry. *The Horse: A Study in Natural History*. London: Kegan Paul, 1891.
Flower, William Henry. "Los museos de historia natural." *Revista del Museo de La Plata* 1 (1891): 2–25.
Flower, William Henry, and Richard Lydekker. *An Introduction to the Study of Mammals Living and Extinct*. London: Adam and Charles Black, 1891.
Fondouce, Paul Cazalis de. "Sur la rencontre de quelques ossements fossiles dans les environs de Durfort (Gard)." *Bulletin de la Société Géologique de France*, ser. 2, 27 (1869): 264–67.
Forgan, Sophie. "Building the Museum: Knowledge, Conflict, and the Power of Place." *Isis* 96, no. 4 (2005): 572–85.
"Fossil Remains in New Holland." *Asiatic Journal* 5 (1831): 26–32.
Fraas, Eberhard. "Wüstenreise eines Geologen in Ägypten." *Kosmos* 3 (1906): 263–69.
Fraas, Oscar. *Vor der Sündfluth! Eine Geschichte der Urwelt*. Stuttgart: Hoffmann'sche Verlags-Buchhandlung, 1866.
Frazer, Persifor. "Life and Letters of Edward Drinker Cope." *American Geologist* 26 (1900): 67–128.
Freeman, Carol. *Paper Tiger: A Visual History of the Thylacine*. Leiden: Brill, 2010.
Fudge, Erica. *Perceiving Animals: Humans and Beasts in Early Modern English Culture*. Urbana: University of Illinois Press, 2002.
Fudge, Erica, ed. *Renaissance Beasts: Of Animals, Humans, and Other Wonderful Creatures*. Urbana: University of Illinois Press, 2004.
Fuller, Errol. *The Passenger Pigeon*. Princeton, NJ: Princeton University Press, 2014.
Gambarotto, Andrea. "Lorenz Oken (1779–1851): *Naturphilosophie* and the Reform of Natural History." *Journal for the History of Science* 50, no. 2 (2017): 329–40.
Gamble, Clive. *Making Deep History: Zeal, Perseverance, and the Time Revolution of 1859*. Oxford: Oxford University Press, 2021.
Garriga, José, ed. *Descripción del esqueleto de un quadrúpedo muy corpulento y raro que se conserva en el Real Gabinete de Historia Natural de Madrid*. Madrid: Imprenta de la Viuda de Don Joaquin Ibarra, 1796.
Gaudant, Jean. "Albert Gaudry (1827–1908) et les 'Enchaînements du Monde Animal.'" *Revue d'histoire des sciences* 44, no. 1 (1991): 117–28.
Gaudry, Albert. *Les ancêtres de nos animaux dans les temps géologiques*. Paris: J.-B. Baillière et Fils, 1888.
Gaudry, Albert. *Animaux fossiles aux environs d'Athènes*. Paris: G. Baillière, 1866.
Gaudry, Albert. *Animaux fossiles et géologie de l'Attique d'après les recherches faites en 1855/56 et en 1860 sous les auspices de l'Académie des Sciences*. 2 vols. Paris: Savy, 1862–1867.
Gaudry, Albert. "L'éléphant de Durfort." In *Centenaire de la Fondation du Muséum d'Histoire Naturelle*, edited by the Professors of the Museum, 327–47. Paris: Imprimerie nationale, 1893.
Gaudry, Albert. *Les enchaînements du monde animal dans les temps géologiques*. 3 vols. Paris: F. Savy, 1878–1990.
Gaudry, Albert. *Essai de paléontologie philosophique: Ouvrage faisant suite aux Enchainements du monde animal dans les temps géologiques*. Paris: Masson, 1896.
Gaudry, Albert. "Fossiles de Patagonie: Dentition de quelques mammifères." *Mémoires de la Société Géologique de France* 31 (1904): 1–27.
Gaudry, Albert. "Lettre de M.A. Gaudry concernant l'envoi en France des fossiles annoncés

dans une précédente communication." *Comptes rendus de l'Académie des Sciences* 51 (1860): 502.

Gaudry, Albert. "Une mission géologique en Grèce." *Revue des deux mondes* 2 (1857): 502–30.

Gaudry, Albert. "Note sur le monte Pentélique et le gisement d'ossements fossiles situé à sa base." *Comptes rendus de l'Académie des Sciences* 38 (1854): 611–13.

Gaudry, Albert. "Note sur les carnassiers fossiles de Pikermi (Grèce)." *Bulletin de la Société Géologique de France* 2 (1861): 527–28.

Gaudry, Albert. "Le Nouveau Musée de Paléontologie." *Revue des deux mondes* 4 (1898): 799–824.

Gaudry, Albert. "Paléontologie: La nouvelle galerie de paléontologie dans le Muséum d'Histoire Naturelle." *Comptes rendus de l'Académie des Sciences* 100 (1885): 698–701.

Gaudry, Albert. *Recherches scientifiques en Orient, entreprises par les ordres du gouvernement, pendant les années 1853–1854: Partie agricole.* Paris: Imprimerie impériale, 1855.

Gaudry, Albert. "Restauration du squelette du *Dinoceras*." *Comptes rendus de l'Académie des Sciences* 118 (1889): 1292.

Gaudry, Albert. "Résultats des recherches faites à Pikermi (Attique), sous les auspices de l'Académie." *Comptes rendus de l'Académie des Sciences* 42 (1856): 291–93.

Gaudry, Albert. "Sur la marche de l'évolution en Patagonie." *Bulletin de la Société géologique de France*, ser. 4, 3 (1908): 473.

Gaudry, Albert. "Sur la position géologique du gisement de Pikermi." *Comptes rendus de l'Académie des Sciences* 51 (1860): 500–502.

Geer, Alexandra van der, Michael Dermitzakis, and John de Vos. "Fossil Folklore from India: The Siwalik Hills and the Mahâbhârata." *Folklore* 119, no. 1 (2008): 71–92.

Geikie, James. *The Great Ice Age and Its Relation to the Antiquity of Man.* London: W. Isbister, 1874.

Geoffroy Saint-Hilaire, Étienne. *Cours de l'histoire naturelle des mammifères.* Paris: Pichon et Didier, 1829.

Geoffroy Saint-Hilaire, Étienne. "Sur le nouveau genre *Sivatherium*, trouvé fossile au bas du versant méridional de l'Himalaya." *Comptes rendus de l'Académie des Sciences* 4 (1837): 53–60.

Geoffroy Saint-Hilaire, Isidore. "Communication sur le même sujet." *Comptes rendus de l'Académie des Sciences* 4 (1837): 429–30.

George, William. *Sir Antonio Brady (1811–1881), Civil Servant, Fossil Collector and Philanthropist of West Ham, Essex.* Barking, UK: W. H. George, 1999.

Gervais, Paul. "Sur une nouvelle collection d'ossements fossiles de mammifères recueillie par M. Fr. Seguin dans la Confédération Argentine." *Comptes rendus de l'Académie des Sciences* 65 (1867): 279–82.

Gervais, Paul. *Zoologie et paléontologie françaises (animaux vertébrés), ou nouvelles recherches sur les animaux vivants et fossiles de la France.* 2 vols. Paris: A. Bertrand, 1848–1852.

Gidley, James. "Tooth Characters and Revision of the North American Species of the Genus *Equus*." *Bulletin of the American Museum of Natural History* 14 (1901): 91–142.

Gidley, James, and Henry Fairfield Osborn. "Revision of the Miocene and Pliocene Equidae of North America." *Bulletin of the American Museum of Natural History* 23 (1907): 865–934.

"Gigantic Paleotherium." *American Journal of Science and Art*, ser. 2, no. 2 (1846): 288–89.

Gilman, Charlotte Perkins. "Similar Cases." *The Nationalist* 2 (1890): 165–66.

Gindhart, Maria. "Fleshing Out the Museum: Fernand Cormon's Painting Cycle for the

New Galleries of Comparative Anatomy, Paleontology, and Anthropology." *Nineteenth-Century Art Worldwide* 7, no. 2 (2008): 55–79.

Girouard, Mark. *Alfred Waterhouse and the Natural History Museum*. London: Natural History Museum, 1999.

Glangeaud, Philippe. "Les nouvelles galeries du Muséum II: Galerie de paléontologie et galerie d'anatomie comparée." *La Nature* (1898): 307–11.

Glauert, Ludwig. "Fossil Remains from Balladonia in the Eucla Division: The Balladonia 'Soak.'" *Records of the Western Australian Museum* 1, no. 2 (1912): 39–46.

Glauert, Ludwig. "The Mammoth Cave (continued)." *Records of the Western Australian Museum* 1, no. 3 (1914): 244–51.

Goldfuß, August. "Osteologische Beiträge zur Kenntnis verschiedener Säugethiere der Vorwelt." *Academia Caesarea Leopoldino-Carolinae Naturae Curiosorum* 10, no. 2 (1821): 453–502.

Goldfuß, August. *Die Umgebungen von Muggendorf: ein Taschenbuch für Freunde der Natur und Altertumskunde*. Erlangen, Germany: J. J. Palm, 1810.

Goode, George Brown. "The Relationships and Responsibilities of Museums." *Science* 2 (1895): 197–209.

Gordon, Elizabeth Oke Buckland. *The Life and Correspondence of William Buckland, D. D., F.R.S.* London: J. Murray, 1894.

Gould, Stephen Jay. "The Case of the Creeping Fox Terrier Clone." *Natural History* 97 (1987): 16–24.

Gould, Stephen Jay. "G. G. Simpson, Paleontology, and the Modern Synthesis." In *The Evolutionary Synthesis: Perspectives on the Unification of Biology*, edited by Ernst Mayr and William Provine, 153–72. Cambridge, MA: Harvard University Press, 1980.

Gould, Stephen Jay. "The Origin and Function of 'Bizarre' Structures: Antler Size and Skull Size in the 'Irish Elk,' *Megaloceros giganteus*." *Evolution* 28, no. 2 (1974): 191–220.

Gould, Stephen Jay. "The Promise of Paleobiology as a Nomothetic, Evolutionary Discipline." *Paleobiology* 6, no. 1 (1980): 96–118.

Grayson, Donald K. *The Establishment of Human Antiquity*. New York: Academic Press, 1983.

Green, Abigail. *Fatherlands: State-Building and Nationhood in Nineteenth-Century Germany*. Cambridge: Cambridge University Press, 2001.

Greer, Kirsten. "Geopolitics and the Avian Imperial Archive: The Zoogeography of Region-Making in the Nineteenth-Century British Mediterranean." *Annals of the Association of American Geographers* 103, no. 6 (2013): 1317–31.

Gregory, John Walter. *The Dead Heart of Australia: A Journey around Lake Eyre in the Summer of 1901–1902*. London: J. Murray, 1906.

Gross, Alan. *Scientific Sublime: Popular Science Unravels the Mysteries of the Universe*. Oxford: Oxford University Press, 2018.

Grove, Richard. *Green Imperialism: Colonial Expansion, Tropical Island Edens and the Origins of Environmentalism, 1600–1860*. Cambridge: Cambridge University Press, 1995.

Gundlach, Bradley. *Process and Providence: The Evolution Question at Princeton, 1845–1929*. Grand Rapids, MI: William B. Eerdmans, 2013.

Haeckel, Ernst. *Natürliche Schöpfungsgeschichte: Gemeinverständliche wissenschaftliche Vorträge über die Entwickelungslehre im Allgemeinen und diejenige von Darwin, Goethe und Lamarck im Besonderen*. Berlin: Reimer, 1874.

Haeckel, Ernst. "Zur Phylogenie der australischen Fauna: Systematische Einleitung." In

Zoologische Forschungsreisen in Australien und dem malayischen Archipel, I. Band: Ceratodus, edited by Richard Semon, i–xxiv. Jena: Gustav Fischer, 1893–1913.

Hagen, Friedrich Heinrich von der, ed. *Der Nibelungen Lied in der Ursprache mit den Lesarten der verschiedenen Handschriften*. Berlin: Julius Eduard Hitzig, 1810.

Hagerman, Christopher. *Britain's Imperial Muse: The Classics, Imperialism, and the Indian Empire, 1784–1914*. Basingstoke, UK: Palgrave Macmillan, 2013.

Hamilakis, Yannis. *The Nation and Its Ruins: Antiquity, Archaeology, and National Imagination in Greece*. Oxford: Oxford University Press, 2007.

Hamilton Couper, James. "On the Geology of a Part of the Sea-Coast of the State of Georgia; with a Description of the Fossils Remains of the Megatherium, Mastodon and Other Contemporaneous Mammalia and Fossil Marine Shells, Found in the Brunswick Canal and at Skidaway Island." In *Memoir on the Megatherium, and Other Extinct Gigantic Quadrupeds of the Coast of Georgia*, edited by William Hodgson, 31–47. New York: Bartlett & Welford, 1846.

Hammond, Michael. "The Expulsion of the Neanderthals from Human Ancestry: Marcellin Boule and the Social Context of Scientific Research." *Social Studies of Science* 12, no. 1 (1982): 1–36.

Hamy, Ernest. *Edouard Lartet: Sa vie et ses travaux*. Brussels: Weissenbruch, 1872.

Harvey, Eleanor. "Founding Landscape: Charles Willson Peale's Exhumation of the Mastodon." *American Art* 31, no. 2 (2017): 40–42.

Harvey, Joy. *Almost a Man of Genius: Clémence Royer, Feminism and Nineteenth-Century Science*. New Brunswick, NJ: Rutgers University Press, 1997.

Hatcher, John Bell. "The Mysterious Mammal of Patagonia, *Grypotherium domesticum*." *Science* 10 (1899): 814–15.

Hatcher, John Bell. *Reports of the Princeton University Expeditions to Patagonia, 1896–1899*. Vol. 1, *Narrative of the Expeditions: Geography of Southern Patagonia*. Princeton, NJ: Princeton University, 1903.

Hedeen, Stanley, and John Mack Faragher. *Big Bone Lick: The Cradle of American Paleontology*. Lexington: University Press of Kentucky, 2011.

Heintzman, Alix. "E Is for Elephant: Jungle Animals in Late Nineteenth-Century British Picture Books." *Environmental History* 19, no. 3 (2014): 553–63.

Hermann, A. "Modern Laboratory Methods in Vertebrate Paleontology." *Bulletin of the American Museum of Natural History* 26 (1909): 283–331.

Herzfeld, Michael. *Ours Once More: Folklore, Ideology, and the Making of Modern Greece*. New York: Berghahn Books, 2020.

Hesketh, Ian. "The Evolutionary Epic." *Victorian Review* 41, no. 2 (2015): 35–39.

Hesketh-Prichard, Hesketh. *Through the Heart of Patagonia*. New York: D. Appleton, 1902.

Heumann, Ina, Holger Stoecker, Marco Tamborini, and Mareike Vennen, eds. *Dinosaurierfragmente: Zur Geschichte der Tendaguru-Expedition und ihrer Objekte, 1906–2018*. Göttingen: Wallstein, 2018.

Hibbert, Samuel. "Additional Contributions towards the History of the *Cervus euryceros*, or Fossil Elk of Ireland." *Edinburgh Journal of Science* 2 (1830): 301–17.

Hodgson, William. *Memoir on the Megatherium, and Other Extinct Gigantic Quadrupeds of the Coast of Georgia*. New York: Bartlett & Welford, 1846.

Hoffstetter, Robert. "Les rôles respectifs de Bru, Cuvier et Garriga dans les premières études concernant *Megatherium*." *Bulletin du Muséum National d'Histoire Naturelle*, ser. 2, 31, no. 6 (1959): 536–45.

Holland, William Jacob. *Twelfth Annual Report of the Director for the Year Ending March 31st, 1909*. Pittsburgh: Carnegie Museum of Natural History, 1909.

Holland, William Jacob, and Olof Peterson. "The Osteology of the Chalicotheroidea: With Special Reference to a Mounted Skeleton of *Moropus elatus* Marsh, Now Installed in the Carnegie Museum." *Memoirs of the Carnegie Museum* 3 (1907–1914): 189–406.

Holten, Birgitte, and Michael Sterll. "The Danish Naturalist Peter Wilhelm Lund (1801–80), Research on Early Man in Minas Gerais." *Luso-Brazilian Review* 37, no. 1 (2000): 33–45.

Holten, Birgitte, and Michael Sterll. "Peter Wilhelm Lund: Life and Work." In *Archaeological and Paleontological Research in Lagoa Santa: The Quest for the First Americans*, edited by Pedro Da-Gloria, Walter Neves, and Mark Hubbe, 11–26. Cham, Switzerland: Springer, 2017.

Howorth, Henry. *The Mammoth and the Flood*. London: Sampson Low, 1887.

Hubber, Des, and Brian Cowley. "Distinct Creation: Early European Images of Australian Animals." *La Trobe Journal* L66 (2000): 3–32.

Hume, William Fraser. *Catalogue of the Geological Museum, Cairo*. Cairo: National Printing Department, 1905.

Hunt, Robert. *The Agate Hills: History of Paleontological Excavations, 1904–1925*. Lincoln: University of Nebraska State Museum, 1984.

Hunter, William. "Observations on the Bones, Commonly Supposed to Be Elephants Bones, Which Have Been Found near the River Ohio in America." *Philosophical Transactions* 58 (1768): 34–45.

Hutchinson, Henry Neville. *Creatures of Other Days*. London: Chapman & Hall, 1894.

Huth, Frederick Henry. *Works on Horses and Equitation: A Bibliographical Record of Hippology*. London: Bernard Quaritch, 1887.

Huxley, Leonard. *Life and Letters of Thomas Henry Huxley*. 2 vols. New York: D. Appleton, 1901.

Huxley, Thomas Henry. *American Addresses: With a Lecture on the Study of Biology*. London: Macmillan, 1877.

Huxley, Thomas Henry. "On the Application of the Laws of Evolution to the Arrangement of the Vertebrata and More Particularly of the Mammalia." *Proceedings of the Zoological Society of London* (1880): 649–62.

Huxley, Thomas Henry. "Palaeontology and the Doctrine of Evolution (1870)." In *Collected Essays*. Vol. 8, *Discourses Biological and Geological*. New York: Macmillan, 1894, 340–88.

Ides, Evert Ysbrants. *Three Years Travels from Moscow Over-Land to China. Thro' Great Ustiga, Siriania, Permia, Sibiria, Daour, Great Tartary, &c. to Peking*. London: W. Freeman, 1706.

Isenberg, Andrew. *The Destruction of the Bison: An Environmental History, 1750–1920*. Cambridge: Cambridge University Press, 2000.

Jacyna, L. Stephen. "'We Are Veritable Animals': The Nineteenth-Century Paris Menagerie as a Site for the Science of Intelligence." In *The History of the Brain and Mind Sciences: Technique, Technology, Therapy*, edited by Stephen Casper and Delia Gavrus, 25–47. Rochester, NY: University of Rochester Press, 2017.

Jaffe, Mark. *The Gilded Dinosaur: The Fossil War between E. D. Cope and O. C. Marsh and the Rise of American Science*. New York: Crown Publications, 2001.

Jameson, Robert. "On the Fossil Bones Found in the Bone-Caves and Bone-Breccia of New Holland." *Edinburgh New Philosophical Journal* 10 (1831): 393.

Jefferson, Thomas. "A Memoir on the Discovery of Certain Bones of a Quadruped of the

Clawed Kind in the Western Parts of Virginia." *Transactions of the American Philosophical Society* 4 (1799): 246–60.
Jeffreys, David, ed. *Views of Ancient Egypt since Napoleon Bonaparte: Imperialism, Colonialism and Modern Appropriations*. Abingdon, UK: Routledge, 2011.
Jennison, Watson. *Cultivating Race: The Expansion of Slavery in Georgia, 1750–1860*. Lexington: University Press of Kentucky, 2012.
Jerrold, William. *How to See the British Museum, in Four Visits*. London: Bradbury and Evans, 1852.
Jones, Ryan Tucker. *Empire of Extinction: Russians and the North Pacific's Strange Beasts of the Sea, 1741–1867*. Oxford: Oxford University Press, 2014.
Jones, Thomas, ed. *Reliquiæ Aquitanicæ: Being Contributions to the Archaeology and Paleontology of Périgord and the Adjoining Provinces of Southern France*. London: Williams & Norgate, 1875.
Jørgensen, Dolly. *Recovering Lost Species in the Modern Age: Histories of Longing and Belonging*. Cambridge, MA: MIT Press, 2019.
Jubilé de M. Albert Gaudry. Paris: Imprimerie générale Lahure, 1902.
Kalof, Linda. *Looking at Animals in Human History*. London: Reaktion Books, 2007.
Kaup, Johann Jacob. "*Deinotherium giganteum*: Eine Gattung der Vorwelt aus der Ordnung der Pachydermen." *Isis* 22 (1829): 401–4.
Kaup, Johann Jacob. *Descriptions d'ossemens fossiles de mammifères inconnus jusqu'à présent, qui se trouvent au Museum Grand-Ducal de Darmstadt*. 5 vols. Darmstadt, Germany: J. G. Heyer, 1832–1839.
Kaup, Johann Jacob. "Sur la place que doit occuper le *Dinotherium* dans l'échelle animale." *Comptes rendus de l'Académie des Sciences* 4 (1837): 527–29.
Kaup, Johann Jacob. "II. Zoologischer Theil." In Kaup and Klipstein, *Beschreibung und Abbildungen*, 1–6.
Kaup, Johann Jacob, and August von Klipstein. *Atlas* Dinotherii gigantei. Darmstadt, Germany: C. F. Will, 1836.
Kaup, Johann Jacob, and August von Klipstein, *Beschreibung und Abbildungen von dem in Rheinhessen aufgefundenen colossalen Schedel des* Dinotherii gigantei *mit geognostischen Mittheilungen über die knochenführenden Bildungen des mittelrheinischen Tertiärbeckens*. Darmstadt, Germany: C. F. Will, 1836.
Kaup, Johann Jacob, and Jean-Baptiste Scholl. *Catalogue des plâtres des ossemens fossiles, qui se trouvent dans le Cabinet d'Histoire Naturelle du Grand Duc de Hesse*. Darmstadt: J. G. Heyer, 1832.
Keene, Melanie. *Science in Wonderland: The Scientific Fairy Tales of Victorian Britain*. Oxford: Oxford University Press, 2015.
Keith, Arthur. "Marcellin Boule: January 1861–July 1942." *Man* 43 (1943): 42–43.
Kerr, Ashley. "From Savagery to Sovereignty: Identity, Politics, and International Expositions of Argentine Anthropology (1878–1892)." *Isis* 108, no. 1 (2017): 62–81.
Kete, Kathleen. *The Beast in the Boudoir: Petkeeping in Nineteenth-Century Paris*. Berkeley: University of California Press, 1994.
Klipstein, August von. "I. Geognostischer Theil." In Kaup and Klipstein, *Beschreibung und Abbildungen*, 1–32.
Knight, David. *Ordering the World a History of Classifying Man*. London: Burnett Books, 1981.
Knipe, Henry. *Evolution in the Past*. London: Herbert and Daniel, 1912.
Knipe, Henry. *Nebula to Man*. London: J. M. Dent, 1905.

Koch, Albert. *Description of the Missourium, or Missouri Leviathan*. Louisville, KY: Prentice and Weissinger, 1841.

Koditschek, Theodore. *Liberalism, Imperialism and the Historical Imagination: Nineteenth Century Visions of Greater Britain*. Cambridge: Cambridge University Press, 2011.

Koditschek, Theodore. "Narrative Time and Racial/Evolutionary Time in Nineteenth-Century British Liberal Imperial History." In *Race, Nation and Empire: Making Histories, 1750 to the Present*, edited by Catherine Hall and Keith McClelland, 36–55. Manchester, UK: Manchester University Press, 2010.

Kohler, Robert. *All Creatures: Naturalists, Collectors, and Biodiversity, 1850–1950*. Princeton, NJ: Princeton University Press, 2006.

Kohler, Robert. *Landscapes and Labscapes: Exploring the Lab-Field Border in Biology*. Chicago: University of Chicago Press, 2002.

Köstering, Susanne. *Natur zum Anschauen: Das Naturkundemuseum des deutschen Kaiserreichs 1871–1914*. Köln: Böhlau, 2003.

Kovalevsky, Vladimir. "On the Osteology of the Hyopotamidae." *Philosophical Transactions of the Royal Society of London* 163 (1873): 19–94.

Krige, John. "Hybrid Knowledge: The Transnational Co-production of the Gas Centrifuge for Uranium Enrichment in the 1960s." *British Journal for the History of Science* 45, no. 3 (2012): 337–57.

Krüger, Tobias. *Discovering the Ice Ages: International Reception and Consequences for a Historical Understanding of Climate*. Leiden: Brill, 2013.

Lang, John Dunmore. "Account of the Discovery of Bone Caves in Wellington Valley, about 210 Miles West from Sydney in New Holland, from *Sydney Gazette* and *New South Wales Advertiser*, 25 May 1830." *Edinburgh New Philosophical Journal* 10 (1831): 364–68.

Lanham, Url. *The Bone Hunters: Heroic Age of Paleontology in the American West*. New York: Dover Publications, 1991.

Lankester, E. Ray. *Extinct Animals*. London: A. Constable, 1905.

Lankester, E. Ray. *Science from an Easy Chair: A Second Series*. London: Adlard & Son, 1912.

Lartet, Edouard. *Notice sur la colline de Sansan, suivie d'une récapitulation des diverses espèces d'animaux vertébrés fossiles*. Auch, France: J. A. Portes, 1851.

Lartet, Edouard. "Nouvelles recherches sur la coexistence de l'homme et des grands mammifères fossiles réputés caractéristiques de la dernière époque géologique." *Annales des sciences naturelles II: Zoologie* 15 (1861): 177–253.

Lartet, Edouard "Sur les migrations anciennes des mammifères de l'époque actuelle." *Comptes rendus de l'Académie des Sciences* 46 (1858): 409–14.

Latour, Bruno. *Science in Action: How to Follow Scientists and Engineers through Society*. Cambridge, MA: Harvard University Press, 1987.

Laurence, Alison. "Pleistocene Park, and Other Designs on Deep Time in the Interwar United States." *Notes and Records: The Royal Society Journal of the History of Science*. published online October 6, 2021. https://doi.org/10.1098/rsnr.2021.0032.

Lawrence, Natalie. "Exotic Origins: The Emblematic Biogeographies of Early Modern Scaly Mammals." *Itinerario* 39, no. 1 (2015): 17–43.

LeConte, Joseph. *Elements of Geology: A Text-book for Colleges and for the General Reader*. New York: D. Appleton, 1883.

LeConte, Joseph. "The Race Problem in the South." In *Man and the State: Studies in Applied Sociology; Popular Lectures and Discussions before the Brooklyn Ethical Association*, 349–402. New York: D. Appleton, 1892.

Lehmann-Nitsche, Robert. "Coexistencia del hombre con un gran desdantado y un equino en las cavernas patagónicas." *Revista del Museo de La Plata* 9 (1899): 455–72.

Leidy, Joseph. "Dr. Leidy's Memoir: On the Fossil Mammalia and Reptilia Collected during the Survey." In *Report of a Geological Survey of Wisconsin, Iowa, and Minnesota: And Incidentally of a Portion of Nebraska Territory*, edited by David Dale Owen, 1:539–72. Philadelphia: Lippincott, Grambo, 1852.

Leidy, Joseph. *The Extinct Mammalian Fauna of Dakota and Nebraska*. Philadelphia: J. P. Lippincott, 1869.

Leidy, Joseph. "On Some New Species of Fossil Mammalia from Wyoming." *Proceedings of the Academy of Natural Sciences* 24 (1872): 167–69.

Levine, Alex, and Adriana Novoa. *¡Darwinistas! The Construction of Evolutionary Thought in Nineteenth Century Argentina*. Leiden: Brill, 2012.

Levit, Gregory, and Lennart Olsson. "'Evolution on Rails': Mechanisms and Levels of Orthogenesis." *Annals for the History and Philosophy of Biology* 11 (2006): 97–136.

Lifschitz, Avi. "The Book of Job and the Sex Life of Elephants: The Limits of Evidential Credibility in Eighteenth-Century Natural History and Biblical Criticism." *Journal of Modern History* 91, no. 4 (2019): 739–75.

Lightman, Bernard. *Victorian Popularizers of Science: Designing Nature for New Audiences*. Chicago: University of Chicago Press, 2007.

Lightman, Bernard, and Bennett Zon, eds. *Victorian Culture and the Origin of Disciplines*. London: Routledge, 2019.

Limoges, Camille. "The Development of the Museum of Natural History of Paris, 1800–1914." In *The Organization of Science and Technology in France, 1808–1914*, edited by Robert Fox and George Weisz, 211–40. Cambridge: Cambridge University Press, 1980.

Lindermayer, Anton von. "Die fossilen Knochenreste in Pikermi in Griechenland." *Correspondenz-Blatt des Zoologisch-Mineralogischen Vereins in Regensburg* 14 (1860): 109–22.

Linnaeus, Carl. *Systema naturæ per regna tria naturae, secundum classes, ordines, genera, species, cum characteribus, differentiis, synonymis, locis. Editio decima, reformata*. 3 vols. Vienna: Typis Ioannis Thomae, 1758.

Lister, Adrian. *Darwin's Fossils: Discoveries That Shaped the Theory of Evolution*. London: Natural History Museum, 2018.

Ljönnberg, Einar. "On a Remarkable Piece of Skin from Cueva Eberhardt, Last Hope Inlet, Patagonia." *Proceedings of the Zoological Society of London* (1900): 379–84.

Ljönnberg, Einar. "On Some Remains of *Neomylodon listae* Ameghino, Brought Home by the Swedish Expedition to Tierra Del Fuego, 1896." In *Wissenschaftliche Ergebnisse der schwedischen Expedition nach den Magellansländern, 1895–1897*. Vol. 2, *Zoologie*, edited by Otto Nordenskjöld, 149–70. Stockholm: P. A. Norstedt & Söner, 1907.

Loomis, F. B. "Momentum in Variation." *American Naturalist* 39 (1905): 839–43.

Lopes, Maria Margaret. "'Scenes from Deep Times': Bones, Travels, and Memories in the Cultures of Nature in Brazil." *História, Ciências, Saúde-Manguinhos* 15, no. 3 (2008): 615–34.

Lopes, Maria Margaret, and Irina Podgorny. "The Shaping of Latin American Museums of Natural History, 1850–1990." *Osiris* 15, no. 1 (2000): 108–18.

Lorimer, Jamie, and Clemens Driessen. "From 'Nazi Cows' to Cosmopolitan 'Ecological Engineers': Specifying Rewilding through a History of Heck Cattle." *Annals of the American Association of Geographers* 106, no. 3 (2016): 631–52.

Lovejoy, Arthur. *The Great Chain of Being: A Study of the History of an Idea*. Cambridge, MA: Harvard University Press, 1936.

Lubbock, John. *Pre-historic Times: As Illustrated by Ancient Remains, and the Manners and Customs of Modern Savages*. London: Williams and Norgate, 1865.

Lucas, Arthur, and William Henry Dudley Le Souef. *The Animals of Australia: Animals, Reptiles and Amphibians*. Melbourne: Whitcombe and Tombs, 1909.

Lull, Richard. "The Evolution of the Horse Family, as Illustrated in the Yale Collections." *American Journal of Science*, ser. 4, no. 23 (1907): 161–82.

Lund, Peter. "Nouvelles observations sur la faune fossile du Brésil: Extraits d'une lettre adressée aux rédacteurs par M. Lund." *Annales des sciences naturelles* 2 (1839): 205–8.

Lydekker, Richard. *A Geographical History of Mammals*. Cambridge: Cambridge University Press, 1896.

Lydekker, Richard. *The Horse and Its Relatives*. London: George Allen, 1912.

Lydekker, Richard. *Life and Rock: A Collection of Zoological and Geological Essays*. London: Universal Press, 1894.

Lydekker, Richard. *Mostly Mammals: Zoological Essays*. London: Hutchinson, 1903.

Lyell, Charles. *Elements of Geology or the Ancient Changes of the Earth and Its Inhabitants as Illustrated by Geological Monuments*. London: J. Murray, 1865.

Lyell, Charles. *The Geological Evidences of the Antiquity of Man*. London: J. Murray, 1863.

Lyell, Charles. "Organic Remains." *Madras Journal of Literature and Science* 7 (1838): 248–50.

Lyell, Charles. *Principles of Geology, Being an Attempt to Explain the Former Changes of the Earth's Surface, by Reference to Causes Now in Operation*. 3 vols. London: John Murray 1830–1833.

Lyons, Henry George. *The Cadastral Survey of Egypt 1892–1907*. Cairo: National Printing Department, 1908.

Lyons, Sherrie. *Species, Serpents, Spirits, and Skulls: Science at the Margins in the Victorian Age*. Albany: SUNY Press, 2010.

MacDonald, Sally, and Michael Rice, eds. *Consuming Ancient Egypt*. Abingdon, UK: Routledge, 2003.

MacFadden, Bruce, Luz Helena Oviedo, Grace Seymour, and Shari Ellis. "Fossil Horses, Orthogenesis, and Communicating Evolution in Museums." *Evolution: Education and Outreach* 5 (2012): 29–37.

Mackness, B. S. "Reconstructing *Palorchestes* (Marsupialia: Palorchestidae)—from Giant Kangaroo to Marsupial 'Tapir.'" *Proceedings of the Linnean Society of New South Wales* 130 (2009): 21–36.

Maier, Gerhard. *African Dinosaurs Unearthed: The Tendaguru Expeditions*. Bloomington: Indiana University Press, 2003.

Manias, Chris. "From Terra Incognita to Garden of Eden: Unveiling the Prehistoric Life of China and Central Asia, 1900–30." In *Treaty Ports in Modern China*, edited by Robert Bickers and Isabella Jackson, 201–19. London: Routledge, 2016.

Manias, Chris. *Race, Science, and the Nation: Reconstructing the Ancient Past in Britain, France and Germany*. London: Routledge, 2013.

Manias, Chris. "Reconstructing an Incomparable Organism: The Chalicothere in Nineteenth and Early-Twentieth Century Paleontology." *History and Philosophy of the Life Sciences* 40, no. 1 (2018): 1–21.

Marchand, Suzanne. *Down from Olympus: Archaeology and Philhellenism in Germany, 1750–1970*. Princeton, NJ: Princeton University Press, 1996.

Marsh, Othniel Charles. *Dinocerata: A Monograph of an Extinct Order of Gigantic Mammals.* Washington, DC: Government Printing Office, 1886.

Marsh, Othniel Charles. "The Discovery of the Cretaceous Mammalia, Part I." *American Journal of Science*, ser. 3, no. 38 (1889): 81–92.

Marsh, Othniel Charles. "The Discovery of the Cretaceous Mammalia, Part II." *American Journal of Science*, ser. 3, no. 38 (1889): 177–80.

Marsh, Othniel Charles. "The Discovery of the Cretaceous Mammalia, Part III." *American Journal of Science*, ser. 3, no. 43 (1892): 249–62.

Marsh, Othniel Charles. "The Fossil Mammals of the Order Dinocerata." *American Naturalist* 7 (1873): 146–53.

Marsh, Othniel Charles. *History and Methods of Paleontological Discovery.* New Haven, CT: Tuttle, Morehouse & Taylor, 1879.

Marsh, Othniel Charles. *Introduction and Succession of Vertebrate Life in America: An Address.* Nashville, TN: Tuttle, Morehouse & Taylor, 1878.

Marsh, Othniel Charles. "Note on Mesozoic Mammalia." *Proceedings of the Academy of Natural Sciences of Philadelphia* 43 (1891): 237–41.

Marsh, Othniel Charles. "On the Dates of Professor Cope's Recent Publications." *American Naturalist* 7, no. 5 (1873): 303–6.

Martin, William. *A Natural History of Quadrupeds, and Other Mammiferous Animals.* London: Whitehead, 1840.

Masson, David. *The British Museum, Historical and Descriptive with Numerous Wood-Engravings.* London: W. and R. Chambers, 1850.

Matthew, William Diller. *The Evolution of the Horse.* New York: American Museum of Natural History, 1903.

Matthew, William Diller. "Exhibit Illustrating the Evolution of the Horse." *American Museum Journal* 8 (1908): 117–22.

Matthew, William Diller. *The Hall of Fossil Vertebrates.* New York: American Museum of Natural History, 1902.

Matthew, William Diller. "Scourge of the Santa Monica Mountains." *American Museum Journal* 16 (1916): 468–72.

Maybury-Lewis, David. "A New World Dilemma: The Indian Question in the Americas." *Bulletin of the American Academy of Arts and Sciences* 46, no. 7 (1993): 44–59.

Mayor, Adrienne. *The First Fossil Hunters Dinosaurs, Mammoths, and Myth in Greek and Roman Times.* Princeton, NJ: Princeton University Press, 2011.

Mayor, Adrienne. *Fossil Legends of the First Americans.* Princeton, NJ: Princeton University Press, 2007.

Mayr, H. "Karl Alfred von Zittel zum 150 Jährigen Geburtstag (25.9.1839–5.1.1904)." *Mitteilungen der Bayerischen Staatssammlung für Paleontologie und historische Geologie* 29 (1993): 7–51.

McClelland, Charles. *State, Society, and University in Germany, 1700–1914.* Cambridge: Cambridge University Press, 1980.

McCook, Stuart. "'It May Be Truth, but It Is Not Evidence': Paul Du Chaillu and the Legitimation of Evidence in the Field Sciences." *Osiris* 11 (1996): 177–97.

McEntee, J. C. "Why Mulligan Is Not Just Another Irish Name: Lake Callabonna, South Australia." In *Aboriginal Placenames.* Vol. 19, *Naming and Re-naming the Australian Landscape*, edited by Harold Koch and Luise Hercus, 251–56. Canberra: ANU Press, 2009.

McKay, John. *Discovering the Mammoth: A Tale of Giants, Unicorns, Ivory, and the Birth of a New Science*. New York: W. W. Norton, 2017.

McNassor, Cathy. *Los Angeles's La Brea Tar Pits and Hancock Park*. Charleston, SC: Arcadia, 2011.

McShane, Clay, and Joel Tarr. *The Horse in the City: Living Machines in the Nineteenth Century*. Baltimore: John Hopkins University Press, 2007.

"Le Mégathérium du Muséum d'Histoire Naturelle." *La Nature* 1 (1873): 305–7.

Michard, Jean-Guy, Alexandre Mille, and Pascal Tassy. *Le secret de l'Archeobelodon: Deux siècles d'enquête sur un fossile mythique*. Paris: Belin, 2015.

Mieg, Juan. *Paseo por el Gabinete de Historia Natural de Madrid, ó descripción sucinta de los principales objetos de zoología que ofrecen las salas de esta interesante colección*. Madrid: M. de Burgos, 1818.

Mille, Alexandre, Jean-Guy Michard, and Pascal Tassy. *Le secret de l'archéobélodon: Deux siècles d'enquête sur un fossile mythique*. Paris: Belin and Muséum National d'Histoire Naturelle, 2015.

Miller, Hugh. *The Testimony of the Rocks: Or, Geology in Its Bearings on the Two Theologies, Natural and Revealed*. Edinburgh: Thomas Constable, 1858.

Milner, Richard. *Charles R. Knight: The Artist Who Saw through Time*. New York: Abrams, 2012.

Minard, Pete. *All Things Harmless, Useful, and Ornamental: Environmental Transformation through Species Acclimatization, from Colonial Victoria to the World*. Chapel Hill: University of North Carolina Press, 2019.

Minard, Pete. "'Making the "Marsupial Lion': Bunyips, Networked Colonial Knowledge Production between 1830–59 and the Description of *Thylacoleo Carnifex*." *Historical Records of Australian Science* 29, no. 2 (2018): 91–102.

Mitchell, Samuel. "Observations on the Teeth of the Megatherium Recently Discovered in the United States, Read Nov. 17, 1823." *Annals of the Lyceum of Natural History of New York* 1 (1824): 58–61.

Mitchell, Thomas. "An Account of the Limestone Caves at Wellington Valley, and of the Situation, near One of Them, Where Fossil Bones Have Been Found." *Proceedings of the Geological Society of London, November 1826 to June 1833* 1 (1834): 321–22.

Mitchell, Thomas. *Three Expeditions into the Interior of Eastern Australia*. 2 vols. London: T & W. Boone, 1839.

Moreno, Francisco. "El Museo de La Plata: Rápida ojeada sobre su fundación y desarrollo." *Revista del Museo de La Plata* 1 (1890–1891): 27–55.

Moreno, Francisco. *Viaje á la Patagonia Austral: Emprendido bajo los auspicios del gobierno nacional, 1876–1877*. 2nd ed. Buenos Aires: Imprenta de la Nacion, 1879.

Moreno, Francisco, and Arthur Smith Woodward. "On a Portion of Mammalian Skin, Named *Neomylodon listai*, from a Cavern near Consuelo Cove, Last Hope Inlet, Patagonia." *Proceedings from the Zoological Society of London* (1899): 144–56.

Morgan, Vincent, and Spencer Lucas. "Notes from Diary—Fayum Trip, 1907." *New Mexico Museum of Natural History and Science* 22 (2002): 1–148.

Morrill, Charles. *The Morrills and Reminiscences*. Chicago: University Publishing, 1918.

Morris, Amy. "Geomythology on the Colonial Frontier: Edward Taylor, Cotton Mather, and the Claverack Giant." *William and Mary Quarterly* 70, no. 4 (2013): 701–24.

Moser, Stephanie. *Ancestral Images: The Iconography of Human Origins*. Ithaca, NY: Cornell University Press, 1998.

Moyal, Ann. *Bright and Savage Land: Scientists in Colonial Australia*. Sydney: HarperCollins, 1987.

Moyal, Ann. *Platypus: The Extraordinary Story of How a Curious Creature Baffled the World*. Baltimore: Johns Hopkins University Press, 2004.

Müller-Wille, Staffan. "Linnaeus and the Four Corners of the World." In *The Cultural Politics of Blood, 1500–1900*, edited by Kimberly Anne Coles, Ralph Bauer, Zita Nunes, and Carla Peterson, 191–209. Basingstoke, UK: Palgrave MacMillan, 2015.

Müller-Wille, Staffan. "Walnuts at Hudson Bay, Coral Reefs in Gotland: The Colonialism of Linnaean Botany." In *Colonial Botany: Science, Commerce, and Politics in the Early Modern World*, edited by Londa Schiebinger and Claudia Swan, 34–48. Philadelphia: University of Pennsylvania Press, 2005.

Müller-Wille, Staffan, and Isabelle Charmantier. "Natural History and Information Overload: The Case of Linnaeus." *Studies in History and Philosophy of Science Part C* 43, no. 1 (2012): 4–15.

Murchison, Charles. "Biographical Sketch." In *Palaeontological Memoirs and Notes of H. Falconer, with a Biographical Sketch of the Author*, edited by Charles Murchison, vol. 1, xxiii–liii. London, 1868.

Murphey, P. C., and Emmett Evanoff. *Stratigraphy, Fossil Distribution, and Depositional Environments of the Upper Bridger Formation (Middle Eocene), Southwestern Wyoming*. Report of Investigations 57. Fort Collins: Wyoming State Geological Survey, 2007.

Musil, Rudolf. "A *Dinotherium* Skeleton from Česká Třebová." *Acta Musei Moraviae* 82 (1997): 105–22.

Nair, Savithri Preetha. "'Eyes and No Eyes': Siwalik Fossil Collecting and the Crafting of Indian Paleontology (1830–1847)." *Science in Context* 18, no. 3 (2005): 359–92.

Nance, Susan. *Entertaining Elephants: Animal Agency and the Business of the American Circus*. Baltimore: John Hopkins University Press, 2013.

Nead, Lynda. *Victorian Babylon: People, Streets and Images in Nineteenth-Century London*. New Haven, CT: Yale University Press, 2000.

Neumann, Roderick P. "Through the Pleistocene: Nature and Race in Theodore Roosevelt's African Game Trails." In *Environment at the Margins: Literary and Environmental Studies in Africa*, edited by Byron Caminero-Santangelo and Garth Myers, 43–72. Athens: University of Ohio Press, 2011.

Nieuwland, Ilja. *American Dinosaur Abroad: A Cultural History of Carnegie's Plaster Diplodocus*. Pittsburgh: University of Pittsburgh Press, 2019.

Nordenskjöld, Otto. "Neue Untersuchungen über *Neomylodon listai*." *Zoologischer Anzeiger* 22 (1899): 335–36.

Nothdurft, William, and Josh Smith. *The Lost Dinosaurs of Egypt*. New York: Random House, 2003.

"La nouvelle galerie de paléontologie au Muséum." *L'année scientifique et industrielle* 9 (1885): 464–67.

"Les nouvelles galeries du Muséum." *L'Anthropologie* 9 (1898): 319–36.

Nyhart, Lynn. *Modern Nature: The Rise of the Biological Perspective in Germany*. Chicago: University of Chicago Press, 2009.

Ochev, Vitalii, and Mikhail Surkov. "The History of Excavation of Permo-Triassic Vertebrates from Eastern Europe." In *The Age of Dinosaurs in Russia and Mongolia*, edited by Michael Benton, Mikhail Shishkin, David Unwin, and Evgenii Kurochkin, 1–16. Cambridge: Cambridge University Press, 2003.

O'Connor, Ralph. *The Earth on Show: Fossils and the Poetics of Popular Science, 1802–1856*. Chicago: University of Chicago Press, 2007.

O'Connor, Ralph. "From the Epic of Earth History to the Evolutionary Epic in Nineteenth-Century Britain." *Journal of Victorian Culture* 14, no. 2 (2009): 207–23.

Olivier-Mason, Joshua. "'These Blurred Copies of Himself': T. H. Huxley, Paul Du Chaillu, and the Reader's Place among the Apes." *Victorian Literature and Culture* 42, no. 1 (2014): 99–122.

Olsen, Penny. *Upside Down World: Early European Impressions of Australia's Curious Animals*. Canberra: National Library Australia, 2010.

Olsen, Penny, and Lynette Russell. *Australia's First Naturalists: Indigenous Peoples' Contribution to Early Zoology*. Canberra: National Library of Australia, 2019.

Orcutt, Mary Logan. "The Discovery in 1901 of the La Brea Fossil Beds." *Historical Society of Southern California Quarterly* 36, no. 4 (1954): 338–41.

Osborn, Henry Fairfield. *The Age of Mammals in Europe, Asia and North America*. New York: Macmillan, 1910.

Osborn, Henry Fairfield. "The Evolution of Mammalian Molars to and from the Tritubercular Type." *American Naturalist* 22, no. 264 (1888): 1067–79.

Osborn, Henry Fairfield. "The Fayûm Expedition of the American Museum." *Science* 25 (1907): 513–16.

Osborn, Henry Fairfield. "The Geological and Faunal Relations of Europe and America during the Tertiary Period and the Theory of the Successive Invasions of an African Fauna." *Science* 11 (1900): 561–74.

Osborn, Henry Fairfield. *The Horse, Past and Present, in the American Museum of Natural History and in the Zoological Park*. New York: Irving Press, 1913.

Osborn, Henry Fairfield. "The Law of Adaptive Radiation." *American Naturalist* 36 (1902): 353–63.

Osborn, Henry Fairfield. "A Memoir upon *Loxolophodon* and *Uintatherium*: Two Genera of the Sub-Order *Dinocerata*." *Contributions from the E. M. Museum of Geology and Archaeology of the College of New Jersey* 1, no. 1 (1881): 3–54.

Osborn, Henry Fairfield. *Men of the Old Stone Age: Their Evolution, Life and Art*. New York: Charles Scribner's Sons, 1915.

Osborn, Henry Fairfield. "On the Structure and Classification of the Mesozoic Mammalia." *Proceedings of the Academy of Natural Sciences of Philadelphia* 39 (1887): 282–92.

Osborn, Henry Fairfield. *Origin and History of the Horse: Address before the New York Farmers. Metropolitan Club, Tuesday Evening, December 19, 1905*. New York, 1905.

Osborn, Henry Fairfield. *Proboscidea: A Monograph of the Discovery, Evolution, Migration and Extinction of the Mastodonts and Elephants of the World*. 2 vols. New York: American Museum of Natural History, 1936–1942.

Osborn, Henry Fairfield. "A Reply to Professor Marsh's 'Note on Mesozoic Mammalia.'" *American Naturalist* 25 (1891): 775–83.

Osborn, Henry Fairfield. "A Review of the Cretaceous Mammalia." *Proceedings of the Academy of Natural Sciences of Philadelphia* 43 (1891): 124–35.

Outram, Dorinda. *Georges Cuvier: Vocation, Science, and Authority in Post-revolutionary France*. Manchester, UK: Manchester University Press, 1984.

Owen, David Dale. *Report of a Geological Survey of Wisconsin, Iowa, and Minnesota: And Incidentally of a Portion of Nebraska Territory*. 2 vols. Philadelphia: Lippincott, Grambo, 1852.

Owen, Richard. "Description of the Fossil Remains of a Mammal (*Hyracotherium leporinum*) and of a Bird (*Lithornis vulturinus*) from the London Clay." *Transactions of the Geological Society of London*, ser. 2, 6 (1841): 203–8.

Owen, Richard. *Description of the Skeleton of an Extinct Gigantic Sloth,* Mylodon robustus. London: R. and J. E. Taylor, 1842.

Owen, Richard. *A History of British Fossil Mammals, and Birds.* London: J. Van Voorst, 1846.

Owen, Richard. *Memoir on the* Megatherium, *or Giant Ground-Sloth of America.* London: Williams and Norgate, 1861.

Owen, Richard. *Monograph of the Fossil Mammalia of the Mesozoic Formations.* London: Palæontographical Society, 1871.

Owen, Richard. "Observations on Certain Fossil Bones from the Collection of the Academy of Sciences of Philadelphia." *Journal of the Academy of Natural Sciences of Philadelphia*, ser. 2, no. 1 (1847–1850): 18–20.

Owen, Richard. *Odontography or, a Treatise on the Comparative Anatomy of the Teeth: Their Physiological Relations, Mode of Development, and Microscopic Structure in the Vertebrate Animals.* 2 vols. London: H. Baillière, 1845.

Owen, Richard. "On Fossil Remains of Equines from Central and South America Referable to *Equus conversidens*, Ow. *Equus tan*, Ow. and *Equus arcidens*, Ow." *Philosophical Transactions of the Royal Society of London* 159 (1869): 559–73.

Owen, Richard. *On the Classification and Geographical Distribution of the Mammalia.* London: Parker, 1859.

Owen, Richard. "On the Discovery of the Remains of a Mastodontoid Pachyderm in Australia." *Annals and Magazine of Natural History* 11 (1843): 7–12.

Owen, Richard. *On the Extent and Aims of a National Museum of Natural History.* London: Saunders, Otley, 1862.

Owen, Richard. *Palaeontology, or, a Systematic Summary of Extinct Animals and Their Geological Relations.* Edinburgh: Adam and Charles Black, 1860.

Owen, Richard. "Report on the Extinct Mammals of Australia." In *Report of the Fourteenth Meeting of the British Association for the Advancement of Science, Held at York in September 1844*, 223–40. London, 1845.

Owen, Richard. "Report on the Missourium Now Exhibiting at the Egyptian Hall, with an Inquiry into the Claims of the Tetracaulodon to Generic Distinction." *Proceedings of the Geological Society of London* 3, no. 2 (1842): 689–95.

Owen, Richard. "Report on the Reptilian Fossils of South Africa: Part II. Description of the Skull of a Large Species of Dicynodon (*D. tigriceps*, Ow.), Transmitted from South Africa by A. G. Bain, Esq." *Transactions of the Geological Society of London*, ser. 2, 7 (1845): 233–40.

Page, David. *The Past and Present Life of the Globe: Being a Sketch in Outline of the World's Life-System.* London: William Blackwood and Sons, 1861.

Palacio-Pérez, Eduardo. "Salomon Reinach and the Religious Interpretation of Paleolithic Art." *Antiquity* 84 (2010): 853–63.

"Palaeontologie—Nouveau genre de didelphe fossile en Auvergne." *L'écho du monde savant et l'Hermès* 1 (1838): 253–54.

Pander, Christian, and Eduard d'Alton. *Das Riesen-Faulthier,* Bradypus giganteus, *abgebildet, beschrieben und mit den verwandten Geschlechtern verglichen.* Bonn: Eduard Weber, 1821.

Parish, Woodbine. *Buenos Ayres and the Provinces of the Rio de La Plata.* London: John Murray, 1838.

Pavlinov, Igor. *Taxonomic Nomenclature: What's in a Name—Theory and History.* Boca Raton, FL: CRC Press, 2022.

Payne, Emma. "Casting a New Canon: Collecting and Treating Casts of Greek and Roman Sculpture, 1850–1939." *Cambridge Classical Journal* 65 (2019): 113–49.

Peale, Rembrandt. *Account of the Skeleton of the Mammoth, a Non-descript Carnivorous Animal of Immense Size, Found in America*. London: E. Lawrence, 1802.

Peale, Rembrandt. *An Historical Disquisition on the Mammoth: or, Great American Incognitum, an Extinct, Immense, Carnivorous Animal, Whose Fossil Remains Have Been Found in North America*. London: E. Lawrence, 1803.

Penny, Hugh Glenn. "German Polycentrism and the Writing of History." *German History* 30, no. 2 (2012): 265–82.

Pentland, Joseph. "Observations on a Collection of Fossil Bones Sent to Baron Cuvier from New Holland." *Edinburgh New Philosophical Journal* 14 (1832–1833): 120–21.

Pentland, Joseph. "On the Fossil Bones of Wellington Valley, New Holland or New South Wales." *Edinburgh New Philosophical Journal* 12 (1832): 301–8.

Pfeifer, Edward. "The Genesis of American Neo-Lamarckism." *Isis* 56, no. 2 (1965): 156–67.

Phillips, Denise. *Acolytes of Nature: Defining Natural Science in Germany, 1770–1850*. Chicago: University of Chicago Press, 2012.

Pickstone, John. "Museological Science? The Place of the Analytical/Comparative in Nineteenth-Century Science, Technology and Medicine." *History of Science* 32 (1994): 111–38.

Pickstone, John. *Ways of Knowing: A New History of Science, Technology and Medicine*. Manchester, UK: Manchester University Press, 2000.

Pietsch, Tamson. *Empire of Scholars: Universities, Networks and the British Academic World, 1850–1939*. Manchester, UK: Manchester University Press, 2013.

Pimentel, Juan. *The Rhinoceros and the Megatherium: An Essay in Natural History*. Cambridge, MA: Harvard University Press, 2017.

Piñero, José M. López. "Juan Bautista Bru (1740–1799) and the Description of the Genus Megatherium." *Journal of the History of Biology* 21, no. 1 (1988): 147–63.

Pledge, Neville. "Fossils of the Lake: A History of the Lake Callabonna Excavations." *Records of the South Australian Museum* 27, no. 2 (1994): 64–77.

Pliny the Elder. *Natural History*. Vol. 3, *Books 8–11*. Translated by H. Rackham. Cambridge, MA: Harvard University Press, 1940.

Plumb, Christopher. "'In Fact, One Cannot See It without Laughing': The Spectacle of the Kangaroo in London, 1770–1830." *Museum History Journal* 3, no. 1 (2010): 7–32.

Podgorny, Irina. *Los Argentinos vienen de los peces: Ensayo de filogenia nacional*. Rosario, Argentina: Beatriz Viterbo, 2021.

Podgorny, Irina. "Bones and Devices in the Constitution of Paleontology in Argentina at the End of the Nineteenth Century." *Science in Context* 18, no. 2 (2005): 249–83.

Podgorny, Irina. "Bureaucracy, Instructions, and Paperwork—The Gathering of Data about the Three Kingdoms of Nature in the Americas, 1770–1815." *Nuevo Mundo, Mundos Nuevos*, February 2019. https://doi.org/10.4000/nuevomundo.75454.

Podgorny, Irina. "El camino de los fósiles: Las colecciones de mamíferos Pampeanos en los museos franceses e ingleses del siglo XIX." *Asclepio* 53, no. 2 (2001): 97–116.

Podgorny, Irina. "The Daily Press Fashions a Heroic Intellectual: The Making of Florentino Ameghino in Late Nineteenth-Century Argentina." *Centaurus* 58 (2016): 166–84.

Podgorny, Irina. *Florentino Ameghino y hermanos*. Buenos Aires: Edhasa, 2021.

Podgorny, Irina. "Fossil Dealers, the Practices of Comparative Anatomy and British Diplomacy in Latin America, 1820–1840." *British Journal for the History of Science* 46, no. 4 (2013): 647–74.

Podgorny, Irina. "Human Origins in the New World? Florentino Ameghino and the Emergence of Prehistoric Archaeology in the Americas (1875–1912)." *PaleoAmerica* 1, no. 1 (2015): 68–80.

Poliquin, Rachel. *The Breathless Zoo: Taxidermy and the Cultures of Longing*. University Park: Pennsylvania State University Press, 2012.

Poskett, James. *Materials of the Mind: Phrenology, Race, and the Global History of Science, 1815–1920*. Chicago: University of Chicago Press, 2019.

Powell, Miles. *Vanishing America: Species Extinction, Racial Peril, and the Origins of Conservation*. Cambridge, MA: Harvard University Press, 2016.

Pratt, Mary Louise. *Imperial Eyes: Travel Writing and Transculturation*. New York: Routledge, 2007.

Prout, Hiram. "Description of a Fossil Maxillary Bone of a Paleotherium, Found near White River." *American Journal of Science and Art*, ser. 2, 3 (1847): 248–50.

Putnam, Walter. "Captive Audiences: A Concert for the Elephants in the Jardin des Plantes." *TDR: The Drama Review* 51, no. 1 (2007): 154–60.

Quirke, Stephen. *Hidden Hands: Egyptian Workforces in Petrie Excavation Archives, 1880–1924*. London: Bristol Classical Press, 2010.

Qureshi, Sadiah. "Dying Americans: Race, Extinction and Conservation in the New World." In *From Plunder to Preservation: Britain and the Heritage of Empire, c.1800–1940*, edited by Astrid Swenson and Peter Mandler, 269–88. Oxford: Oxford University Press, 2013.

Rader, Karen, and Victoria Cain. *Life on Display: Revolutionizing U.S. Museums of Science and Natural History in the Twentieth Century*. Chicago: University of Chicago Press, 2014.

Rainger, Ronald. *An Agenda for Antiquity: Henry Fairfield Osborn and Vertebrate Paleontology at the American Museum of Natural History, 1890–1935*. Tuscaloosa: University of Alabama Press, 1991.

Rains Wallace, David. *Beasts of Eden: Walking Whales, Dawn Horses, and Other Enigmas of Mammal Evolution*. Berkeley: University of California Press, 2004.

Rains Wallace, David. *The Bonehunters' Revenge: Dinosaurs, Greed, and the Greatest Scientific Feud of the Gilded Age*. Boston: Houghton Mifflin, 2000.

Raj, Kapil. *Relocating Modern Science: Circulation and the Construction of Knowledge in South Asia and Europe, 1650–1900*. Basingstoke, UK: Palgrave MacMillan, 2007.

Ramaswamy, Sumathi. *The Lost Land of Lemuria: Fabulous Geographies, Catastrophic Histories*. Berkeley: University of California Press, 2005.

Ratcliff, Jessica. "The East India Company, the Company's Museum, and the Political Economy of Natural History in the Early Nineteenth Century." *Isis* 107, no. 3 (2016): 495–517.

Rauch, Alan. "The Sins of Sloths: The Giant Ground Sloth as a Paleontological Parable." In *Victorian Animal Dreams: Representations of Animals in Victorian Literature and Culture*, edited by Martin Danahay and Deborah Morse, 215–28. Aldershot, UK: Ashgate, 2007.

Regal, Brian. *Henry Fairfield Osborn: Race, and the Search for the Origins of Man*. Burlington, VT: Ashgate, 2002.

Reich, Mike, and Alexander Gehler. "Giant's Bones and Unicorn Horns." *Georgia Augusta: Research Magazine of the University of Göttingen* 8 (2011): 44–50.

Reid, Donald Malcolm. "The Egyptian Geographical Society: From Foreign Laymen's Society to Indigenous Professional Association." *Poetics Today* 14, no. 3 (1993): 539–72.

Reid, Donald M. "French Egyptology and the Architecture of Orientalism: Deciphering the Façade of Cairo's Egyptian Museum." In *Franco-Arab Encounters: Studies in the Memory of David C. Gordon*, edited by L. Car Brown and Matthew S. Gordon, 35–69. Beirut: American University of Beirut, 1996.

Reid, Donald. *Whose Pharaohs? Archaeology, Museums, and Egyptian National Identity from Napoleon to World War I.* Berkeley: University of California Press, 2002.

Reinach, Salomon. "L'art et la magie à propos des peintures et des gravures de l'âge du renne." *L'Anthropologie* 14 (1903): 257–66.

Reybrouck, David van. *From Primitives to Primates: A History of Ethnographic and Primatological Analogies in the Study of Prehistory*. Leiden: Sidestone Press, 2012.

Richards, Robert. *The Romantic Conception of Life: Science and Philosophy in the Age of Goethe.* Chicago: University of Chicago Press, 2002.

Rieppel, Lukas. "Albert Koch's Hydrarchos Craze: Credibility, Identity, and Authenticity in Nineteenth-Century Natural History." In *Science Museums in Transition: Cultures of Display in Nineteenth-Century Britain and America*, edited by Carin Berkowitz and Bernard Lightman, 139–62. Pittsburgh: University of Pittsburgh Press, 2017.

Rieppel, Lukas. *Assembling the Dinosaur: Fossil Hunters, Tycoons, and the Making of a Spectacle.* Cambridge, MA: Harvard University Press, 2019.

Rieppel, Lukas. "Plaster Cast Publishing in Nineteenth-Century Paleontology." *History of Science* 53, no. 4 (2015): 456–91.

Rieppel, Lukas. "Prospecting for Dinosaurs on the Mining Frontier: The Value of Information in America's Gilded Age." *Social Studies of Science* 45, no. 2 (2015): 161–86.

Rieppel, Olivier. "Othenio Abel (1875–1946) and 'the Phylogeny of the Parts.'" *Cladistics* 29, no. 3 (2013): 328–35.

Rigal, Laura. *The American Manufactory: Art, Labor, and the World of Things in the Early Republic.* Princeton, NJ: Princeton University Press, 2001.

Riggs, Christina. *Photographing Tutankhamun: Archaeology, Ancient Egypt, and the Archive.* London: Routledge, 2018.

Riper, A. Bowdoin van. *Men among the Mammoths: Victorian Science and the Discovery of Human Prehistory*. Chicago: University of Chicago Press, 1993.

Ritvo, Harriet. *The Animal Estate: The English and Other Creatures in the Victorian Age.* Cambridge, MA: Harvard University Press, 1987.

Ritvo, Harriet. *Noble Cows and Hybrid Zebras: Essays on Animals and History*. Charlottesville: University of Virginia Press, 2010.

Ritvo, Harriet. *The Platypus and the Mermaid, and Other Figments of the Classifying Imagination*. Cambridge, MA: Harvard University Press, 1997.

Robin, Libby. *How a Continent Created a Nation*. Sydney: University of New South Wales Press, 2007.

Robinson, Phillip. "Looking for Enigmas in the Forest: A Chronicle of Travelers, Trappers and Researchers." In *The Pygmy Hippo Story: West Africa's Enigma of the Rainforest*, edited by Phillip Robinson, Knut Hentschel, and Gabriella Flacke, 24–44. Oxford: Oxford University Press, 2017.

Roosevelt, Theodore. *African Game Trails: An Account of the African Wanderings of an American Hunter-Naturalist*. New York: Charles Scribner's Sons, 1910.

Roosevelt, Theodore. *A Book-Lover's Holidays in the Open*. New York: Charles Scribner's Sons, 1916.

Rossi, Paolo. *The Dark Abyss of Time: The History of the Earth and the History of Nations from Hooke to Vico*. Chicago: University of Chicago Press, 1985.

Roth, Santiago. "Beobachtungen über Entstehung und Alter der Pampasformation in Argentinien." *Zeitschrift der Deutschen Geologischen Gesellschaft* 40 (1888): 375–464.

Roth, Santiago. "Descripción de los restos encontrados en la caverna de Última Esperanza." *Revista del Museo de La Plata* 9 (1899): 421–53.

Rothfels, Nigel. *Elephant Trails: A History of Animals and Cultures*. Baltimore: John Hopkins University Press, 2021.

Rowley-Conwy. Peter, *From Genesis to Prehistory: The Archaeological Three Age System and Its Contested Reception in Denmark, Britain, and Ireland*. Oxford: Oxford University Press, 2007.

Rudwick, Martin. *Bursting the Limits of Time: The Reconstruction of Geohistory in the Age of Revolution*. Chicago: University of Chicago Press, 2005.

Rudwick, Martin. *Earth's Deep History: How It Was Discovered and Why It Matters*. Chicago: University of Chicago Press, 2014.

Rudwick, Martin. "The Emergence of a Visual Language for Geological Science 1760–1840." *History of Science* 14, no. 3 (1976): 149–95.

Rudwick, Martin, ed. *Georges Cuvier, Fossil Bones, and Geological Catastrophes: New Translations and Interpretations of the Primary Texts*. Chicago: University of Chicago Press, 1997.

Rudwick, Martin. "Georges Cuvier's Paper Museum of Fossil Bones." *Archives of Natural History* 27, no. 1 (2000): 51–68.

Rudwick, Martin. *The Great Devonian Controversy: The Shaping of Scientific Knowledge among Gentlemanly Specialists*. Chicago: University of Chicago Press, 1985.

Rudwick, Martin. *The Meaning of Fossils: Episodes in the History of Palaeontology*. Chicago: University of Chicago Press, 1985.

Rudwick, Martin. *Scenes from Deep Time: Early Pictorial Representations of the Prehistoric World*. Chicago: University of Chicago Press, 1992.

Rudwick, Martin. *Worlds before Adam: The Reconstruction of Geohistory in the Age of Reform*. Chicago: University of Chicago Press, 2008.

Rupke, Nicolaas. *The Great Chain of History: William Buckland and the English School of Geology, 1814–48*. Oxford: Oxford University Press, 1983.

Rupke, Nicolaas. *Richard Owen: Biology without Darwin*. Chicago: University of Chicago Press, 2009.

Rupke, Nicolaas. "The Road to Albertopolis: Richard Owen (1804–92) and the Founding of the British Museum of Natural History." In *Science, Politics and the Public Good: Essays in Honour of Margaret Gowing*, edited by Nicolaas Rupke, 63–89. London: Palgrave Macmillan, 1988.

Saha, Jonathan. *Colonizing Animals: Interspecies Empire in Myanmar*. Cambridge: Cambridge University Press, 2021.

Saha, Jonathan. "Colonizing Elephants: Animal Agency, Undead Capital and Imperial Science in British Burma." *British Journal for the History of Science Themes* 2 (2017): 169–89.

Schaffer, Simon. "The Nebular Hypothesis and the Science of Progress." In *History, Humanity and Evolution: Essays for John C. Greene*, edited by James Moore, 131–64. Cambridge: Cambridge University Press, 1989.

Schaffer, Simon, Lissa Roberts, Kapil Raj, and James Delbourgo, eds. *The Brokered World: Go-Betweens and Global Intelligence, 1770–1820*. Sagamore Beach, MA: Science History, 2009.

Schama, Simon. *Landscape and Memory*. New York: Vintage Books, 1996.

Scheffel, Joseph Victor. *Gaudeamus!* Stuttgart: Adolf Bonz, 1883.

Schiebinger, Londa. "Why Mammals Are Called Mammals: Gender Politics in Eighteenth-Century Natural History." *American Historical Review* 98, no. 2 (1993): 382–411.
Schlosser, Max. "Beiträge zur Kenntnis der oligozänen Landsäugetiere aus dem Fayum: Ägypten." *Beiträge zur Paläontologie und Geologie Österreich-Ungarns und des Orients* 24 (1911): 51–167.
Schlosser, Max. *Führer durch die Münchener Paläontologische Staatssammlung*. Munich: Selbstverlag des Konservatoriums der paläontologischen Staatssammlung, 1912.
Schmidt, F. A. *Petrefactenbuch, oder allgemeine und besondere Versteinerungskunde mit Berücksichtigung der Lagerungs-Verhältnisse, besonders in Deutschland*. Stuttgart: Krais & Hoffmann, 1855.
Schmitt, Cannon. *Darwin and the Memory of the Human: Evolution, Savages, and South America*. Cambridge: Cambridge University Press, 2009.
Schmitt, Stéphane. "From Eggs to Fossils: Epigenesis and Transformation of Species in Pander's Biology." *International Journal of Developmental Biology* 49, no. 1 (2005): 1–8.
Schnapp, Alain. *La conquête du passé*. Paris: LGF, 1998.
Schnitter, Claude. "Le développement du Muséum National d'Histoire Naturelle de Paris au cours de la seconde moitié du XIX siècle: 'Se transformer ou périr.'" *Revue d'histoire des sciences* 49, no. 1 (1996): 53–98.
Schuchert, Charles. *Biographical Memoir of Othniel Charles Marsh, 1831–1899*. Washington, DC: National Academy of Sciences, 1939.
Schuller, Kyla. *The Biopolitics of Feeling: Race, Sex, and Science in the Nineteenth Century*. Durham, NC: Duke University Press Books, 2018.
Schuller, Kyla. "The Fossil and the Photograph: Red Cloud, Prehistoric Media, and Dispossession in Perpetuity." *Configurations* 24, no. 2 (2016): 229–61.
Schwartz, Vanessa. *Spectacular Realities: Early Mass Culture in Fin-de-Siècle Paris*. Berkeley: University of California Press, 1998.
Sclater, Philip. "On the General Geographical Distribution of the Members of the Class Aves." *Journal of the Proceedings of the Linnean Society* 2 (1858): 130–45.
Sclater, William, and Philip Sclater. *The Geography of Mammals*. London: K. Paul, Trench, Trübner, 1899.
Scott, William Berryman. "Development of American Palæontology." *Proceedings of the American Philosophical Society* 66 (1927): 409–29.
Scott, William Berryman. *A History of Land Mammals in the Western Hemisphere*. New York: Macmillan, 1913.
Scott, William Berryman. "Paleontology as a Morphological Discipline." *Science* 4 (1896): 177–88.
Scott, William Berryman. *Some Memories of a Paleontologist*. Princeton, NJ: Princeton University Press, 1939.
Scott, William Berryman. *The Theory of Evolution, with Special Reference to the Evidence upon Which It Is Founded*. New York: Macmillan, 1917.
Secord, James. "Global Geology and the Tectonics of Empire." In *Worlds of Natural History*, edited by Helen Anne Curry, Nicholas Jardine, James Secord, and Emma Spary, 401–17. Cambridge: Cambridge University Press, 2018.
Secord, James. "Knowledge in Transit." *Isis* 95, no. 4 (2004): 654–72.
Seeley, Harry Govier. "Croonian Lecture: On *Pareiasaurus bombidens* (Owen), and the Significance of Its Affinities to Amphibians, Reptiles, and Mammals (Abstract)." *Proceedings of the Royal Society of London* 42 (1887): 337–42.

Seeley, Harry Govier. *Dragons of the Air: An Account of Extinct Flying Reptiles*. London: Methuen, 1901.

Seeley, Harry Govier. "Note on the Skeleton of *Pareiasaurus baini*." *Geological Magazine*, decade 4, 2 (1895): 1–3.

Seeley, Harry Govier. "Researches on the Structure, Organization, and Classification of the Fossil Reptilia. VII, Further Observations on *Pareiasaurus*." *Philosophical Transactions of the Royal Society of London* 183 (1893): 311–70.

Sellers, C. C. *Mr. Peale's Museum: Charles Willson Peale and the First Popular Museum of Natural Science and Art*. New York: W. W. Norton, 1980.

Sellier, Charles. *Curiosités du vieux Montmartre: Les carrières à plâtre*. Paris: J. Kugelmann, 1893.

Semonin, Paul. *American Monster: How the Nation's First Prehistoric Creature Became a Symbol of National Identity*. New York: NYU Press, 2000.

Semonin, Paul. "Empire and Extinction: The Dinosaur as a Metaphor for Dominance in Prehistoric Nature." *Leonardo* 30, no. 3 (1997): 171–82.

Sepkoski, David. *Catastrophic Thinking: Extinction and the Value of Diversity from Darwin to the Anthropocene*. Chicago: University of Chicago Press, 2020.

Sepkoski, David. *Rereading the Fossil Record: The Growth of Paleobiology as an Evolutionary Discipline*. Chicago: University of Chicago Press, 2012.

Sera-Shriar, Efram, ed. *Historicizing Humans: Deep Time, Evolution, and Race in Nineteenth-Century British Sciences*. Pittsburgh: University of Pittsburgh Press, 2018.

Seurat, M. "Les nouvelles galeries du Muséum d'Histoire Naturelle, à Paris." *Le génie civil* 33, no. 3 (1898): 37–40.

Shapin, Steven. "The Invisible Technician." *American Scientist* 77, no. 6 (1989): 554–63.

Sheets-Pyenson, Susan. *Cathedrals of Science: The Development of Colonial Natural History Museums during the Late Nineteenth Century*. Kingston, ON: McGill-Queen's University Press, 1988.

Shindler, Karolyn. *Discovering Dorothea: The Life of the Pioneering Fossil-Hunter Dorothea Bate*. London: HarperCollins, 2005.

Shteir, Ann. "Botany in the Breakfast Room: Women and Early Nineteenth-Century British Plant Study." In *Uneasy Careers and Intimate Lives: Women in Science, 1789–1979*, edited by Pnina Abir-Am and Dorinda Outram, 31–44. New Brunswick, NJ: Rutgers University Press, 1987.

Simons, John. *Obaysch: A Hippopotamus in Victorian London*. Sydney: Sydney University Press, 2019.

Simpson, George Gaylord, and H. Tobien. "The Rediscovery of Peale's Mastodon." *Proceedings of the American Philosophical Society* 98, no. 4 (1954): 279–81.

Sivasundaram, Sujit. *Nature and the Godly Empire: Science and Evangelical Mission in the Pacific, 1795–1850*. Cambridge: Cambridge University Press, 2005.

Sloan, Phillip. "John Locke, John Ray, and the Problem of the Natural System." *Journal of the History of Biology* 5, no. 1 (1972): 1–53.

Smith, Julia. *Slavery and Rice Culture in Low Country Georgia, 1750–1860*. Knoxville: University of Tennessee Press, 1991.

Sollas, William Johnson. *Ancient Hunters and Their Modern Representatives*. London: Macmillan, 1911.

Solounias, Nikos. "Mammalian Fossils of Samos and Pikermi. Part 2, Resurrection of a Classic Turolian Fauna." *Annals of the Carnegie Museum* 59, no. 8 (1981): 231–70.

Somerset, Richard. "Textual Evolution." *Translator* 17, no. 2 (2011): 255–74.
Sommer, Marianne. "Ancient Hunters and Their Modern Representatives: William Sollas's (1849–1936) Anthropology from Disappointed Bridge to Trunkless Tree and the Instrumentalisation of Racial Conflict." *Journal of the History of Biology* 38, no. 2 (2005): 327–65.
Sommer, Marianne. *History Within: The Science, Culture, and Politics of Bones, Organisms, and Molecules*. Chicago: University of Chicago Press, 2016.
Sommer, Marianne. "Mirror, Mirror on the Wall: Neanderthal as Image and 'Distortion' in Early 20th-Century French Science and Press." *Social Studies of Science* 36, no. 2 (2006): 207–40.
Sommer, Marianne. "The Romantic Cave? The Scientific and Poetic Quests for Subterranean Spaces in Britain." *Earth Sciences History* 22, no. 2 (2003): 172–208.
Sommer, Marianne. "Seriality in the Making: The Osborn-Knight Restorations of Evolutionary History." *History of Science* 48, no. 3/4 (2010): 461–82.
Sorignet, A. *La cosmogonie de la Bible devant les sciences perfectionnées*. Paris: Gaume frères, 1854.
Spary, Emma. *Utopia's Garden: French Natural History from Old Regime to Revolution*. Chicago: University of Chicago Press, 2000.
Speece, Darren, and Paul Sutter. *Defending Giants: The Redwood Wars and the Transformation of American Environmental Politics*. Seattle: University of Washington Press, 2019.
Spiro, Jonathan. *Defending the Master Race: Conservation, Eugenics, and the Legacy of Madison Grant*. Burlington: University of Vermont Press, 2008.
Stahl, M. "Nouveau procédé pour la solidification des substances friables." *Comptes rendus de l'Académie des Sciences* 58 (1864): 1052–54.
Star, Susan, and James Griesemer. "Institutional Ecology, 'Translations' and Boundary Objects: Amateurs and Professionals in Berkeley's Museum of Vertebrate Zoology, 1907–39." *Social Studies of Science* 19, no. 3 (1989): 387–420.
"Stated Meeting, Sept. 20, 1872." *Proceedings of the American Philosophical Society* 12 (1873): 514–16.
Statistical Register of the Colony of Western Australia for 1914 and Previous Years. Perth: Frederick William Simpson, Government Printer, 1916.
Stein, Barbara. *On Her Own Terms: Annie Montague Alexander and the Rise of Science in the American West*. Berkeley: University of California Press, 2001.
Stepan, Nancy. *The Idea of Race in Science: Great Britain, 1800–1960*. Hamden, CT: Archon Books, 1982.
Stepan, Nancy. *Picturing Tropical Nature*. London: Reaktion Books, 2001.
Sternberg, Charles. *The Life of a Fossil Hunter*. New York: H. Holt, 1909.
Stevenson, Alice. *Scattered Finds: Archaeology, Egyptology and Museums*. London: UCL Press, 2019.
Stirling, Edward Charles. "The Recent Discovery of Fossil Remains at Lake Calabonna, South Australia I." *Nature* 50 (1894): 184–88.
Stirling, Edward Charles. "The Recent Discovery of Fossil Remains at Lake Calabonna, South Australia II." *Nature* 50 (1894): 206–11.
Stirling, Edward Charles. "Reconstruction of Diprotodon from the Calabonna Deposits, South Australia." *Nature* 76 (1907): 543–44.
Stromer von Reichenbach, Ernst. "Bericht über eine von den Privatdozenten Dr. Max Blanckenhorn und Dr. Ernst Stromer von Reichenbach ausgeführte Reise nach Ae-

gypten." *Sitzungsberichte der Mathematich-physikalischen Classe der Königlichen Bayerischen Akademie der Wissenschaften* 32 (1903): 341–52.

Stromer von Reichenbach, Ernst. "Eine geologische Forschungsreise in die Libysche Wüste." *Bericht der Senckenbergsichen Naturforschenden Gesellschaft in Frankfurt-am-Main* (1904): 109–10.

Stromer von Reichenbach, Ernst. "Richard Markgraf und seine Bedeutung für die Erforschung der Wirbeltierpaläontologie Ägyptens." *Centralblatt für Mineralogie, Geologie und Paläontologie* 11 (1916): 287–88.

Sutter, Paul. "The World with Us: The State of American Environmental History." *Journal of American History* 100, no. 1 (2013): 94–119.

Swart, Sandra. "O Is for Okapi." In *Animalia: An Anti-Imperial Bestiary for Our Times*, edited by Antoinette Burton and Renisa Mawani, 131–44. Durham, NC: Duke University Press, 2020.

Swart, Sandra. "Writing Animals into African History." *Critical African Studies* 8, no. 2 (2016): 95–108.

Swart, Sandra. "Zombie Zoology: History and Reanimating Extinct Animals." In *The Historical Animal*, edited by Susan Nance, 54–72. Syracuse, NY: Syracuse University Press, 2015.

Tamborini, Marco. "'If the Americans Can Do It, So Can We': How Dinosaur Bones Shaped German Paleontology." *History of Science* 54, no. 3 (2016): 225–56.

Tamborini, Marco. "The Reception of Darwin in Late Nineteenth-Century German Paleontology as a Case of Pyrrhic Victory." *Studies in History and Philosophy of Science Part C* 66 (2017): 37–45.

Tamborini, Marco. "Die Wurzeln der idiographischen Paläontologie: Karl Alfred von Zittels Praxis und sein Begriff des Fossils." *NTM Zeitschrift für Geschichte der Wissenschaften, Technik und Medizin* 23, no. 3–4 (2015): 117–42.

Tassy, Pascal. *L'évolution au Muséum, Albert Gaudry*. Paris: Editions matériologiques, 2020.

Thackray, John, and Bob Press. *The Natural History Museum: Nature's Treasurehouse*. London: Natural History Museum, 2001.

Thelin, John. *A History of American Higher Education*. Baltimore: John Hopkins University Press, 2019.

Thomson, Keith. "The 'Great-Claw' and the Science of Thomas Jefferson." *Proceedings of the American Philosophical Society* 155, no. 4 (2011): 394–403.

Thomson, Keith. *The Legacy of the Mastodon: The Golden Age of Fossils in America*. New Haven, CT: Yale University Press, 2008.

Tilesius, Wilhelm Gottlieb. *On the Mammoth or Fossil Elephant, Found in the Ice, at the Mouth of the River Lena, in Siberia*. London: William Phillips, 1819.

Todd, David. *A Velvet Empire: French Informal Imperialism in the Nineteenth Century*. Princeton, NJ: Princeton University Press, 2021.

Toews, John Edward. *Becoming Historical: Cultural Reformation and Public Memory in Early Nineteenth-Century Berlin*. Cambridge: Cambridge University Press, 2004.

Tolmachoff, I. P. "The Carcasses of the Mammoth and Rhinoceros Found in the Frozen Ground of Siberia." *Transactions of the American Philosophical Society* 23 (1929): i–74.

Topham, Jonathan R. *Reading the Book of Nature: How Eight Best Sellers Reconnected Christianity and the Sciences on the Eve of the Victorian Age*. Chicago: University of Chicago Press, 2022.

Torcelli, Alfredo, ed. *Obras completas y correspondencia cientifica de Florentino Ameghino*. 24 vols. La Plata: Taller de impresiones oficiales, 1913–1936.

Tournouër, André. "Notes sur la Patagonie." *Bulletin de la Société Linnéenne de la Seine-Maritime* 1 (1914): 10–12.

Tournouër, André. "Recherches paléontologiques en Patagonie." *Comptes rendus de l'Académie des Sciences de Paris* 135 (1902): 540–43.

Tournouër, André. "Sur le *Neomylodon* et l'hyimché des indiens Tehuelches." *Bulletin du Muséum National d'Histoire Naturelle* 6 (1900): 343–44.

Trautmann, Thomas. *Aryans and British India.* Berkeley: University of California Press, 1997.

Tyrrell, Brian. "Bred for the Race: Thoroughbred Breeding and Racial Science in the United States, 1900–1940." *Historical Studies in the Natural Sciences* 45, no. 4 (2015): 549–76.

Tyrrell, Ian. *Crisis of the Wasteful Nation: Empire and Conservation in Theodore Roosevelt's America.* Chicago: University of Chicago Press, 2015.

Urgell, Guiomar de. *Arte en el Museo de La Plata: Pintura.* La Plata, Argentina: Fundación Museo de La Plata, 1995.

Vetter, Jeremy. "Cowboys, Scientists, and Fossils: The Field Site and Local Collaboration in the American West." *Isis* 99, no. 2 (2008): 273–303.

Vetter, Jeremy. *Field Life: Science in the America West during the Railroad Era.* Pittsburgh: University of Pittsburgh Press, 2016.

Vetter, Jeremy. "Wallace's Other Line: Human Biogeography and Field Practice in the Eastern Colonial Tropics." *Journal of the History of Biology* 39, no. 1 (2006): 89–123.

Vos, Paula de. "Natural History and the Pursuit of Empire in Eighteenth-Century Spain." *Eighteenth-Century Studies* 40, no. 2 (2007): 209–39.

Wagner, Andreas. "Sitzung der mathematisch-physikalischen Klasse am 9. December 1837." *Gelehrte Anzeigen* 58 (1838): 466–69.

Wagner, Andreas. "Ueber fossile Ueberreste von einem Affenschädel und andern Saügthieren aus Griechenland." *Gelehrte Anzeigen* 8 (1839): 305–11.

Wagner, Andreas. "Urweltliche Säugthier-Ueberreste aus Griechenland." *Abhandlungen der Bayerischen Akademie der Wissenschaften* 5 (1847): 373–464.

Wagner, Andreas. "Zur Berichtigung einiger Angaben des Dr Lindermayer in dessen Aufsatze über die fossilen Knochenreste von Pikermi." *Sitzungsberichte der Königlichen Bayerischen Akademie der Wissenschaften zu München* (1860): 647–55.

Wagner, Andreas, and Roth, Johannes. "Die fossilen Knochenüberreste von Pikermi in Griechenland." *Abhandlungen der Bayerischen Akademie der Wissenschaften* 7 (1854): 371–464.

Walcott, Charles, and S. P. Langley. "Correspondence Relating to Collections of Vertebrate Fossils Made by the Late Professor O. C. Marsh." *Science* 11 (1900): 21–24.

Walker, Henry. "A Day's Elephant Hunting in Essex." *Essex Field Club, Transactions* 1 (1881): 27–58.

Walker, James. *Lakota Belief and Ritual.* Lincoln: University of Nebraska Press, 1991.

Wallace, Alfred Russel. *The Geographical Distribution of Animals, with a Study of the Relations of Living and Extinct Faunas as Elucidating the Past Changes of the Earth's Surface.* 2 vols. London, 1876.

Ward, Henry. *Catalogue of Casts of Fossils, from the Principal Museums of Europe and America, with Short Descriptions and Illustrations.* Rochester: Benton & Andrews, 1866.

Ward, Henry. "Los museos Argentinos." *Revista del Museo de La Plata* 1 (1891): 145–51.

Ward, Henry. *Notice of the* Megatherium cuvieri*: The Giant Fossil Ground-Sloth of South America.* Rochester NY: University of Rochester, 1864.

Wells, H. G. *Tales of Space and Time.* London: Macmillan, 1906.

Western Australian Museum and Art Gallery. *Guide to the Contents of the Western Australian Museum and Art Gallery, with a List of the Western Australian Marsupials and Birds.* Perth: Ames & Heller, 1900.

Western Australian Museum and Art Gallery. *Report of the Committee for the Year 1897–8*. Perth: Richard Pether, 1898.

Wheeler, Walter. "The Uintatheres and the Cope-Marsh War." *Science* 131 (1960): 1171–76.

White, Mark John. *William Boyd Dawkins and the Victorian Science of Cave Hunting: Three Men in a Cavern*. Barnsley, UK: Pen & Sword Books, 2017.

Williamson, George. *The Longing for Myth in Germany: Religion and Aesthetic Culture from Romanticism to Nietzsche*. Chicago: University of Chicago Press, 2004.

Winterer, Caroline. *American Enlightenments: Pursuing Happiness in the Age of Reason*. New Haven, CT: Yale University Press, 2016.

Wistar, Caspar. "A Description of the Bones Deposited, by the President, in the Museum of the Society, and Represented in the Annexed Plates." *Transactions of the American Philosophical Society* 4 (1799): 526–31.

Witton, Mark, Darren Naish, and John Conway. "State of the Paleoart." *Paleontologia Electronica* 17, no. 3 (2014): 1–10.

Wiwjorra, Ingo. *Der Germanenmythos: Konstruktion einer Weltanschauung in der Altertumsforschung des 19. Jahrhunderts*. Darmstadt, Germany: Wissenschaftliche Buchgesellschaft, 2006.

Woodward, Bernard. "Fossil Marsupials of Western Australia." *Records of the Western Australian Museum* 1, no. 1 (1910): 9–10.

Woodward, Henry. "Excursion to Ilford and Visit to Sir Antonio Brady's Museum, June 17th, 1871." *Proceedings of the Geological Society of London* 2 (1872): 273–74.

Woodward, Henry. *A Guide to the Exhibition Galleries of the Department of Geology and Palæontology in the British Museum (Natural History)*. London: Printed by order of the Trustees of the British Museum (Natural History), 1881.

Woodward, Henry. "How the Skull of the Mammoth Was Got Out of the Brick Earth at Ilford." *Geological Magazine*, decade 1, 2 (1865): 93–94.

Woodward, Henry. "On the Curvature of the Tusks in the Mammoth, *Elephas primigenius* (Blumenbach)." *Geological Magazine*, decade 1, 5 (1868): 540–43.

Wylie, Caitlin Donahue. *Preparing Dinosaurs: The Work behind the Scenes*. Cambridge, MA: MIT Press, 2021.

Yanni, Carla. *Nature's Museums: Victorian Science and the Architecture of Display*. Baltimore: Johns Hopkins University Press, 1999.

Young, Robert. *Darwin's Metaphor: Nature's Place in Victorian Culture*. Cambridge: Cambridge University Press, 1985.

Yusoff, Kathryn. *A Billion Black Anthropocenes or None*. Minneapolis: University of Minnesota Press, 2018.

Yuval-Naeh, Naomi. "Cultivating the Carboniferous: Coal as a Botanical Curiosity in Victorian Culture." *Victorian Studies* 61, no. 3 (2019): 419–45.

Zalensky, Vladimir. *Prjevalsky's Horse* (Equus prejewalskii *Pol.)*. London: Hurst and Blackett, 1907.

Zimmerman, Virginia. *Excavating Victorians*. Albany: State University of New York Press, 2007.

Zimmermann, W. F. A. *Die Wunder der Urwelt: Eine populäre Darstellung der Geschichte der Schöpfung und des Urzustandes unseres Weltkörpers*. Berlin: Gustav Hempel, 1861.

Ziolkowski, Theodore. *German Romanticism and Its Institutions*. Princeton, NJ: Princeton University Press, 1992.

Zittel, Karl Alfred von. *Aus der Urzeit: Bilder aus der Schöpfungsgeschichte*. Munich: Oldenbourg, 1875.

Zittel, Karl Alfred von. *Geschichte der Geologie und Paläontologie bis Ende des 19. Jahrhunderts.* Munich: Oldenbourg, 1899.

Zittel, Karl Alfred von. *Grundzüge der Paläontologie (Paläzoologie).* Munich: Oldenbourg, 1895.

Zittel, Karl Alfred von. *Handbuch der Palaeontologie.* 5 vols. Munich: Oldenbourg, 1876–1893.

Zittel, Karl Alfred von. "Museums of Natural History in the United States." *Science* 3 (1884): 191–96.

Zizzamia, Daniel. "Making the West Malleable: Coal, Geohistory, and Western Expansion, 1800–1920." PhD dissertation, Montana State University, 2015.

Zizzamia, Daniel. "Restoring the Paleo-West: Fossils, Coal, and Climate in Late Nineteenth-Century America." *Environmental History* 24, no. 1 (2019): 130–56.

INDEX